PHYSICS OF HIGH TEMPERATURE PLASMAS

Second Edition

PHYSICS OF
HIGH TEMPERATURE
PLASMAS

SECOND EDITION

George Schmidt

Department of Physics
Stevens Institute of Technology
Hoboken, New Jersey

ACADEMIC PRESS   New York San Francisco London   1979

A Subsidiary of Harcourt Brace Jovanovich, Publishers

ACADEMIC PRESS, INC.
111 Fifth Avenue, New York, New York 10003

United Kingdom Edition published by
ACADEMIC PRESS, INC. (LONDON) LTD.
24/28 Oval Road, London NW1 7DX

Library of Congress Cataloging in Publication Data

Schmidt George, (Date)
    Physics of high temperature plasmas, second edition.

    First ed. published in 1966 under title: Physics of
high temperature plasmas, an introduction.

    1. High temperature plasmas. I. Title.
QC718.5.H5S35   1979   530.4'4   79–6945
ISBN 0–12–626660–3

PRINTED IN THE UNITED STATES OF AMERICA

79 80 81 82   9 8 7 6 5 4 3 2 1

Contents

Preface to Second Edition

During the years that have elapsed between the first and the present edition of this volume, plasma physics has grown into a mature discipline. Linear waves and instabilities are well understood and many of them have been experimentally observed. The principal advance has been in the understanding of a wide range of nonlinear phenomena, with the notable exception of the elusive subject of strong turbulence.

Significant advances have been made in the principal application of plasma physics: the controlled production of thermonuclear energy. In addition to the classical approach of magnetic containment, inertial confinement fusion, using laser or particle beams, is making vigorous progress. It may appear to the superficial observer that the slow but steady progress of magnetic containment fusion is based on a Darwinian model; produce mutations on existing machines and let the fittest survive. These mutations are, however, far from random; they are based on the intuition derived from the knowledge of the principles of plasma physics and their manifestations in fusion devices. The building of such an intuitive understanding of plasma behavior is among the principal aims of this volume. Hence mathematical descriptions of phenomena are supplemented by physical pictures emphasizing the underlying physics.

Many topics that were not discussed in the first edition have been added. These include drift waves, parametric instabilities, solitons, trapped particle orbits in Tokomaks, and tearing modes. The chapter on linear waves and instabilities in uniform plasmas has been expanded and split into two, and a new chapter on nonlinear waves has been added. The detailed reference lists of the first edition have been dropped and replaced by general references at the end of the volume.

In selecting the material to be covered, an attempt has been made to distinguish the basic and important from the merely fashionable. To what extent the author has succeeded in these endeavors will have to be judged by the reader.

Many useful comments on the first edition have been received, which have been of considerable help in correcting inaccuracies, while preparing the second edition. Particular thanks are due to Dr. K. Nishikawa, Dr. P. Noerdlinger, and Dr. T. Stringer for thoughtful comments.

Preface to First Edition

This book had its origin in a graduate course in plasma physics taught by the author over the years, first in 1957–1958 at the Israel Institute of Technology and after 1958 at the Stevens Institute of Technology. During this period, both the material contained in the course and the method of presentation have undergone a substantial evolution with our growing understanding of plasma phenomena and the range of applicability and the limitations of various plasma models. This volume has benefited equally from the author's research experience in the field and from his efforts in the classroom to communicate to the students various aspects of this rapidly growing and complex field.

In spite of the conceptual simplicity of the physics of classical plasmas, more often than not their behavior in the laboratory defied the theorist's predictions. The reason for this is to be found not in the theorist's limited knowledge of mathematical methods, but in the difficulty in sorting out the really important aspects in an experimental situation and using the model best suited for the case. The necessity of determining whether to use MHD or particle models, whether to use zero or finite ion gyroradii, and whether the laws derived for infinite uniform plasmas can apply for finite nonuniform plasmas are examples of the predicament of the physicist who tries to understand what goes on in the laboratory. These considerations led to a heavy emphasis on the physical understanding of plasma behavior and the meaning of various models used to describe it.

The book was written primarily for graduate students taking a course in plasma physics as well as researchers who want to become acquainted with the field, but it is hoped that people already engaged in plasma research, particularly experimentalists, will also benefit from reading it.

The book is based on a two-semester course with classical mechanics and electromagnetic theory as prerequisites. It can also be used for a condensed one-term course, omitting Sections 2-9, 5-3, 5-5, 6-4 to 6-7, 7-3, 8-3, 8-5,* and some of the more involved proofs, without loss of continuity.

The book deals with high temperature fully ionized plasmas. Collisions are considered rare; the effects of multiple small angle scattering are discussed in the last chapter. The subject of magnetohydrodynamics of conductive fluids is confined to Chapters IV and V. Most of the remaining material deals with the collisionless many-particle apsects of plasma physics.

Each chapter is followed by a brief summary and a large number of exercises. The reader is urged to work out at least a few in each chapter. No specific references are made in the body of the text to the scientific papers on which the material is partly based; these and recommended further reading can be found in the references listed at the end of each chapter. These lists contain only books and papers which have appeared in the scientific literature; they are by no means complete.

The author is much indebted to many colleagues at Stevens Institute of Technology and other institutions for discussions and helpful criticism. The book owes much to a large number of students whose questions, or sometimes simply whose worried look during the explanation of a particular point, indicated lack of clarity or an occasional gap in logic. Several students also helped with the preparation of the lecture notes; the work of J. Sinnis and G. Halpern was particularly useful. The book never would have been completed without the unfailing encouragement of my wife Kati to whom I am most thankful.

* Note that section numbers have been changed in the second edition.

PHYSICS OF HIGH TEMPERATURE PLASMAS

Second Edition

I

Introduction

The heating of a solid or liquid substance leads to *phase transitions*. Molecules or atoms with sufficient energy to overcome the binding potential will evaporate. At temperatures high enough to impart this energy to almost every particle, the substance becomes a gas. It is characteristic of phase transitions that at a fixed pressure they occur at a constant temperature. The amount of energy that must be fed into the system at this temperature to bring about the transition is the *latent heat*.

Further heating of a gas results in additional transitions. For example, a molecular gas dissociates gradually into an atomic gas if the thermal energy of some particles exceeds the molecular binding energy. An even more drastic change takes place as soon as the temperature of the gas is high enough so that some electrons can overcome the atomic binding energy. With increasing temperature, more and more atoms get stripped of their electrons until the gas becomes a mixture of freely moving electrons and nuclei. We shall call this fully ionized substance a *plasma*. Although the transition from a gas to a plasma takes place gradually and is therefore not a phase transition in the thermodynamic sense, plasma is often referred to as a "fourth state of matter."

The investigation of the behavior of plasmas has great importance for the understanding of our universe. Although our earth consists mainly of the first three states of matter, this cannot be said about most of the stars and interstellar matter. All but a tiny fraction of the universe is plasma.

The full scope of possible earthly application of plasmas cannot even be estimated as yet. One of the most important possibilities to be seen at present is the application of the hot plasma as thermonuclear fuel. Nuclear fusion reactions resulting in the release of considerable energy take place in plasmas composed of certain light elements (D, T, etc.) at temperatures of the order of 10 to 100 million degrees Kelvin. This is the energy source of many stars (including the sun), and man is attempting to use similar processes for his own use. The hydrogen bomb applies thermonuclear fusion in a hot plasma for energy production in an uncontrolled explosive form, and there are hopes of finding ways to produce peaceful power by means of controlled thermonuclear fusion.

The latter problem poses many challenging questions. As the extremely hot plasma—necessary to obtain fusion reactions—would cause every material in close contact to evaporate, it must be confined by some kind of field. The plasma in the sun, for instance, is confined by its own gravitational field. This is evidently impossible to achieve on an earthly scale; the most probable answer is confinement by an electromagnetic field. Two questions arise immediately: 1. Do field configurations exist where the internal pressure of the plasma is counterbalanced by electromagnetic forces (equilibrium configuration)? 2. If such configurations were found, how do they behave if the equilibrium is slightly perturbed; are there any stable confined equilibria? The first question can be answered positively, but the second one (not less vital than the first) has not yet been settled.

In the following, while attempting to investigate the laws of plasma physics in general, emphasis will be placed on those phenomena which are likely to prove important for the realization and operation of thermonuclear machines rather than to applications in astrophysics.

Gravitational forces are much smaller than electromagnetic ones on an earthly scale. Therefore we shall deal mainly with electromagnetic forces. Similarly, for conceivable controlled thermonuclear applications, the momentum of the particles is high and the density low enough to keep the de Broglie wavelengths of particles well below the mean particle distance. Except for some cases of particle collisions (the close collisions), quantum effects are therefore negligible.

The complete system of equations, describing the behavior of a plasma under these conditions, can be presented in a straightforward way. The electromagnetic field inside and outside the plasma is completely defined by Maxwell's equations:

$$\nabla \times \mathbf{H} = \mathbf{J} + \frac{\partial \mathbf{D}}{\partial t} \tag{1-1}$$

$$\nabla \times \mathbf{E} = -\frac{\partial \mathbf{B}}{\partial t} \tag{1-2}$$

$$\nabla \cdot \mathbf{D} = \rho \tag{1-3}$$

$$\nabla \cdot \mathbf{B} = 0 \tag{1-4}$$

and the additional conditions

$$\mathbf{B} = \mu \mathbf{H} \tag{1-5}$$

and
$$\mathbf{D} = \varepsilon\mathbf{E} \tag{1-6}$$

Inside the plasma the particles move in vacuum, and therefore $\mu = \mu_0$ and $\varepsilon = \varepsilon_0$ (the mks system is used). If the charge $\rho(\mathbf{r},t)$ and current density $\mathbf{J}(\mathbf{r},t)$ are given, the electromagnetic field as a function of space and time is uniquely defined, provided the initial and boundary conditions are known. In a plasma, however, the charge and current densities are unknown, as the particles moving in the field (to be determined from the solution of Maxwell's equations) give rise to charge accumulations and currents

If, on the other hand, the electromagnetic fields were known, the equations of motion of each particle could be calculated, and the resulting charge and current densities computed. The equation of motion of a nonrelativistic particle with charge q_i and mass m_i in an $\mathbf{E}$ (electric) and $\mathbf{B}$ (magnetic) field is

$$m_i\ddot{\mathbf{r}}_i = q_i(\mathbf{E} + \dot{\mathbf{r}}_i \times \mathbf{B}) \tag{1-7}$$

If the total number of plasma particles is N, we have N equations of this type. With known $\mathbf{E}(\mathbf{r},t)$ and $\mathbf{B}(\mathbf{r},t)$ fields, these equations (with given initial conditions) can also be uniquely solved. To "plug back" the results of these solutions into (1-1) to (1-4), we express the plasma charge and current density in the form

$$\rho_{\text{pl}} = \frac{\sum\limits_{\Delta V} q_i}{\Delta V} \tag{1-8}$$

and

$$\mathbf{J}_{\text{pl}} = \frac{\sum\limits_{\Delta V} \dot{\mathbf{r}}_i q_i}{\Delta V} \tag{1-9}$$

where the summation is carried out over a "suitably chosen" small volume element ΔV. As the positions and velocities of particles as a function of time are given as solutions of (1-7), ρ_{pl} and $\mathbf{J}_{\text{pl}}$ in (1-8) and (1-9) can be computed.

Obviously the set of equations (1-1) to (1-9) is complete. The solutions are self-consistent in the sense that the particle motions obtained create the appropriate electromagnetic fields necessary to produce just the particle motions with which one started (Fig. 1-1).

An important approximation is hidden behind (1-8) and (1-9). Since we are dealing with point charges, ρ and $\mathbf{J}$ should be described by δ functions. In other words, carrying out the $\lim\limits_{\Delta V \to 0}$ transition in these equation we find either nothing or a single particle in our volume element. If, however, one keeps ΔV big enough to contain a fairly large number of particles, we obtain "smooth" functions for ρ and $\mathbf{J}$, which are suitable for analytical cal-culations. This is physically equivalent to "smearing out" the point particles and forgetting about their individuality. When looking for interactions of individual particles, such as collisions, more refined expressions shall be invoked.

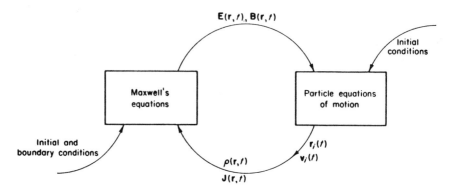

FIG. 1–1. Self–consistent plasma equations.

In addition to the plasma, known charges and currents might be present, associated with external conductors. In this case,

$$\rho = \rho_{pl} + \rho_{ext} \tag{1-10}$$

and

$$\mathbf{J} = \mathbf{J}_{pl} + \mathbf{J}_{ext} \tag{1-11}$$

Owing to the very large number of equations (1–7), it is practically inconceivable to carry out the above–outlined program of solution, even with the fastest computers available. We have the choice of approximate solution of the equations, or of finding precise solution for simplified models resembling (more or less) the real plasma.

Motion of Charged Particles in Electromagnetic Fields

2-1. The Static Magnetic Field

For one particle moving in a magnetic field with velocity **v**, (1-7) reduces to

$$m \frac{d\mathbf{v}}{dt} = q(\mathbf{v} \times \mathbf{B}) \tag{2-1}$$

As the force is perpendicular to the velocity, no work is done by the magnetic field. Indeed, the scalar multiplication of (2-1) by **v** yields

$$m\mathbf{v} \frac{d\mathbf{v}}{dt} = \frac{d}{dt} (\tfrac{1}{2}mv^2) = 0 \tag{2-2}$$

showing that the kinetic energy of the particle, in an arbitrary magnetic field, is a constant of motion.

Let us restrict ourselves for the moment to the special case where the magnetic field lines are straight and parallel (but the field is not necessarily uniform). Denoting vector components parallel to the field with the subscript $\parallel$ and those perpendicular to it with $\perp$, we obtain

$$\mathbf{v} = \mathbf{v}_{\parallel} + \mathbf{v}_{\perp} \tag{2-3}$$

and (2-1) becomes

$$\frac{d\mathbf{v}_{\parallel}}{dt} + \frac{d\mathbf{v}_{\perp}}{dt} = \frac{q}{m} (\mathbf{v}_{\perp} \times \mathbf{B}) \tag{2-4}$$

since $\mathbf{v}_{\parallel} \times \mathbf{B}$ vanishes. Equation (2-4) splits into a $\parallel$-component equation and a $\perp$-component equation

$$\frac{d\mathbf{v}_{\parallel}}{dt} = 0 \qquad (\mathbf{v}_{\parallel} = \text{const}) \tag{2-5}$$

and

$$\frac{d\mathbf{v}_{\perp}}{dt} = \frac{q}{m} (\mathbf{v}_{\perp} \times \mathbf{B}) \tag{2-6}$$

Since the right side of (2-6) is perpendicular to $\mathbf{v}_\perp$, the left side is a centripetal acceleration. It can be written

$$\frac{v_\perp^2}{r^2}(-\mathbf{r}) = \frac{q}{m}(\mathbf{v}_\perp \times \mathbf{B}) \tag{2-7}$$

where $\mathbf{r}$ is the local radius of curvature of the particle path (Fig. 2-1). Its

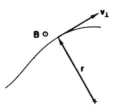

FIG. 2-1. Particle moving in a magnetic field of straight and parallel field lines. The field intensity varies in the plane perpendicular to $\mathbf{B}$.

value is, from (2-7),

$$r = \left|\frac{mv_\perp}{qB}\right| \tag{2-8}$$

In the special case of a uniform magnetic field, $B = $ const, and considering the constancy of $v_\perp$ from (2-2) and (2-5), the radius of curvature

$$R = \left|\frac{mv_\perp}{qB}\right| \tag{2-9}$$

is also a constant. In a uniform magnetic field, therefore, the particle moves in a circle with the so-called *cyclotron* or *gyroradius* R in the perpendicular plane, while it moves with a constant velocity along the field lines. The resulting path is a *helix*.

The angular frequency of the circular motion is

$$\omega_c = \frac{v_\perp}{R} = \frac{q}{m}B \tag{2-10}$$

which is often called the *cyclotron frequency*, as its constancy for nonrelativistic velocities ($m = $ const) enables the operation of the cyclotron. For $v_\parallel = 0$ (which does not represent any restriction, just a suitable choice of coordinate system moving with the particle in the parallel direction), the particle moves in a circular path, giving rise to a magnetic field of its own. The time average of this field, over many gyration periods, is that of a ring current with the intensity

$$I = \frac{q\omega_c}{2\pi} = \frac{1}{2\pi}\frac{q^2B}{m} \tag{2-11}$$

The corresponding magnetic moment is

$$\mu_m = I\pi R^2 = \frac{1}{2}\frac{q^2 R^2 B}{m} \tag{2-12}$$

Denoting the magnetic flux surrounded by the path by ϕ, (2-12) becomes

$$\mu_m = \frac{1}{2\pi}\frac{q^2}{m}\phi \tag{2-13}$$

which shows that the magnetic moment is proportional to the flux enclosed. Inserting R from (2-9) into (2-12) leads to another form of the magnetic moment,

$$\mu_m = \frac{\frac{1}{2}mv_\perp^2}{B} \tag{2-14}$$

or, using (2-9) again, we obtain

$$\mu_m = |\tfrac{1}{2}qv_\perp R| \tag{2-15}$$

The latter is often put into vector form,

$$\mathbf{\mu}_m = \tfrac{1}{2}q\mathbf{R} \times \mathbf{v}_\perp \tag{2-16}$$

Note that the direction of μ_m does not depend on the sign of particle charge in a given field.

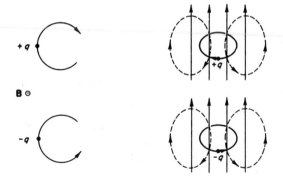

FIG. 2-2. The gyrating particle in the magnetic field generates a field like that of a diamagnetic dipole.

The magnetic field generated by a ring current at a distance much larger than R is similar to that of a magnetic dipole with the same moment. Figure 2-2 shows that the magnetic field associated with both a $+q$ and $-q$ particle moving in a uniform magnetic field opposes the external field inside the

path, corresponding to a diamagnetic dipole. This is the source of the diamagnetic properties of plasmas to be studied later.

We are now going to investigate the motion of a charged particle in a
uniform magnetic field with an additional constant force present. The
equation of motion then becomes

$$\frac{d\mathbf{v}}{dt} = \frac{q}{m}(\mathbf{v} \times \mathbf{B}) + \frac{\mathbf{F}}{m} \qquad (2\text{-}17)$$

This splits again into component equations:

$$\frac{d\mathbf{v}_\parallel}{dt} = \frac{\mathbf{F}_\parallel}{m} \qquad (2\text{-}18)$$

and

$$\frac{d\mathbf{v}_\perp}{dt} = \frac{q}{m}(\mathbf{v}_\perp \times \mathbf{B}) + \frac{\mathbf{F}_\perp}{m} \qquad (2\text{-}19)$$

Equation (2-18) represents a constant acceleration along the field line. The
external force term in (2-19) results in a drift velocity perpendicular to both
the magnetic field and $\mathbf{F}$. As shown in Fig. 2-3, the particle accelerated by

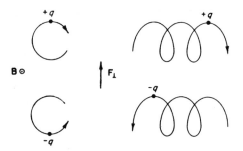

FIG. 2-3. A constant force $\mathbf{F}_\perp$ acting on a particle gyrating in a uniform magnetic field
results in a drift motion perpendicular to $\mathbf{F}_\perp$ and $\mathbf{B}$.

this force gains velocity, which in turn increases the radius of curvature
according to (2-8). Reaching the turning point, the particle moves backward
on a symmetrical path, now decelerated by the force, with decreasing radius
of curvature to the opposite turning point.

Equation (2-19) can be resolved by introducing the drift velocity $\mathbf{w}^D$ and
writing

$$\mathbf{v}_\perp = \mathbf{w}^D + \mathbf{u} \qquad (2\text{-}20)$$

We shall now show that a suitable choice of the drift velocity, namely,

$$\mathbf{w}^D = \frac{1}{q}\frac{\mathbf{F}_\perp \times \mathbf{B}}{B^2} \qquad (2\text{-}21)$$

"transforms away" the force term in (2-19). With the substitution of (2-20) and (2-21), (2-19) becomes

$$\frac{d\mathbf{u}}{dt} = \frac{1}{m}\left[\frac{(\mathbf{F}_\perp \times \mathbf{B}) \times \mathbf{B}}{B^2}\right] + \frac{q}{m}(\mathbf{u} \times \mathbf{B}) + \frac{\mathbf{F}_\perp}{m} \tag{2-22}$$

The first term on the right side can be written

$$\frac{(\mathbf{F}_\perp \times \mathbf{B}) \times \mathbf{B}}{B^2} = \frac{\mathbf{B}(\mathbf{F}_\perp \cdot \mathbf{B}) - \mathbf{F}_\perp(\mathbf{B} \cdot \mathbf{B})}{B^2} = -\mathbf{F}_\perp \tag{2-23}$$

Substitution in (2-22) leads to cancellation of the force term. The remainder is simply

$$\frac{d\mathbf{u}}{dt} = \frac{q}{m}(\mathbf{u} \times \mathbf{B}) \tag{2-24}$$

which means that the particle motion in a coordinate system moving with velocity $\mathbf{w}^D$ is governed entirely by the magnetic field and therefore moves on a circular path. The constant drift velocity superposed on this motion yields a *cycloid*, such as the one in Fig. 2-3. Note that the drift velocity depends on the particle charge.

We conclude by establishing the following rule: The motion of a charged particle in a uniform magnetic field, under the influence of an external force, can be described as the superposition of a gyration around the so-called *guiding center* with the cyclotron frequency and the motion of this guiding center. The guiding-center motion does not follow the laws of particle mechanics; it responds differently to the external force parallel and perpendicular to **B**. It is accelerated in the parallel direction according to (2-18), as if no magnetic field were present, while it drifts in the perpendicular plane with the constant velocity $\mathbf{w}^D$ as given by (2-21).

2-2. The Guiding-Center Approximation; Dipole-Like Motion

Using the conclusion of the previous section, we are now going to attack several more complicated cases. The combination of a homogeneous magnetic field and an electric field leads, for instance, to the additional force

$$\mathbf{F} = q\mathbf{E} \tag{2-25}$$

and the drift velocity

$$\mathbf{w}^E = \frac{\mathbf{E} \times \mathbf{B}}{B^2} \tag{2-26}$$

which is independent of the charge and mass as well. The electric drift is identical for every particle; consequently the electric field can be entirely transformed away. The special theory of relativity shows that this can indeed be done. The Lorentz transformation from a coordinate system, where a homogeneous electric and a magnetic field are present, to another one, which moves with velocity $\mathbf{w}^E$ ($w^E \ll c$) with respect to the first, transforms away the electric field, leaving the magnetic field unchanged (see Exercise 2-1).

A homogeneous gravitational field, however, with gravitational acceleration $\mathbf{g}$, gives rise to a force

$$\mathbf{F} = m\mathbf{g} \tag{2-27}$$

and the drift velocity

$$\mathbf{w}^g = \frac{m}{q} \frac{\mathbf{g} \times \mathbf{B}}{B^2} \tag{2-28}$$

depends on the m/q ratio. The gravitational field cannot, therefore—in this context—be transformed away.

An important example is the motion of a charged particle in a slightly inhomogeneous magnetic field, "slightly" meaning that the variation of the magnetic field inside the particle orbit is small compared to the magnitude of the field. If $\mathbf{B}_0$ is the field at the guiding center and $\mathbf{r}$ represents the momentary particle position in the guiding-center coordinate system, the magnetic field at the particle can be expressed by the Taylor expansion,

$$\mathbf{B(r)} = \mathbf{B}_0 + (\mathbf{r} \cdot \nabla_0)\mathbf{B} + \cdots \tag{2-29}$$

where ∇_0 means that the differentiation is to be performed at point 0. [Actually the momentary guiding center also varies slightly during a gyration period, while we hold point 0 fixed for this time (see Fig. 2-4)].

FIG. 2-4. Particle motion in a slightly nonuniform magnetic field.

In our case the higher-order terms can be neglected, and, in addition,

$$|\mathbf{B}_0| \gg |(\mathbf{r} \cdot \nabla_0)\mathbf{B}| \tag{2-30}$$

is true; thus the magnetic field "felt" by the particle differs but little from that prevailing at the guiding center. The particle path differs little, therefore, from a helix corresponding to $\mathbf{B} = \text{const}$, or in the case of $v_\parallel = 0$ from a circle. The latter case will be assumed in the following calculation.

In this approximation the motion of the particle can be described in the following way. The ring current represented by its dipole moment is located in an inhomogeneous magnetic field. The force exerted by the field inhomogeneity makes the particle drift as described in the previous section. The equation of motion (2-1) becomes, when inserting (2-29),

$$\frac{d\mathbf{v}}{dt} = \frac{q}{m} \{[\mathbf{v} \times \mathbf{B}_0] + [\mathbf{v} \times (\mathbf{r} \cdot \nabla_0)\mathbf{B}]\} \tag{2-31}$$

As the last term is a small first-order one compared to the first zero-order term, we write the velocity as a superposition:

$$\mathbf{v} = \mathbf{v}_0 + \mathbf{v}_1 \tag{2-32}$$

where $\mathbf{v}_0$ is the solution of the zero-order equation

$$\frac{d\mathbf{v}}{dt} = \frac{q}{m} [\mathbf{v} \times \mathbf{B}_0] \tag{2-33}$$

and $\mathbf{v}_1$ is a perturbation of the first order. Since we neglect second-order terms, the last term of (2-31) can be written in the form

$$\frac{q}{m} [\mathbf{v}_0 \times (\mathbf{r} \cdot \nabla_0)\mathbf{B}] \tag{2-34}$$

Similarly, up to the first order we can put $\mathbf{r} = \mathbf{R}$ (the cyclotron radius corresponding to $\mathbf{B}_0$), and (2-31) becomes

$$\frac{d\mathbf{v}}{dt} = \frac{q}{m} (\mathbf{v} \times \mathbf{B}_0) + \frac{q}{m} [\mathbf{v}_0 \times (\mathbf{R} \cdot \nabla_0)\mathbf{B}] \tag{2-35}$$

The second right-hand term constitutes the external force of (2-17). It is, however, not a constant, since it depends on the momentary particle position $\mathbf{R}$. We calculate the average of this quantity over a gyration period

$$\mathbf{F} = \langle q\mathbf{v}_0 \times (\mathbf{R} \cdot \nabla_0)\mathbf{B}\rangle \tag{2-36}$$

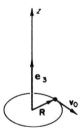

Fig. 2-5. Illustration to aid computation of $\mathbf{F}$.

In a local cylindrical coordinate system with the z coordinate pointing in the $\mathbf{B}_0$ direction (Fig. 2-5), (2-36) becomes

$$\mathbf{F} = \left\langle q\mathbf{v}_0 \times R \frac{\partial_0}{\partial r} \mathbf{B} \right\rangle \tag{2-37}$$

where ∂_0 again means that the differentiation has to be performed at point 0. Of the three components of $\mathbf{B}$, B_θ is parallel to $\mathbf{v}_0$ and therefore gives no contribution to $\mathbf{F}$, while B_r contributes to $\mathbf{F}_\parallel$ and B_z to $\mathbf{F}_\perp$:

$$\mathbf{F}_\parallel = \left\langle q\mathbf{v} \times \mathbf{R} \frac{\partial B_r}{\partial r} \right\rangle \tag{2-38}$$

and

$$\mathbf{F}_\perp = \left\langle q\mathbf{v} \times \mathbf{R} \frac{\partial B_z}{\partial r} \mathbf{e}_3 \right\rangle \tag{2-39}$$

where $\mathbf{e}_3$ is the unit vector in the z direction, and the 0 index has been omitted.

Observing that $\mathbf{v} \times \mathbf{R}$ is a constant vector pointing in the z direction, (2-38) becomes

$$\mathbf{F}_\parallel = q\mathbf{v} \times \mathbf{R} \left\langle \frac{\partial B_r}{\partial r} \right\rangle = 2\mu_m \left\langle \frac{\partial B_r}{\partial r} \right\rangle \mathbf{e}_3 \tag{2-40}$$

where (2-16) has been used. From $\nabla \cdot \mathbf{B} = 0$ follows

$$\frac{\partial B_r}{\partial r} + \frac{B_r}{r} + \frac{\partial B_z}{\partial z} + \frac{1}{r}\frac{\partial B_\theta}{\partial \theta} = 0 \tag{2-41}$$

By averaging over a gyration period the last term vanishes:

$$\left\langle \frac{1}{r}\frac{\partial B_\theta}{\partial \theta} \right\rangle = \frac{1}{2\pi r}\oint \frac{1}{r}\frac{\partial B_\theta}{\partial \theta}\, r\, d\theta = 0 \tag{2-42}$$

since B is single-valued. Furthermore, in the $r \to 0$ limit

$$\lim_{r \to 0} \frac{B_r}{r} = \frac{\partial B_r}{\partial r} \tag{2-43}$$

since at $r = 0$, $B_r = 0$. By inserting in (2-41), this leads to

$$2\left\langle \frac{\partial B_r}{\partial r} \right\rangle = -\left\langle \frac{\partial B_z}{\partial z} \right\rangle = -\left(\frac{\partial B_z}{\partial z} \right) \tag{2-44}$$

The parallel force becomes, therefore,

$$\mathbf{F}_\parallel = -\mu_m \left(\frac{\partial B_z}{\partial z} \right) \mathbf{e}_3 \tag{2-45}$$

Equation (2-45) can be put into a coordinate-independent form,

$$\mathbf{F}_\parallel = -\frac{\mu_m}{B}\left[(\mathbf{B} \cdot \nabla)\mathbf{B} \right]_\parallel \tag{2-46}$$

The evaluation of $\mathbf{F}_\perp$ also requires some computation. The vectors $\mathbf{v}_0$, $\mathbf{R}$, and $\mathbf{e}_3$ form a rectangular coordinate system with $\mathbf{v}_0 \times \mathbf{e}_3 = -(\mathbf{R}/R)v_0$

(Fig. 2-5). Equation (2-39) may therefore be written

$$\mathbf{F}_\perp = -qv\left\langle \mathbf{R}\,\frac{\partial B_z}{\partial r}\right\rangle \tag{2-47}$$

For the evaluation of the $\langle\ \rangle$ we set up a two-dimensional Cartesian system in the perpendicular plane (Fig. 2-6) with

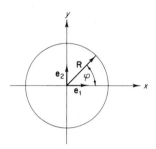

FIG. 2-6. Illustration to aid evaluation of $\mathbf{F}_\perp$.

$$\mathbf{R} = R\cos\varphi\,\mathbf{e}_1 + R\sin\varphi\,\mathbf{e}_2 \tag{2-48}$$

$$x = r\cos\varphi, \qquad y = r\sin\varphi \tag{2-49}$$

and

$$\frac{\partial}{\partial r} = \frac{dx}{dr}\frac{\partial}{\partial x} + \frac{dy}{dr}\frac{\partial}{\partial y} = \cos\varphi\,\frac{\partial}{\partial x} + \sin\varphi\,\frac{\partial}{\partial y} \tag{2-50}$$

For convenience the coordinate system can be chosen so as to make, for example, $\partial B_z/\partial y = 0$. Taking into account that $\langle\sin\varphi\cos\varphi\rangle = 0$ and $\langle\cos^2\varphi\rangle = \frac{1}{2}$, (2-47) becomes

$$\mathbf{F}_\perp = -qvR\mathbf{e}_1\langle\cos^2\varphi\rangle\frac{\partial B_z}{\partial x} = -\frac{qvR}{2}\frac{\partial B_z}{\partial x}\mathbf{e}_1 = -\mu_m\frac{\partial B_z}{\partial x}\mathbf{e}_1 \tag{2-51}$$

Since $(B_x)_0 = (B_y)_0 = 0$ one may write

$$\frac{\partial B_z}{\partial x} = \frac{1}{2B_z}\frac{\partial B_z{}^2}{\partial x} = \frac{1}{2B}\frac{\partial B^2}{\partial x} \tag{2-52}$$

Equation (2-51) therefore finally reduces to

$$\mathbf{F}_\perp = -\frac{\mu_m}{2B}\nabla_\perp B^2 \tag{2-53}$$

The vector identity [from (A-12)]

$$(\nabla\times\mathbf{B})\times\mathbf{B} = (\mathbf{B}\cdot\nabla)\mathbf{B} - \nabla\left(\frac{B^2}{2}\right) \tag{2-54}$$

yields, for the parallel components,

$$[(\mathbf{B} \cdot \nabla)\mathbf{B}]_{\parallel} = \left[\nabla \frac{B^2}{2}\right]_{\parallel} \tag{2-55}$$

The force components $\mathbf{F}_{\parallel}$ and $\mathbf{F}_{\perp}$ can thus be reduced to

$$\mathbf{F} = -\frac{\mu_m}{B} \nabla \frac{B^2}{2} = -\mu_m \nabla B \tag{2-56}$$

or, using (2-54), the force can be rewritten in terms of the magnetic moment,

$$\mathbf{F} = (\nabla \times \mathbf{B}) \times \frac{\mathbf{B}}{B} \mu_m - \left(\frac{\mu_m}{B} \mathbf{B} \cdot \nabla\right)\mathbf{B}$$

$$= \boldsymbol{\mu}_m \times (\nabla \times \mathbf{B}) + (\boldsymbol{\mu}_m \cdot \nabla)\mathbf{B} \tag{2-57}$$

the usual expression for the force acting on a small ring current placed in an inhomogeneous magnetic field. Note that it differs in the $\boldsymbol{\mu}_m \times (\nabla \times \mathbf{B})$ term from the force acting on a dipole.

As a result of $\mathbf{F}_{\parallel}$, the guiding center is accelerated according to (2-18):

$$m \frac{d\mathbf{v}_{\parallel}}{dt} = -\frac{\mu_m}{B}[(\mathbf{B} \cdot \nabla)\mathbf{B}]_{\parallel} = -\frac{\mu_m}{B}\left[\nabla \frac{B^2}{2}\right]_{\parallel} \tag{2-58}$$

The force and the acceleration point in the direction of decreasing field strength, as expected from a diamagnetic dipole. In the special case of the so-called *mirror field* of Fig. 2-7, an axially symmetric magnetic field con-

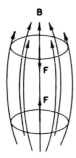

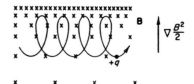

FIG. 2-7. Forces acting on the guiding center in a mirror field.

FIG. 2-8. Particle drift due to the field gradient perpendicular to $\mathbf{B}$.

figuration with lines of force constricting at two points along the axis, the particle is reflected from both mirrors and is trapped in the central region. $\mathbf{F}_{\perp}$ causes the guiding center to drift, with the velocity

$$\mathbf{w}^B = -\frac{1}{q} \frac{\mu_m}{B} \frac{\nabla(B^2/2) \times \mathbf{B}}{B^2} \tag{2-59}$$

This drift is perpendicular to the field and to the field gradient (Fig. 2-8). Note that this drift is charge-dependent, unlike the electrical drift. Positive and negative particles therefore drift in opposite directions, giving rise to a current.

2-3. The Guiding-Center Approximation; Inertial Forces

The arbitrary assumption of $v_{\parallel} = 0$ has been made in the previous calculation. In case of a uniform magnetic field this choice was justified by introducing a coordinate system moving with the guiding center with velocity $v_{\parallel}$.

This procedure can be followed again for more general fields with two conditions kept in mind:

1. The guiding-center approximation is justified only if the conditions of its applicability, e.g., (2-30), are now valid in the moving system. In addition, it must be considered that the coordinate system moving with velocity $v_{\parallel}$ in the laboratory system finds itself in varying fields on its journey. It is required, therefore, that the rate of change of the field should be small during a gyration period as measured in the moving system.

2. It should also be considered that a coordinate system gliding along a line of force is not necessarily an inertial system. Bent field lines give rise to inertial forces.

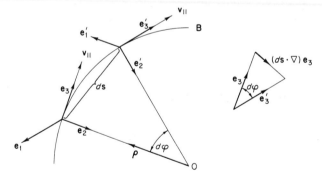

FIG. 2-9. Calculation of the centrifugal drift for a particle gyrating about a curved field line.

The latter can be derived from the equation of motion introducing the guiding-center coordinate system with unit vectors e_i (Fig. 2-9) gliding along a line of force with the velocity $w_{\parallel}$. The particle velocity in the rest frame may be written

$$\mathbf{v} = \mathbf{e}_i v_i \qquad (2\text{-}60)$$

where the usual summation convention over common indices has been adopted. The left side of the equation of motion (2-17) now becomes

$$\frac{d\mathbf{v}}{dt} = \mathbf{e}_i \frac{dv_i}{dt} + v_i \frac{d\mathbf{e}_i}{dt} = \mathbf{e}_i \frac{dv_i}{dt} + v_i(\mathbf{w}_\| \cdot \nabla)\mathbf{e}_i \tag{2-61}$$

The first right-hand term contains the usual parallel and perpendicular accelerations in the moving system, while the second term represents the centripetal acceleration seen in the moving system. The latter term when transferred to the right side of the equation of motion (2-17) gives rise to a centrifugal force

$$\mathbf{F}^c = -mv_i(\mathbf{w}_\| \cdot \nabla)\mathbf{e}_i \tag{2-62}$$

$(\mathbf{w}_\| \cdot \nabla)\mathbf{e}_i$ is a constant vector at each point, while v_1 and v_2 oscillate because of the gyration of $\mathbf{v}$. Hence, the time average coming from $i = 1$ and $i = 2$ over a gyration period vanishes, while the third component yields

$$\mathbf{F}^c = -mw_\|(\mathbf{w}_\| \cdot \nabla)\mathbf{e}_3 = -mw_\|^2(\mathbf{e}_3 \cdot \nabla)\mathbf{e}_3 \tag{2-63}$$

From similar triangles, $ds/\rho = ds(\mathbf{e}_3 \cdot \nabla)\mathbf{e}_3/1$; hence $\mathbf{F}^c$ is indeed the centrifugal force

$$\mathbf{F}^c = mw_\|^2 \frac{\boldsymbol{\rho}}{\rho^2} \tag{2-64}$$

where ρ is the local radius of curvature of the field line. $\mathbf{F}^c$ can also be expressed in terms of the magnetic field from (2-63):

$$\mathbf{F}^c = -\frac{mw_\|^2}{B^2}[(\mathbf{B} \cdot \nabla)\mathbf{B}]_\perp \tag{2-65}$$

since $\mathbf{B} = B\mathbf{e}_3$.

$\mathbf{F}^c$ is obviously a perpendicular vector, and it gives rise to the centrifugal drift

$$\mathbf{w}^c = \frac{m}{q} \frac{w_\|^2}{B^4}[\mathbf{B} \times (\mathbf{B} \cdot \nabla)\mathbf{B}] \tag{2-66}$$

which lies in the perpendicular plane.

If volume currents are not present, $\nabla \times \mathbf{B} = 0$ in (2-54) and the two different drifts expressed in (2-59) and (2-66) add up to

$$\mathbf{w} = \frac{m}{qB^4}\left(w_\|^2 + \frac{v_\perp^2}{2}\right)\left[\mathbf{B} \times \nabla \frac{B^2}{2}\right] \tag{2-67}$$

A more general problem can now be attacked—the motion of a charged particle under the combined influence of inhomogeneous and time-dependent magnetic and electric fields. For the applicability of the guiding-center approximation, the former requirements should hold again in the moving guiding-center coordinate system.

It is required, furthermore, that the rate of change of any field quantity and its gradient, measured in the guiding-center system, should be small during a gyration period. The change of any field quantity $\mathbf{A}$ in this system is due to the time variation of fields and to the motion of the system through inhomogeneous fields. The inequality required can therefore be written

$$\left|\frac{d_0\mathbf{A}}{dt}\right| = \left|\frac{\partial_0\mathbf{A}}{\partial t} + (\mathbf{w}_0 \cdot \nabla_0)\mathbf{A}\right| \ll |\omega_c \mathbf{A}_0| \tag{2-68}$$

where subscripts refer again to the guiding center and ω_c is the cyclotron frequency.

The equation of motion is, again,

$$\frac{d\mathbf{v}}{dt} = \frac{q}{m}(\mathbf{E} + \mathbf{v} \times \mathbf{B}) \tag{2-69}$$

The zero-order motion of the guiding center is given by

$$\mathbf{w}^0 = \mathbf{w}^E + \mathbf{w}_\| \tag{2-70}$$

since all perpendicular drifts—with the exception of $\mathbf{w}^E$—depend on the space or time variation of the fields and consequently are of higher order. One may write

$$\mathbf{v} = \mathbf{w}^0 + \mathbf{u} \tag{2-71}$$

where $\mathbf{u}$ includes the gyration about the guiding center as well as the first-order drifts. The equation of motion now becomes

$$\frac{d\mathbf{u}}{dt} = \frac{q}{m}(\mathbf{u} \times \mathbf{B}) + \frac{q}{m}\mathbf{E}_\| - \frac{d\mathbf{w}^0}{dt} \tag{2-72}$$

since $\mathbf{w}_\| \times \mathbf{B} = 0$ and $\mathbf{w}^E \times \mathbf{B} = -\mathbf{E}_\perp$. The first two terms of this equation represent a particle gyrating in a slightly nonuniform magnetic field, measured in a frame moving with velocity $\mathbf{w}^0$. In addition, one finds two force terms: a parallel electric force accelerating the guiding center along the magnetic field, and an inertial force which gives rise to the inertial drift

$$\mathbf{w}^i = -\frac{m}{qB^2}\left(\frac{d\mathbf{w}^0}{dt} \times \mathbf{B}\right) \tag{2-73}$$

This includes, of course, the centrifugal drift (2-66). (The parallel part of the inertial force serves only to balance the parallel force terms in the accelerated frame.)

Our results can be summarized as follows: The guiding center of a particle

placed in magnetic and electric fields is accelerated along the magnetic field
line as given by

$$\left(\frac{d\mathbf{w}}{dt}\right)_{\parallel} = \frac{q}{m}\,\mathbf{E}_{\parallel} - \frac{\mu_m}{mB}\,[(\mathbf{B}\cdot\nabla)\mathbf{B}]_{\parallel} \tag{2-74}$$

The guiding center drifts in the perpendicular direction as

$$\mathbf{w}_{\perp} = \mathbf{w}^E + \frac{\mu_m}{qB^3}\left[\mathbf{B}\times\nabla\frac{B^2}{2}\right] + \frac{m}{qB^2}\,[\mathbf{B}\times\dot{\mathbf{w}}^0] \tag{2-75}$$

where the first term is the dominant zero-order quantity.

Equation (2-74) is often expressed in terms of $w_{\parallel}$, the guiding-center speed
in the parallel direction, from

$$\frac{dw_{\parallel}}{dt} = \frac{d}{dt}(\mathbf{w}\cdot\mathbf{e}_3) = (\dot{\mathbf{w}})_{\parallel} + \mathbf{w}\cdot\dot{\mathbf{e}}_3 \tag{2-76}$$

where $\mathbf{e}_3 = \mathbf{B}/B$. When neglecting higher-order terms, the last term can be
written as

$$(\mathbf{w}^E + \mathbf{w}_{\parallel})\cdot\dot{\mathbf{e}}_3 = \mathbf{w}^E\cdot\dot{\mathbf{e}}_3$$

$$= \mathbf{w}^E\cdot\left[\frac{\partial\mathbf{e}_3}{\partial t} + (\mathbf{w}^E + \mathbf{w}_{\parallel})\cdot\nabla\mathbf{e}_3\right] \tag{2-77}$$

Inserting (2-76) and (2-77) into (2-74) leads to the alternative form of (2-74),

$$\frac{dw_{\parallel}}{dt} = \frac{q}{m}\,E_{\parallel} - \frac{\mu_m}{mB}\,[(\mathbf{B}\cdot\nabla)\mathbf{B}]_{\parallel} + \mathbf{w}^E\cdot\left[\frac{\partial\mathbf{e}_3}{\partial t} + (\mathbf{w}^E + \mathbf{w}_{\parallel})\cdot\nabla\mathbf{e}_3\right] \tag{2-78}$$

If the initial velocity of the particle motion and the space and time variation
of the electric and magnetic field are given, (2-74) and (2-75) can be integrated,
provided that the variation of μ_m is known. It will be shown that in the
approximation considered, μ_m is a constant of motion.

2-4. The Guiding-Center Approximation; Constancy of the Magnetic Moment

We have not yet considered the motion of the particle in the guiding-
center coordinate system. Although it was assumed that in this approximation
the particle path in the guiding-center system is nearly circular, nothing has
been said about the variation of the radius R in times much longer than the
gyration period. Instead of considering the change in the cyclotron radius,
we can just as well describe the change of the magnetic moment μ_m, which

by knowing the magnetic field $\mathbf{B}(\mathbf{r},t)$ gives a complete description of the particle orbit up to second-order terms. Furthermore, following (2-74) and (2-75), knowledge of the variation of μ_m is required for the solution of the equations of motion for the guiding center.

Multiplying the particle equation of motion by $\mathbf{v}$, the rate of change of particle energy becomes

$$\frac{d}{dt}\left(\tfrac{1}{2}mv^2\right) = q\mathbf{v} \cdot \mathbf{E} \tag{2-79}$$

The particle velocity can be represented as the sum of the guiding-center velocity and gyration velocity:

$$\mathbf{v} = \mathbf{w} + \mathbf{v}_\perp \tag{2-80}$$

Squaring (2-80) yields

$$v^2 = w^2 + v_\perp{}^2 + 2\mathbf{w} \cdot \mathbf{v}_\perp \tag{2-81}$$

Since $\mathbf{w}$ is a slow variable and hence nearly a constant vector during a gyration period, while $\mathbf{v}_\perp$ rotates, the last quantity on the right side is (after averaging) of the first order, and its time derivative can be neglected as a second-order term. The rate of change of the gyration kinetic energy now becomes from (2-79) to (2-81),

$$\frac{d}{dt}\left(\tfrac{1}{2}mv_\perp{}^2\right) = q\langle \mathbf{v}_\perp \cdot \mathbf{E}\rangle + q\mathbf{w} \cdot \mathbf{E} - \frac{d}{dt}\left(\tfrac{1}{2}mw^2\right) \tag{2-82}$$

The first right-hand term is the increment of the work performed on the gyrating particle by the electric field. This can be averaged over the gyration period,

$$\langle q\mathbf{v}_\perp \cdot \mathbf{E}\rangle = \frac{q}{T}\int \mathbf{E} \cdot \mathbf{v}_\perp \, dt = \frac{q}{T}\oint \mathbf{E} \cdot d\mathbf{s} \tag{2-83}$$

where $T = 2\pi/\omega_c$ and $d\mathbf{s} = \mathbf{v}_\perp \, dt$ is the line element of a circle of radius R and not of the actual particle path (see Fig. 2-10). Although $\mathbf{E}$ is the value

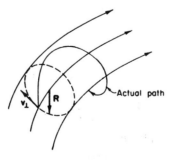

Fig. 2-10. Particle moving in an electromagnetic field. The dominant motion is gyration around the guiding center.

of the electric field the particle encounters on its actual path at time t, one may convince oneself with the help of (2-68) that in our approximation this can be replaced by the corresponding **E** value on the circle. Using the induction law and the definition of μ_m one obtains

$$\langle q\mathbf{v}_\perp \cdot \mathbf{E}\rangle = \frac{q}{2\pi}\,\omega_c R^2 \pi \frac{\partial B}{\partial t} = \tfrac{1}{2}qv_\perp R \frac{\partial B}{\partial t} = \mu_m \frac{\partial B}{\partial t} \tag{2-84}$$

With help of (2-84), (2-82) yields

$$\frac{d}{dt}(\tfrac{1}{2}mv_\perp{}^2) = \mu_m \frac{\partial B}{\partial t} + \mathbf{w}\cdot(q\mathbf{E} - m\dot{\mathbf{w}}) \tag{2-85}$$

The last term can be evaluated by using (2-75),

$$\mathbf{w}\cdot(q\mathbf{E} - m\dot{\mathbf{w}})$$

$$= \left[\frac{\mathbf{E}\times\mathbf{B}}{B^2} + \frac{\mu_m}{qB^3}\left(\mathbf{B}\times\nabla\frac{B^2}{2}\right) + \frac{m}{qB^2}(\mathbf{B}\times\dot{\mathbf{w}}) + \mathbf{w}_\parallel\right]\cdot[q\mathbf{E} - m\dot{\mathbf{w}}] \tag{2-86}$$

Performing the multiplications, two terms vanish because of orthogonality and two others cancel, reducing (2-86) to

$$\frac{\mu_m}{qB^3}\left[\mathbf{B}\times\nabla\frac{B^2}{2} + \frac{qB^3}{\mu_m}\mathbf{w}_\parallel\right]\cdot[q\mathbf{E} - m\dot{\mathbf{w}}]$$

$$= \frac{\mu_m}{B}\frac{\mathbf{E}\times\mathbf{B}}{B^2}\cdot\nabla\frac{B^2}{2} + \frac{m\mu_m}{qB^3}(\mathbf{B}\times\dot{\mathbf{w}})\cdot\nabla\frac{B^2}{2} + \frac{\mu_m}{B}\mathbf{w}_\parallel\cdot(\mathbf{B}\cdot\nabla)\mathbf{B} \tag{2-87}$$

where triple products have been regrouped, and in the last term use has been made of (2-74).

Owing to (2-55),

$$\frac{\mathbf{w}_\parallel}{B}\cdot(\mathbf{B}\cdot\nabla)\mathbf{B} = \frac{\mathbf{w}_\parallel}{B}\cdot\nabla\frac{B^2}{2} = \mathbf{w}_\parallel\cdot\nabla B \tag{2-88}$$

The right side of (2-87) now reduces to

$$\mu_m\left[\mathbf{w}^E + \frac{m}{qB^2}(\mathbf{B}\times\dot{\mathbf{w}}) + \mathbf{w}_\parallel\right]\cdot\nabla B = \mu_m\mathbf{w}\cdot\nabla B \tag{2-89}$$

where use has been made of (2-75) and the identity

$$\left(\mathbf{B}\times\nabla\frac{B^2}{2}\right)\cdot\nabla B = 0 \tag{2-90}$$

Equation (2-85) becomes, finally,

$$\frac{d}{dt}(\tfrac{1}{2}mv_\perp{}^2) = \mu_m\left[\frac{\partial B}{\partial t} + (\mathbf{w}\cdot\nabla)B\right] = \mu_m \frac{dB}{dt} \tag{2-91}$$

From the definition of the magnetic moment (2-14), however,

$$\frac{d}{dt}(\tfrac{1}{2}mv_\perp^2) = \frac{d}{dt}(B\mu_m) = \mu_m\frac{dB}{dt} + B\frac{d\mu_m}{dt} \tag{2-92}$$

From comparison of the last two equations follows the important result

$$\boxed{\frac{d\mu_m}{dt} = 0} \tag{2-93}$$

The magnetic moment of the particle in the guiding-center system is a constant of motion (within the approximation used).

The perpendicular component of the initial velocity vector yields the constant μ_m immediately, which inserted in (2-74) and (2-75) renders them soluble for $\mathbf{w}(t)$.

It should be mentioned that from the form of μ_m given in (2-13) the constancy of the magnetic flux enclosed by the particle path follows also. The magnetic flux is "frozen into" the particle path. We shall frequently meet similar rules later on.

The constancy of the magnetic moment enables one to describe the motion of the guiding center with the help of the following model. Consider a particle of charge q, mass m, and magnetic moment μ_m. If placed in a non-uniform magnetic field, such a particle experiences the force $\mathbf{F} = -\mu_m \nabla B$. As a result of this force, the particle is accelerated along the field line with the acceleration $\dot{v}_\parallel = F_\parallel/m$, while it drifts across the field with the velocity $q^{-1}\mathbf{F}_\perp \times \mathbf{B}/B^2$. This is also the motion of the guiding center of a particle with the same charge and mass and orbital magnetic moment μ_m. In the description of a plasma it is sometimes sufficient to follow the motion of the guiding centers of the particles. This is often called a *guiding-center plasma.*

Constants of motion are very useful tools in particle kinetics. We illustrate this in two examples.

From (2-91) and (2-93) it follows that the kinetic energy of the gyration is proportional to the magnetic field. If, for example, the field increases in time, or (with the help of electric fields) we force the particle to drift into a stronger magnetic field, this energy part increases accordingly. This sometimes means only a transformation of "drift energy" into gyration energy, but an increase of the total kinetic energy can also frequently be achieved (e.g., by increasing a homogeneous magnetic field).

As another example we discuss the mirror geometry (Fig. 2-11) as a means of confining charged particles. For the construction of the mirror field, the knowledge of the turning point is desirable. This could be obtained by

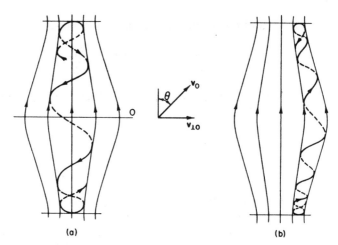

FIG. 2-11. Particle motion in a mirror geometry: (a) guiding center on axis, (b) guiding center off axis.

solving the equations of motion for the guiding center,

$$\frac{d\mathbf{w}}{dt} = -\frac{\mu_m}{mB}\left[(\mathbf{B}\cdot\nabla)\mathbf{B}\right]_{\parallel} \tag{2-94}$$

and

$$\mathbf{w}_\perp = 0 \tag{2-95}$$

if the guiding center is on the axis.

Instead of doing this, we can simply write down the constants of motion for the symmetry plane and a turning point. From the constancy of the magnetic moment,

$$\left(\frac{\frac{1}{2}mv_\perp{}^2}{B}\right)_0 = \left(\frac{\frac{1}{2}mv_\perp{}^2}{B}\right)_{\text{turn}} \tag{2-96}$$

Another constant of motion for a static magnetic field is the kinetic energy,

$$[\tfrac{1}{2}mv^2]_0 = [\tfrac{1}{2}mv^2]_{\text{turn}} \tag{2-97}$$

Denoting the angle between the velocity vector and the field by θ, one obtains

$$[\sin\theta]_0 = \left[\frac{v_\perp}{v}\right]_0 \tag{2-98}$$

and

$$[\sin\theta]_{\text{turn}} = 1 \tag{2-99}$$

The combination of the last four equations leads to

$$[\sin^2\theta]_0 = \frac{B_0}{B_{\text{turn}}} \tag{2-100}$$

Hence all particles with the same θ at the symmetry plane will be reflected from the same B_{turn} field, independent of their velocity. The ratio of the maximum field B_{max} to B_0 (for a given mirror geometry) is usually called the *mirror ratio*. Obviously only particles with $B_{turn} \leqq B_{max}$ will be contained. For the critical angle one obtains

$$[\sin^2\theta_{crit}]_0 = \frac{B_0}{B_{max}} \tag{2-101}$$

Particles with $\theta_0 < \theta_{crit}$ will escape.

The motion of particles with guiding centers outside the symmetry axis is somewhat more complicated (Fig. 2-11b). The $\mathbf{w}_\perp$ drift component is now

$$\mathbf{w}_\perp = \frac{\mu_m}{qB^3}\left[\mathbf{B} \times \nabla \frac{B^2}{2}\right] + \frac{m}{qB^2}[\mathbf{B} \times \dot{\mathbf{w}}^0] \tag{2-102}$$

For an axially symmetric field, $\mathbf{B}$, $\nabla(B^2/2)$ and $\mathbf{w}^0$ lie in the rz plane; hence $\mathbf{w}_\perp$ is an azimuthal vector which causes the guiding center to rotate slowly around the symmetry axis, while (2-100) still holds for the reflection from a "mirror."

2-5. Adiabatic Invariants

We found that the magnetic moment of a particle in a nearly uniform, slowly varying magnetic field is a constant of motion in our approximation. Such quantities, where the degree of "constancy" depends on restriction such as the "slowness" of the variation of some parameters of the motion, are the *adiabatic invariants*.

The classical example of an adiabatic invariant is due to Lorentz and Einstein, who considered the behavior of a pendulum under the slow variation of the length of its suspending thread. It turned out that if the variation of the length of the thread is slow enough (the pendulum swings many times before an appreciable change in the frequency v takes place), the ratio W/v is a constant, where W is the energy of the pendulum. This has an important meaning in quantum theory, where $W = nhv$. In this case, the adiabatic invariant describes the constancy of the number of quanta associated with the oscillator.

Similarly we found that when field quantities as measured in the particle system are subject to certain restrictions—prescribing essentially slowness of variation—the magnetic moment

$$\mu_m = \frac{W_\perp}{B} = \frac{q}{m}\frac{W_\perp}{\omega_c} \tag{2-103}$$

or $W_\perp/\omega_c$ is a constant of motion.

Let us investigate now a somewhat more general problem, that of one-dimensional motion of a particle in the potential field $V(x)$, which is subject to slow variation in time (Fig. 2-12a). If the potential is held fixed, the

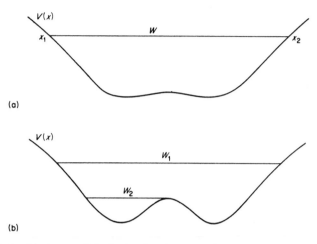

(a)

(b)

FIG. 2-12. Particle in an adiabatically changing potential well. In (b) the particle with the energy W_2 is no longer adiabatic.

equation of motion is

$$m \frac{d^2x}{dt^2} + \frac{dV}{dx} = 0 \tag{2-104}$$

which yields, after multiplication by dx/dt and integration, the energy integral

$$\frac{m}{2}\left(\frac{dx}{dt}\right)^2 + V(x) = W \tag{2-105}$$

The period of oscillation can be expressed from (2-105):

$$T = \oint dt = \oint \frac{dx}{[(2/m)(W - V)]^{1/2}} \tag{2-106}$$

We now define the action integral:

$$J = \oint m \frac{dx}{dt} \, dx = \oint [2m(W - V)]^{1/2} \, dx \tag{2-107}$$

To investigate the particle motion in a potential field which changes its shape slowly (compared to T), we express the potential as a function of a set of parameters λ_i in addition to x, making the replacement

$$V(x) \rightarrow V(x, \lambda_1 \cdots \lambda_n) \tag{2-108}$$

The λ's are slowly varying functions of time, changing the shape of the potential "slowly."

We will show that the action integral J is an adiabatic invariant. We are looking for the total time derivative

$$\frac{dJ}{dt} = \frac{d}{dt} 2 \int_{x_1}^{x_2} [2m(W - V)]^{1/2} \, dx \qquad (2\text{-}109)$$

where x_1 and x_2 are the turning points. Owing to the "slowness" of the process, we assume that W is still an integral of motion during a period of oscillation but that it varies slowly over many periods. This means that the λ's are fixed when evaluating the integral in (2-109), while the turning points and the integrand are time (and λ) dependents in the long run. Keeping this in mind we carry out the differentiation in (2-109):

$$\frac{1}{2} \frac{dJ}{dt} = [2m(W - V)]_{x_2}^{1/2} \frac{dx_2}{dt} - [2m(W - V)]_{x_1}^{1/2} \frac{dx_1}{dt}$$

$$+ \int_{x_1}^{x_2} \frac{\partial}{\partial t} [2m(W - V)]^{1/2} \, dx \qquad (2\text{-}110)$$

Since $W - V$ vanishes at both turning points, the first two terms on the right side are zero. The integrand in the third term can be evaluated:

$$\frac{\partial}{\partial t} [2m(W - V)]^{1/2} = \frac{1}{[(2/m)(W - V)]^{1/2}} \left(\frac{\partial T}{\partial t} - \frac{\partial W}{\partial t} \right) \qquad (2\text{-}111)$$

Furthermore,

$$\frac{\partial V}{\partial t} = \sum_i \frac{\partial V}{\partial \lambda_i} \frac{d\lambda_i}{dt} \qquad (2\text{-}112)$$

and using (2-105),

$$\frac{\partial W}{\partial t} = -\dot{x} \frac{\partial V}{\partial x} + \sum_i \frac{\partial V}{\partial \lambda_i} \frac{d\lambda_i}{dt} \qquad (2\text{-}113)$$

where use has been made of (2-104). Inserting (2-112) and (2-113) into (2-111) and using (2-105), one obtains

$$\frac{\partial}{\partial t} [2m(W - V)]^{1/2} = -\frac{\partial V}{\partial x} \qquad (2\text{-}114)$$

Equation (2-110) finally yields the result

$$\frac{dJ}{dt} = -\oint \frac{\partial V}{\partial x} \, dx = 0 \qquad (2\text{-}115)$$

Note that in addition to the "slowness" assumption, we tacitly assumed during the derivation that the turning points vary continuously. For instance, an emergence of a potential peak, as shown in Fig. 2-12b, trapping the particle at a certain stage of the process on the left or right side, must be excluded, however slow the time variation of V. Similarly, the inverse process, which leads to escaping particles from a region of previous confinement, cannot be considered either.

An alternative derivation of the adiabatic invariant property of the action integral illustrates its physical meaning from a statistical viewpoint. Instead of looking at a single system, let us consider now a great number of systems—an ensemble—each one consisting of the same varying potential $V(Q, \lambda_1 \cdots \lambda_n)$ but containing a particle with different energy and phase. Q is the generalized coordinate, while the canonical momentum is P. We represent this ensemble in the two-dimensional QP phase space, where the members of the ensemble populate a certain area (Fig. 2-13). Holding the

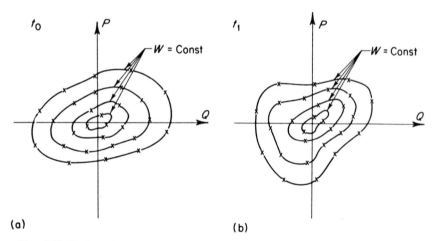

FIG. 2-13. Trajectories in the phase plane: (a) at t_0, (b) after adiabatic deformation of the potential well.

λ's fixed for the moment, and considering again only periodic motion, each point in phase space moves around on a closed $W = \text{const}$ line. The variation of V as a function of the λ's deforms the $W = \text{const}$ lines of Fig. 2-13a into those of Fig. 2-13b, but because of the "slowness" of the process, points lying originally on a $W(P,Q) = \text{const}$ line and differing in phase will always have nearly equal energy, although this varies in time. Furthermore, while the deformation of V deforms the family of the closed $W = \text{const}$ lines (nested one inside the other), no "crossing" of lines occurs. This follows from the unambiguity of the equations of motion, that is, given initial values of Q and P determine the motion. Hence, even after distorting the curves (slowly), they remain nested in the initial order.

On the other hand, according to Liouville's equation, the points in phase space behave as an incompressible fluid: they occupy a constant area in the course of their motion. This can be applied, of course, to phase points nested inside the energy curve of any given system of the ensemble. Consequently the phase surface ·

$$\oint P \, dQ = J \qquad (2\text{-}116)$$

corresponding to any one of the systems is a constant of motion.

We have arrived again at the constancy of the action variable, but now in a somewhat generalized form, as Q and P are general (one-dimensional) coordinates. The trapping and escaping of particles, resulting from breaking up and opening phase curves, must, of course, again be avoided.

We are now in a position to look at some applications. For the case of the pendulum or harmonic oscillator with slowly varying parameters, it is easy to show that the constancy of J is really equivalent to the constancy of W/ω (see Exercise 2-7).

We have seen that a charged particle moving in a slowly varying (in space and time) magnetic field moves nearly in a circle if viewed from the proper coordinate system. The circular motion can be regarded as a superposition of two linear harmonic oscillators, while the field variations at the momentary particle position can be described with the help of the slowly varying λ's. The adiabatic invariant of the oscillators becomes the well-known magnetic moment invariant of the gyrating particle as seen from (2-103).

The constancy of the magnetic moment enables one to derive the second or "longitudinal" adiabatic invariant. Let us consider a particle traveling between two magnetic mirrors. In addition to the gyration around the guiding center, the particle will oscillate between the mirrors. In the absence of parallel electric fields, the parallel component of the guiding-center velocity obeys the equation of motion (2-74):

$$m \frac{dv_\parallel}{dt} + \frac{\partial}{\partial s} (\mu_m B) = 0 \qquad (2\text{-}117)$$

where $\partial/\partial s$ denotes differentiation along a field line. With

$$V(s) = \mu_m B \qquad (2\text{-}118)$$

(2-117) reduces to (2-104). If there is a change in the fields, much slower than the oscillation period between the mirrors, the action variable

$$J = \oint [2m(\mu_m B_{\text{turn}} - \mu_m B)]^{1/2} \, ds \qquad (2\text{-}119a)$$

or

$$\oint [B_{\text{turn}} - B]^{1/2}\, ds = \text{invariant} \qquad (2\text{-}119\text{b})$$

Since the variation of B is known, this equation yields the turning points on the field lines (see, e.g., Exercise 2-10).

The slow variation in the field quantities can be caused either by the particle drifting to neighboring field lines in a static field or by slight time variation of the field itself. Consider, for example, a particle trapped between two mirrors, and increase the magnetic field in time. We have seen previously that the constancy of μ_m results in an increase of the perpendicular kinetic energy proportional to B. The longitudinal invariant can also be written

$$\oint v_{\parallel}\, ds = \text{invariant} \qquad (2\text{-}119\text{c})$$

It follows that by reducing the length of the confinement region the particle gains longitudinal kinetic energy. This corresponds to a heating effect by longitudinal magnetic compression.

The constancy of both the magnetic moment and the longitudinal invariant can be used to arrive at the third or "flux" invariant. Consider a static azimuthally symmetric mirror field. The guiding center of a particle oscillates between the mirrors, while the path drifts slowly around in the azimuthal direction, sweeping out a barrel-like surface. If the magnetic field deviates slightly from azimuthal symmetry (as, e.g., the earth's magnetic field does), or it varies slowly with time, this surface will not exactly close. We may ask what (besides μ_m and J remaining constant) happens to this surface during many azimuthal rotations of the particle path. It turns out that the magnetic flux enclosed is also an invariant. An example of this behavior is discussed in Exercise 2-11.

Note that the particle gyration around the guiding center is fast, the oscillation of the guiding center between the mirrors slower, and the rotation of the path the slowest process. Therefore, with given spatial and time variations μ_m is the "most" and the flux the "least" invariant quantity.

If in the Hamiltonian representation of a system (see the next section), all but one of the coordinates are almost "ignorable" and the remaining one is "almost periodic," the action variable is again an adiabatic invariant. For instance, the two-dimensional motion of a ball being reflected from the walls of a channel with slightly varying width (Fig. 2-14) can be treated that way. The x coordinate is almost ignorable and the y motion nearly periodic. We take x for λ, write the potential as $V(y,\lambda)$ and obtain for the action variable

$$v_y l = l(v^2 - v_x^{\,2})^{1/2} = \text{inv} \qquad (2\text{-}120)$$

since the kinetic energy is conserved. The turning point is where $v_x = 0$;

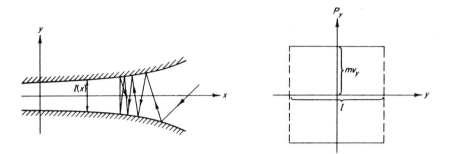

FIG. 2-14. Motion of a ball in a channel of slightly varying width.

therefore,

$$l_{\text{turn}} = \frac{\text{inv}}{v} \qquad (2\text{-}121)$$

The significance of this formula in plasma physics is in the treatment of particle motion in the so-called *cusp geometry* (Fig. 2-15). There particles

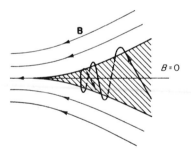

FIG. 2-15. Particle motion near a point cusp.

move on straight lines in the interior field-free region, while they are reflected entering the magnetic field. If the cyclotron radius in the field is small compared to the dimensions of the field-free region, (2-121) can be used to determine the turning point. As in the mirror geometry, some of the particles (those with $\text{inv}/v < d_{\text{min}}$) can escape. Loss-rate formulas in line cusps can be derived easily from (2-121) (see Exercise 2-8).

2–6. Particle Motion in Fields with Spatial Symmetry; the Hamiltonian Method

In cases where the guiding-center approximation is not applicable, the general equation of motion needs to be solved. With given electromagnetic fields and initial conditions this constitutes no difficulty in principle although tedious numerical or graphical methods usually have to be invoked. If, however, the field configuration bears certain forms of symmetry, the Hamiltonian formalism yields constants of motion, providing information with respect to the general character of the motion. In the following we illustrate this method by some examples.

The nonrelativistic Lagrangian of a particle in an electromagnetic field is

$$L = \tfrac{1}{2}mv^2 + q(\mathbf{A} \cdot \mathbf{v}) - qV \qquad (2\text{-}122)$$

where $\mathbf{A}$ and V are the vector and scalar potentials, related to the field strengths by

$$\mathbf{B} = \nabla \times \mathbf{A} \qquad (2\text{-}123)$$

and

$$\mathbf{E} = -\nabla V - \frac{\partial \mathbf{A}}{\partial t} \qquad (2\text{-}124)$$

The Hamiltonian is obtained from the Lagrangian by

$$\mathscr{H} = P_i \dot{Q}_i - L \qquad (2\text{-}125)$$

where P_i is the canonical momentum, defined as

$$P_i = \frac{\partial L}{\partial \dot{Q}_i} \qquad (2\text{-}126)$$

Q_i is the canonical coordinate, the dots denote time derivatives, and the summation convention for common indices is again adopted.

The equations of motion are obtained from the canonical equations

$$\dot{Q}_i = \frac{\partial \mathscr{H}}{\partial P_i} \qquad (2\text{-}127)$$

and

$$\dot{P}_i = -\frac{\partial \mathscr{H}}{\partial Q_i} \qquad (2\text{-}128)$$

If $\mathscr{H}$ does not depend on the coordinate Q_k then

$$\dot{P}_k = 0 \qquad (2\text{-}129)$$

yielding P_k as a constant of motion; Q_k is often called a *cyclic or ignorable coordinate*.

For fields with planar symmetry (e.g., $\partial/\partial Q_3 = 0$ for every field quantity), we introduce Cartesian coordinates as canonicals. The velocity vector is now $v_i = \dot{Q}_i$ and the Lagrangian becomes

$$L = \tfrac{1}{2}m\dot{Q}_i\dot{Q}_i + qA_i\dot{Q}_i - qV \qquad (2\text{-}130)$$

while the canonical momentum becomes, from (2-126),

$$P_i = m\dot{Q}_i + qA_i \; ; \qquad \dot{Q}_i = \frac{P_i - qA_i}{m} \tag{2-131}$$

For the Hamiltonian the $\dot{Q}_i$'s have to be expressed in terms of the P_i's. Therefore,

$$\mathcal{H} = P_i \frac{P_i - qA_i}{m} - \frac{(P_i - qA_i)(P_i - qA_i)}{2m} - \frac{qA_i(P_i - qA_i)}{m} + qV$$

$$= \frac{(P_i - qA_i)^2}{2m} + qV \tag{2-132}$$

If one of the coordinates (e.g., $i = 3$) is cyclic, we write (2-132) in the form

$$\mathcal{H} = \frac{(P_1 - qA_1)^2}{2m} + \frac{(P_2 - qA_2)^2}{2m} + \psi \tag{2-133}$$

where

$$\psi = \frac{(P_3 - qA_3)^2}{2m} + qV \tag{2-134}$$

is the effective potential, a known space-time function, because P_3 is a constant. The equations of motion, as derived from (2-133), reduce to those of two-dimensional motion in the magnetic field defined by A_1 and A_2 with the effective potential ψ.

Equation (2-133) can be further reduced for the special case of straight and parallel magnetic field lines (pointing in the z direction), and with y also cyclic (Fig. 2-16). Let the magnetic field $\mathbf{B}(0,0,B_z)$ be an arbitrary function of x. It is easy to see that (2-123) can be satisfied with $\mathbf{A}(0,A_y(x),0)$. (A_y represents the magnetic flux crossing the strip of unit width marked in Fig. 2-16a.)

For this case the Hamiltonian assumes the simple form

$$\mathcal{H} = \frac{P_x^2}{2m} + \psi \tag{2-135}$$

where

$$\psi = \frac{(P_y - qA_y)^2}{2m} + \frac{P_z^2}{2m} + qV \tag{2-136}$$

Equation (2-135) represents the one-dimensional particle motion in the potential field ψ. Figure 2-16b shows the variation of the effective potential for $V = 0$. Note that P_y and P_z, and consequently ψ, are different for different particles. The first term of (2-136) vanishes for vanishing $\dot{y}$ according to (2-131).

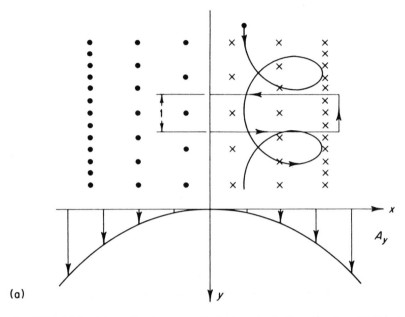

(a)

FIG. 2-16. (a) Particle motion in a magnetic field varying in the x direction. (b) Construction of the effective potential. (c) Effective potential with a superimposed uniform electric field in the x direction.

For static fields $\mathscr{H}$ is also a constant of motion. Its value determines the turning planes x_1 and x_2 of the particle in the effective potential trough. The particle drifts along the y axis perpendicular to $\mathbf{B}$ and ∇B, the motion going over to the drift of (2-59) in the appropriate limiting case.

The addition of the electric potential $V(x)$ [corresponding to an electric field $\mathbf{E}(E_x,0,0)$] results only in a deformation of the effective potential (Fig. 2-16c).

For a system of axial symmetry ($\partial/\partial\varphi = 0$), the introduction of cylindrical coordinates (r,z,φ) proves advantageous. The components of the velocity vector are now $\mathbf{v}(\dot{r},\dot{z},r\dot{\varphi})$ and the Lagrangian becomes

$$L = \tfrac{1}{2}m(\dot{r}^2 + \dot{z}^2 + r^2\dot{\varphi}^2) + q(A_r\dot{r} + A_z\dot{z} + A_\varphi r\dot{\varphi}) - qV \qquad (2\text{-}137)$$

For the canonical momenta one obtains

$$P_r = \frac{\partial L}{\partial \dot{r}} = m\dot{r} + qA_r; \qquad\qquad \dot{r} = \frac{P_r - qA_r}{m} \qquad (2\text{-}138)$$

$$P_z = \frac{\partial L}{\partial \dot{z}} = m\dot{z} + qA_z; \qquad\qquad \dot{z} = \frac{P_z - qA_z}{m} \qquad (2\text{-}139)$$

$$P_\varphi = \frac{\partial L}{\partial \dot{\varphi}} = mr^2\dot{\varphi} + qA_\varphi r; \qquad\qquad \dot{\varphi} = \frac{P_\varphi - qA_\varphi r}{mr^2} \qquad (2\text{-}140)$$

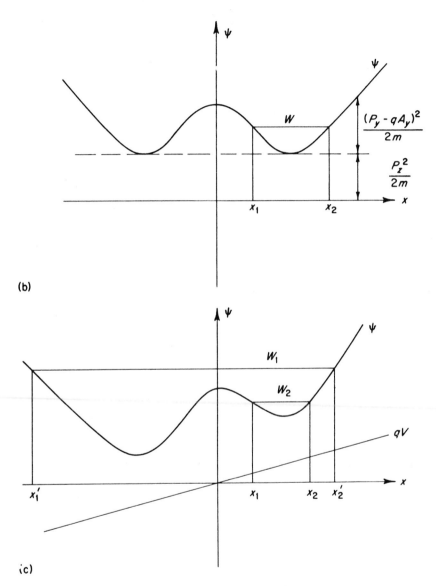

(b)

(c)

Using (2-125) again, the Hamiltonian assumes the form

$$\mathcal{H} = \frac{(P_r - qA_r)^2}{2m} + \frac{(P_z - qA_z)^2}{2m} + \frac{(P_\varphi - qrA_\varphi)^2}{2mr^2} + qV \qquad (2\text{-}141)$$

For a "mirror-type" magnetic field configuration, (2-123) can be satisfied by the proper choice of $A_\varphi(r,z)$ with $A_r = A_z = 0$. The vector potential and the flux are again closely related. From (2-123) it follows that the flux enclosed by a circle of radius r is $\phi = 2\pi r A_\varphi$.

We set

$$\mathcal{H} = \frac{P_r^2}{2m} + \frac{P_z^2}{2m} + \psi \tag{2-142}$$

with

$$\psi = \frac{1}{2m}\left(\frac{P_\varphi - qrA_\varphi}{r}\right)^2 + qV \tag{2-143}$$

As $P_\varphi = \text{const}$, ψ is known, and the particle motion can be described as a two-dimensional one in the rz plane. We first consider the $V = 0$ case.

The minimum line of the effective potential trough is defined by

$$\frac{\partial \psi}{\partial r} = \frac{P_\varphi - qrA_\varphi}{mr} \frac{r(-qA_\varphi - qr\,\partial A_\varphi/\partial r) - (P_\varphi - qrA_\varphi)}{r^2} = 0 \tag{2-144}$$

One solution is

$$\frac{P_\varphi - qrA_\varphi}{r} = 0 \tag{2-145}$$

consequently $\psi = 0$. For finite r (2-145) yields $\dot\phi = 0$ from (2-140). This can be satisfied, evidently, only for particle paths that form loops such as those in Fig. 2-17a.

A stationary field again yields a constant $\mathcal{H}$, and the turning points r_1 and r_2 can be plotted easily. Figure 2-17b represents a cross section of the potential trough at a $z = \text{const}$ plane. Note that $rA_\varphi = \text{const}$ as a function of z corresponds to a $\phi = \text{const}$ surface defined by a family of **B** lines. Consequently the bottom line of the potential trough—defined by $qrA_\varphi = P_\varphi$ —follows a B line. Note that in this case ϕ and P_φ have the same sign.

If, however, the particle encircles the axis as in Fig. 2-17c, (2-145) cannot be satisfied; $\dot\phi$ is never zero. The bottom of the potential trough corresponds then to the other solution of (2-144), namely,

$$qr^2\frac{\partial A_\varphi}{\partial r} + P_\varphi = 0 \tag{2-146}$$

Substituting in (2-140) and using (2-123) yields

$$mr\dot\phi = -qr\frac{\partial A_\varphi}{\partial r} - qA_\varphi = -qrB_z \tag{2-147}$$

or

$$r = R = \left|\frac{mv_\varphi}{qB_z}\right| \tag{2-148}$$

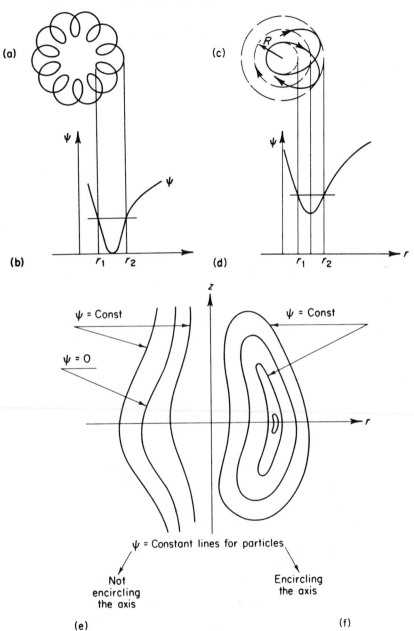

FIG. 2-17. Particle motion in a magnetic field with axial symmetry: (a) off-axis particle, (b) effective potential of off-axis particle, (c) particle encircling the axis, (d) effective potential well, (e) $\psi = $ const lines for off-axis particles, (f) $\psi = $ const lines for particles encircling the axis.

This criterion for the minimum line of the potential trough corresponds, by virtue of (2-9), to a particle moving in a circle with the cyclotron radius corresponding to $B_z(R)$. Such a path is degenerate, and the two turning points coincide. Other particles, with the same P_φ but more energy, oscillate between the turning points r_1 and r_2, corresponding to $\mathcal{H} = \psi$ (see Fig. 2-17c and d).

Some important differences between the potential troughs of Fig. 2-17b and d should be noted:

1. Contrary to the previous case, the bottom line of the potential trough of particles encircling the central line does *not* follow a field line. While a **B** line is still described by the $rA_\varphi = $ const equation, this does not necessarily coincide with the criterion of (2-146).

2. While for the "off-axis" particles the bottom line of the potential trough maintains the constant level $\psi_{min} = 0$, for particles encircling the axis it has a slope. For a mirror geometry, for instance, this slope is an upward one, approaching larger fields in the *mirror regions*. Thus particles encircling the axis with a certain P_φ are strictly unable to escape if their energy is smaller than $(\psi_{min})_{max}$ corresponding to the *mirror throat* (see Fig. 2-17e and f). The same cannot be said in general about the off-axis particles. The guiding-center approximation does not give an account of this delicate effect (see Fig. 2-18).

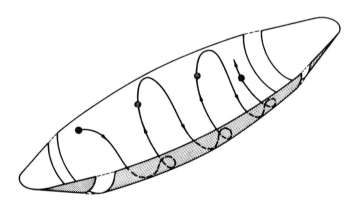

FIG. 2-18. Particle confined in a potential trough.

The inclusion of the electric potential term in (2-143) may improve the conditions for confinement. ψ_{min} is now no longer zero for the off-axis case and the proper application of the electric field results in increasing ψ_{min} approaching the mirrors (Fig. 2-19). This is utilized in various thermonuclear devices where a radial electric field is superimposed on the magnetic mirror field (see Exercise 2-13).

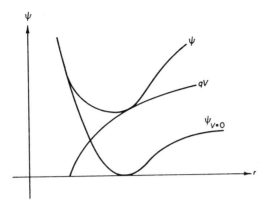

FIG. 2-19. Effective potential in the presence of a radial electric field.

The applicability of the Hamiltonian method is, of course, not restricted to time-independent problems. The discussion of the betatron, for example, follows similar lines. The effective potential has a shape like that plotted in Fig. 2-17d, increasing in time, with a corresponding increase in the particle energy (see Exercise 2-19).

2-7. Nonadiabatic Particle Motion in Axially Symmetric Fields

One is now in the position to use the "strict" constants of motion to derive general laws for particles moving in axially symmetric fields. The Hamiltonian method is particularly valuable when one is faced with a non-adiabatic problem, where the guiding-center approximation breaks down completely, like in cases, frequently encountered in applications, where the particle moves through vanishing magnetic fields either in space or in time. Clearly in such instances $\mu_m \to \infty$, and the adiabatic invariant is not even approximately conserved.

Consider a magnetic field configuration of axial symmetry where the field lines lie in the rz planes (e.g., a mirror geometry, dipole field, etc.). In the guiding-center approximation, a particle moving along a field line will drift slowly to other field lines. This drift, however, turns out to be a purely azimuthal one; hence the guiding center remains on the same *flux surface*, defined by rotating a field line around the axis of symmetry.

If the magnetic field varies slowly in time, while retaining its symmetry, an azimuthal electric field is set up which gives rise to an $\mathbf{E} \times \mathbf{B}$ drift. It can be shown (see Exercise 2-9) that this drift just suffices to keep the particle on the same (contracting or expanding) flux surface.

As a first example we are going to show that the particle "sticks to its flux surface" under much more general conditions.

Consider first a static magnetic field of axial symmetry and $B_\varphi = 0$. Let us place a particle in the $z = z_0$ plane of such a field and make it an off-axis particle by a proper choice of P_φ satisfying (2-145) at some $r = R(z_0)$. For a given P_φ there might be one such R, as in the case of a mirror geometry, or more, as for a dipole field, where there are two. Such a particle will move in the r direction between the turning points $r_1(z)$ and $r_2(z)$ and drift in the φ and z directions, while staying always around the bottom line $R = R(z)$ of the potential trough. This line, however, is determined by

$$qRA_\varphi(R) = P_\varphi \tag{2-149}$$

or

$$\phi(R) = 2\pi RA_\varphi(R) = \text{const} \tag{2-150}$$

where $\phi(R)$ is the magnetic flux inside the circle of radius R. Thus the bottom line of the potential trough defines a flux surface. Hence there exists a general flux conservation law which holds strictly, however rapid the variation of the magnetic field might be:

I.   *For an off-axis particle whose trajectory is enveloped by the surfaces $r_1(z,\varphi)$ and $r_2(z,\varphi)$, there is always a flux surface $R(z,\theta)$ confined between these surfaces.*

Consider next magnetic fields varying in time. The rapidity of this variation need not be restricted, but we demand that it takes place in a time interval $T_1 < t < T_2$, the fields being stationary before T_1 and after T_2. Since P_φ is a strict constant of motion, the same arguments can be used as before to prove the following flux conservation law:

II.   *For off-axis particles at T_1 and T_2, the flux surfaces $R(z,\varphi,T_1)$ and $R(z,\varphi,T_2)$ contain the same flux.*

For an illustration of this law we consider the case of the homogeneous magnetic field $\mathbf{B} = B_1\mathbf{k}$, with the vector potential $A_\varphi = (B_1/2)r$. For an off-axis particle, at some $r = R$,

$$P_\varphi = q\frac{B_1}{2}R^2 \tag{2-151}$$

The effective potential ψ is easily constructed (Fig. 2-20a). A particle with the canonical momentum P_φ and energy E will oscillate between the turning points r_1 and r_2. The real motion in the $r\varphi$ plane is, of course, circular (Fig. 2-20b). The distance of the guiding center from the axis is $r_g \neq R$.

Now consider B as a function of time. At $t = T_1$ it is B_1 and it settles down at $t = T_2$ to the value B_2. The effective potential ψ_2 can again be constructed with $R_2 = (2P_\theta/qB_2)^{1/2}$ pointing to the "same flux line" as before, or, more precisely, the circle of radius R_2 contains the same flux of field B_2 as the circle of radius R_1 did with B_1.

To construct the particle orbit at T_2 we need to know E_2, but this requires detailed knowledge of the field variation between T_1 and T_2 and the phase of the particle at T_1. The final orbits are presented in two simple cases in Fig. 2-20c.

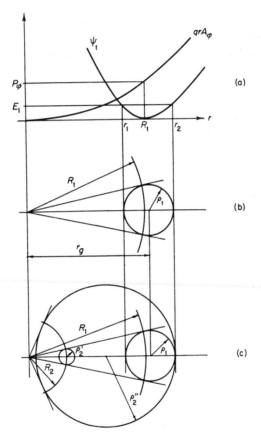

Fɪɢ. 2-20. Illustration of the general flux conservation law. The particle moves in a spatially uniform magnetic field changing in time arbitrarily in the interval $T_1 < t < T_2$.

1. The field variation is adiabatic. Since in this case the magnetic moment is conserved, $\rho_2'/\rho_1 = R_2/R_1$.

2. The field changes instantaneously from B_1 to B_2 at the time the particle is at r_2. Now $\rho_2'' \gg \rho_2'$, the magnetic moment is increased considerably, but the flux conservation law II still prevails.

The question arises whether conservation laws similar to I and II do exist for particles encircling the axis. We have seen, however, that the bottom line of the potential trough defined by (2-146) is not a flux line, and it is easy to see that it does not conserve flux either. A well-known example is the betatron, where the flux leaks freely into the particle orbit when the field rises, and it would leak out again during the second quarter cycle.

The motion of a particle in a field, which goes through zero in time, is clearly nonadiabatic. Consider, for example, a spatially uniform field $B_1 = B$ which is reversed in time to $B_2 = -B$. An off-axis particle follows the expanding flux surface for a while, but as the field becomes zero this shifts to infinity, and disappears completely after the field is reversed. Since P_φ remains the same and qA_φ changes sign during the process, the particle is bound to encircle the axis as soon as the field passes through zero. (Of course, an originally "encircling" particle follows the inverse fate.) The vector potential after reversal is

$$A_{\varphi 2} = \frac{r}{2} B_2 = -\frac{r}{2} B \qquad (2\text{-}152)$$

Again denoting the radius of the initial potential bottom by R, one obtains for P_φ,

$$P_\varphi = qRA_\varphi(R) = q \frac{R^2}{2} B \qquad (2\text{-}153)$$

The bottom of the new potential trough for the particle in the reversed field r_2 is given by (2-146). Using (2-152) and (2-153), (2-146) yields

$$-\frac{qr_2{}^2}{2} B + q \frac{R^2}{2} B = 0 \qquad (2\text{-}154)$$

or

$$r_2 = R \qquad (2\text{-}155)$$

The bottom lines of the two potential troughs coincide. The new well is, however, lifted to the altitude

$$\psi_{\min} = \frac{1}{2m} \left(\frac{P_\varphi + q(R^2/2)B}{R} \right)^2 = \frac{q^2 R^2 B^2}{2m} \qquad (2\text{-}156)$$

Even if we know nothing about the details—how the field reversal took place (provided axial symmetry was maintained)—it is certain that the particle cannot have less energy than $\psi_{\min}$. If, for example, the particle had no energy before field reversal, this is the minimum energy acquired in the process. Even the reversal of moderate fields may impart a very considerable amount of energy to particles, particularly electrons (see Exercise 2-23).

The particle trajectory is plotted in Fig. 2-21. When the magnetic field

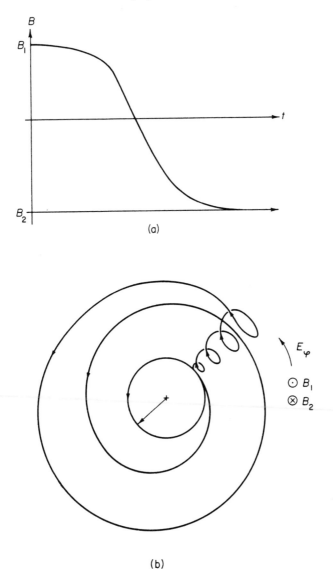

(a)

(b)

FIG. 2-21. Particle trajectory in a uniform field changing sign: (a) magnetic field as function of time, (b) trajectory of particle.

decreases, the particle gyrates with an ever-increasing radius of curvature, drifting radially outward roughly with the $\mathbf{w}^E$ drift. When the magnetic field goes through zero, there is an inflection point in the trajectory and the radius of curvature changes sign. Note that although the magnetic field changes sign, the induced electric field does not, and the particle picks up energy continuously from the electric field after field reversal, while spiralling inward in the increasing magnetic field.

In the cusped geometry, two coils with opposing currents are mounted coaxially, producing a magnetic field like the one sketched in Fig. 2-22a. A

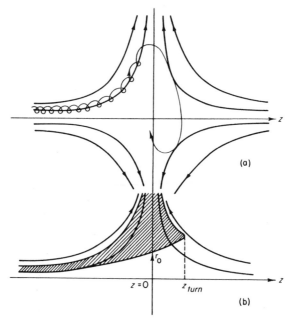

FIG. 2-22. Particle motion in a cusped geometry. The shaded region is accessible for a particle with a given energy and P_φ but otherwise unknown initial conditions.

particle injected along a field line follows the flux surface until it passes the median plane, where both B_z and A_φ change sign. On the other side of this plane the particle is forced to encircle the axis. Since the field is static in addition to P_φ, the particle energy is also a constant.

If these two constants of motion are known, one finds that the particle motion is limited to restricted regions of space. A typical case is shown in Fig. 2-22b, where the particle cannot be outside the shaded region. Some characteristic points on the boundary can be easily calculated. If the particle energy is E, the shortest distance the particle can approach the axis in the $z = 0$ plane r_0 can be determined from

$$E = \frac{1}{2m} \frac{P_\varphi^2}{r_0^2}$$

(2-157)

since in this plane $A_\varphi = 0$. If in the uniform field region inside the coil the particle moves axially,

$$P_\varphi = q\rho A_\varphi = q\frac{\rho^2}{2}B_0 \tag{2-158}$$

where the 0 index refers to the uniform field region and ρ is the injection radius. From the constancy of the particle velocity v and (2-157) and (2-158), one obtains

$$r_0 = \frac{1}{2}\frac{q\rho^2 B_0}{mv} = \frac{1}{2}\left(\frac{\rho}{r_c}\right)\rho \tag{2-159}$$

where r_c is the cyclotron radius corresponding to the uniform magnetic field and v.

The maximum distance a particle can travel axially is z_t, which is not necessarily the turning point for the particle—rather it is the furthest of possible turning points for the given constants of motion. At z_t all the particle energy is azimuthal or "potential energy"; therefore,

$$\psi(z_t, R) = \tfrac{1}{2}mv^2 \tag{2-160}$$

or

$$mv = \frac{P_\theta}{R} - qA(z_t, R) \tag{2-161}$$

where $R(z)$ is obtained by solving (2-146) for the given $A(r,z)$.

All particles for which a point z_t exists are either trapped or reflected, and only those with no real value of z_t can possibly be transmitted, although not all of them are. The minimum condition for possible transmission is that z_t is in the uniform field on the right-hand coil. Here $A(r) = -A_0(r)$ and $B(r) = -B_0(r)$. It follows from (2-146) and (2-158) that

$$-q\frac{R^2}{2}B_0 + q\frac{\rho^2}{2}B_0 = 0 \tag{2-162}$$

Hence

$$R = \rho \tag{2-163}$$

in analogy with (2-154) and (2-155). This is not surprising, since from a co-ordinate system attached to the particle, the problem of transmission from a uniform field to an opposite one agrees with the problem of field reversal in time.

The critical condition now follows from (2-161):

$$v_{\text{crit}} = \frac{q\rho^2 B_0}{2\rho m} + q\frac{\rho}{2m}B_0 = \frac{q\rho B_0}{m} \tag{2-164}$$

where the relationship with (2-156) is apparent. Particles injected into a cusped field of intensity B_0 at the "throat," at a radius ρ, have a chance to pass through the other throat only if their velocity exceeds v_{crit}. It also follows that under otherwise similar conditions the critical velocity is much higher for electrons than it is for ions.

2-8. Static Magnetic and Time-Varying Electric Fields

In various applications one needs to study the motion of charged particles in a uniform static magnetic and uniform but time-varying electric field. We shall investigate two limiting cases.

(a) *The variation of the electric field is much slower than the cyclotron frequency.*

In this case the guiding-center approximation can be applied. The motion along magnetic field lines is trivial. The motion across the field can be obtained from (2-75):

$$\mathbf{w}_\perp = \frac{\mathbf{E} \times \mathbf{B}}{B^2} + \frac{m}{qB^2} \left[\mathbf{B} \times \frac{d}{dt} \frac{\mathbf{E} \times \mathbf{B}}{B^2} \right] \qquad (2\text{-}165)$$

This can be further simplified in our case to

$$\mathbf{w}_\perp = \frac{\mathbf{E} \times \mathbf{B}}{B^2} + \frac{m \, \dot{\mathbf{E}}_\perp}{q \, B^2} = \mathbf{w}^E + \mathbf{w}^P \qquad (2\text{-}166)$$

The term $\mathbf{w}^P$ is the so-called *polarization drift*. With an increasing electric field, positive particles drift in the direction of the electric field and negative particles in the opposite direction, both picking up energy during the process. This energy is needed to provide for the additional drift energy $\frac{1}{2}m(w^E)^2$ (see Exercise 2-27). Similarly, decreasing the electric field results in an energy loss for the particles. Note that the electric drift $\mathbf{w}^E$ itself is perpendicular to $\mathbf{E}$ and hence does not lead to energy exchange between field and particle (on the average).

(b) *The frequency of the electric field is comparable to the cyclotron frequency.*

Assuming a periodic time variation for the electric field in the form

$$\mathbf{E}(t) = \mathbf{E}e^{i\omega t} \qquad (2\text{-}167)$$

the equation of motion becomes

$$\frac{d\mathbf{v}}{dt} = \frac{q}{m} \left(\mathbf{E}e^{i\omega t} + \mathbf{v} \times \mathbf{B} \right) \qquad (2\text{-}168)$$

We try the "Ansatz" of decomposing the velocity vector into two parts:

$$\mathbf{v} = \mathbf{v}_M + \mathbf{v}_E e^{i\omega t} \qquad (2\text{-}169)$$

where v_M contains no time variation with the angular frequency ω. Substituting (2-169) into (2-168) one obtains

$$\frac{d\mathbf{v}_M}{dt} + i\omega\mathbf{v}_E e^{i\omega t} = \frac{q}{m}(\mathbf{E}e^{i\omega t} + \mathbf{v}_M \times \mathbf{B} + \mathbf{v}_E \times \mathbf{B}e^{i\omega t}) \qquad (2\text{-}170)$$

This equation separates into two parts. The terms which do not contain the ω periodicity,

$$\frac{d\mathbf{v}_M}{dt} = \frac{q}{m}\mathbf{v}_M \times \mathbf{B} \qquad (2\text{-}171)$$

lead again to gyration with the cyclotron frequency, while the terms containing the "forced" oscillation yield

$$\left(i\omega + \frac{q}{m}\mathbf{B} \times \right)\mathbf{v}_E = \frac{q}{m}\mathbf{E} \qquad (2\text{-}172)$$

To solve (2-172) for $\mathbf{v}_E$, we multiply by the conjugate operator $(i\omega - (q/m)\mathbf{B}\times)$ and obtain

$$\left(\frac{q^2}{m^2}B^2 - \omega^2\right)\mathbf{v}_E - \frac{q^2}{m^2}(\mathbf{B} \cdot \mathbf{v}_E)\mathbf{B} = \frac{q}{m}\left(i\omega - \frac{q}{m}\mathbf{B} \times\right)\mathbf{E} \qquad (2\text{-}173)$$

This equation immediately yields for the parallel velocity component

$$\mathbf{v}_{E\parallel} = -\frac{i}{\omega}\frac{q}{m}\mathbf{E}_\parallel \qquad (2\text{-}174)$$

and for the perpendicular component

$$\mathbf{v}_{E\perp} = \frac{(q/m)(i\omega - \boldsymbol{\omega}_c \times)\mathbf{E}_\perp}{\omega_c^2 - \omega^2} \qquad (2\text{-}175)$$

where the cyclotron frequency vector $\boldsymbol{\omega}_c = (q/m)\mathbf{B}$ has been introduced.

Equation (2-174) represents an oscillation along the magnetic lines of force with the frequency ω, lagging 90 degrees behind the imposed electric field. To analyze the physical meaning of the motion in the perpendicular plane, we decompose the oscillating electric field vector into two rotating components by putting

$$\mathbf{E}_\perp = \frac{1}{2}\left(\mathbf{E}_\perp + i\frac{\boldsymbol{\omega}_c \times \mathbf{E}}{\omega_c}\right) + \frac{1}{2}\left(\mathbf{E}_\perp - i\frac{\boldsymbol{\omega}_c \times \mathbf{E}}{\omega_c}\right) \qquad (2\text{-}176)$$

$$= \mathbf{E}_L + \mathbf{E}_R$$

where the index L represents rotation to the left and R rotation to the right, looking in the direction of the $\boldsymbol{\omega}_c$ vector. Substituting this decomposed form of $\mathbf{E}_\perp$ into (2-175) the operation can easily be performed for both terms.

For $\mathbf{E}_L$ one obtains

$$\tfrac{1}{2}(i\omega - \boldsymbol{\omega}_c \times)\left(\mathbf{E}_\perp + i\,\frac{\boldsymbol{\omega}_c \times \mathbf{E}}{\omega_c}\right) = \tfrac{1}{2}i(\omega + \omega_c)\mathbf{E}_\perp - \frac{1}{2}\left(1 + \frac{\omega}{\omega_c}\right)\boldsymbol{\omega}_c \times \mathbf{E}$$

$$= \tfrac{1}{2}i(\omega + \omega_c)\left(\mathbf{E}_\perp + i\,\frac{\boldsymbol{\omega}_c \times \mathbf{E}}{\omega_c}\right)$$

$$= i(\omega + \omega_c)\mathbf{E}_L \qquad (2\text{-}177)$$

In a similar fashion, one finds

$$(i\omega - \boldsymbol{\omega}_c \times)\mathbf{E}_R = i(\omega - \omega_c)\mathbf{E}_R \qquad (2\text{-}178)$$

From (2-177) and (2-178) it follows that both $\mathbf{E}_L$ and $\mathbf{E}_R$ are eigenvectors of the complex operator appearing in (2-175). Hence $\mathbf{v}_{E\perp}$ also splits into a vector

$$\mathbf{v}_L = \frac{q}{m}\,\frac{i}{\omega_c - \omega}\,\mathbf{E}_L \qquad (2\text{-}179)$$

rotating to the left, and another,

$$\mathbf{v}_R = \frac{q}{m}\,\frac{-i}{\omega + \omega_c}\,\mathbf{E}_R \qquad (2\text{-}180)$$

rotating to the right. The left component is ahead, the right component lagging behind the driving field. For a positive ion ω_c is positive, while it is negative for an electron. Therefore, as $\omega \to |\omega_c^+|$ there is a resonance between an ion and the left component of the electric field, while at $\omega \to |\omega_c^-|$ an electron resonates with the field component rotating to the right.

Equations (2-174), (2-179), and (2-180) can be summarized in the tensor form

$$\mathbf{v} = \overset{\leftrightarrow}{v}\mathbf{E} \qquad (2\text{-}181a)$$

where $\overset{\leftrightarrow}{v}$ is the mobility tensor. In the rotating system $\overset{\leftrightarrow}{v}$ is diagonal and (2-181a) reads

$$\begin{pmatrix} v_L \\ v_R \\ v_\parallel \end{pmatrix} = \frac{q}{m}\,i \begin{pmatrix} \dfrac{1}{\omega_c - \omega} & 0 & 0 \\ 0 & -\dfrac{1}{\omega_c + \omega} & 0 \\ 0 & 0 & -\dfrac{1}{\omega} \end{pmatrix} \begin{pmatrix} E_L \\ E_R \\ E_\parallel \end{pmatrix} \qquad (2\text{-}181b)$$

In the stationary Cartesian system with the magnetic field pointing in the z direction, (2-181a) can be transformed with the help of (2-174) and (2-175).

It reads

$$
\begin{pmatrix} v_x \\ \\ v_y \\ \\ v_z \end{pmatrix} = \frac{q}{m} \begin{pmatrix} \dfrac{i\omega}{\omega_c^2 - \omega^2} & \dfrac{\omega_c}{\omega_c^2 - \omega^2} & 0 \\ \\ -\dfrac{\omega_c}{\omega_c^2 - \omega^2} & \dfrac{i\omega}{\omega_c^2 - \omega^2} & 0 \\ \\ 0 & 0 & -\dfrac{i}{\omega} \end{pmatrix} \begin{pmatrix} E_x \\ \\ E_y \\ \\ E_z \end{pmatrix} \tag{2-181c}
$$

2-9. High-Frequency Fields; Oscillation-Center Approximation

The guiding-center approximation dealt with slowly varying fields. A similar approximation procedure can be applied to the opposite limiting case, that of a very high frequency field. We start again by discussing the zero-order problem

Given an electric field with a harmonic field variation in time and uniform in space, the equation of motion of a particle reduces to

$$\ddot{\mathbf{r}} = \frac{q}{m}\, \mathbf{E} e^{i\omega t} \tag{2-182}$$

which can be readily solved by the substitution

$$\mathbf{r} = \mathbf{r}_0 e^{i\omega t} + \mathbf{C}_1 t + \mathbf{C}_2 \tag{2-183}$$

yielding

$$\mathbf{r} = -\frac{q}{m\omega^2}\, \mathbf{E} e^{i\omega t} + \mathbf{C}_1 t + \mathbf{C}_2 \tag{2-184}$$

Now we turn again to the case where the electric field is slightly nonuniform and an oscillating magnetic field is also present. The equation of motion now becomes

$$\ddot{\mathbf{r}} = \frac{q}{m}\, [\mathbf{E}(\mathbf{r}) + \dot{\mathbf{r}} \times \mathbf{B}(\mathbf{r})] e^{i\omega t} \tag{2-185}$$

Equation (2-185) reduces to (2-182) in the zero-order approximation if
1. In the Taylor expansion of the electric field,

$$\mathbf{E}(\mathbf{r}) = \mathbf{E}(\mathbf{r}_0) + (\mathbf{r}_1 \cdot \nabla_0)\mathbf{E} + \cdots \tag{2-186}$$

the first term is the dominating one. $\mathbf{r}_0$ is the center of oscillation and $\mathbf{r}_1 = \mathbf{r} - \mathbf{r}_0$ is an oscillating vector. A similar rule must hold for the magnetic field also.

The oscillation center, moving with the velocity $\dot{\mathbf{r}}_0$, does not carry the particle into appreciably different regions of the field during an oscillation period, or

$$\frac{(\dot{\mathbf{r}}_0 \cdot \nabla)\mathbf{E}}{\omega} \ll E \tag{2-187}$$

2. The second right-hand term in (2-185) is small compared to the other terms. This is equivalent to requiring that

$$\omega_c = \frac{q}{m} B \ll \omega \tag{2-188}$$

and also that $\dot{r}_0 \lesssim \dot{r}_1$, since the left side of (2-185) is of the order of $\omega\dot{r}_1$, while the magnetic term is only $\sim \omega_c \dot{r}$.

With these assumptions (2-185) can be expanded into

$$\ddot{\mathbf{r}}_0 + \ddot{\mathbf{r}}_1 = \frac{q}{m}[\mathbf{E}_0 + \underbrace{(\mathbf{r}_1 \cdot \nabla_0)\mathbf{E}}_{0} + \underbrace{\dot{\mathbf{r}}_0 \times \mathbf{B}_0 + \dot{\mathbf{r}}_1 \times \mathbf{B}_0}_{1}$$
$$\underbrace{\phantom{\ddot{\mathbf{r}}_0}}_{1}\ \underbrace{\phantom{\ddot{\mathbf{r}}_1}}_{0}$$
$$+ \underbrace{\dot{\mathbf{r}}_0 \times (\mathbf{r}_1 \cdot \nabla_0)\mathbf{B} + \dot{\mathbf{r}}_1 \times (\mathbf{r}_1 \cdot \nabla_0)\mathbf{B}}_{2}] \tag{2-189}$$

where $\mathbf{E}_0$ and $\mathbf{B}_0$ stand for $\mathbf{E}(\mathbf{r}_0,t)$ and $\mathbf{B}(\mathbf{r}_0,t)$, respectively. The order of each term is marked in the formula. In the zero order, (2-189) reduces to (2-182), yielding an oscillatory solution for $\mathbf{r}_1$. The constants in (2-184) now become slowly varying time functions, and we include the $\mathbf{C}_1 t + \mathbf{C}_2$ term in $\mathbf{r}_0$. Thus

$$\mathbf{r}_1 = -\frac{q}{m\omega^2} \mathbf{E}_0 \tag{2-190}$$

As in the guiding-center approximation, we want to compute the time-average value of the drift velocity $\dot{\mathbf{r}}_0$. Here, of course, the averaging has to be performed over an oscillation period. This yields for the first-order equation

$$\langle \ddot{\mathbf{r}}_0 \rangle = \frac{q}{m} [\langle (\mathbf{r}_1 \cdot \nabla_0)\mathbf{E} \rangle + \langle \dot{\mathbf{r}}_0 \times \mathbf{B}_0 \rangle + \langle \dot{\mathbf{r}}_1 \times \mathbf{B}_0 \rangle] \tag{2-191}$$

As $\dot{\mathbf{r}}_0$ is a slowly varying function ($\ddot{\mathbf{r}}_0$ is of the first order) and $\langle \mathbf{B}_0 \rangle = 0$, the second right-hand term is of the second order and can be neglected. Furthermore, realizing that $\langle \ddot{\mathbf{r}}_0 \rangle = \ddot{\mathbf{r}}_0$, (2-191) becomes

$$\ddot{\mathbf{r}}_0 = \frac{q}{m} [\langle (\mathbf{r}_1 \cdot \nabla_0)\mathbf{E} \rangle + \langle \dot{\mathbf{r}}_1 \times \mathbf{B}_0 \rangle] \tag{2-192}$$

We now substitute the zero-order solution for $\mathbf{r}_1$ from (2-190) in (2-192) to obtain

$$\ddot{\mathbf{r}}_0 = -\frac{q^2}{m^2\omega^2} \langle (\mathbf{E}_0 \cdot \nabla_0)\mathbf{E} \rangle - \frac{q^2}{m^2\omega^2} \langle \dot{\mathbf{E}}_0 \times \mathbf{B}_0 \rangle \tag{2-193}$$

The vector analytical identity, (2-54) can now be applied, and

$$\ddot{\mathbf{r}}_0 = \frac{q^2}{m^2\omega^2}\langle \mathbf{E}_0 \times (\nabla_0 \times \mathbf{E})\rangle - \frac{q^2}{m^2\omega^2}\left\langle \nabla_0 \frac{E^2}{2}\right\rangle - \frac{q^2}{m^2\omega^2}\langle \dot{\mathbf{E}}_0 \times \mathbf{B}_0\rangle \quad (2\text{-}194)$$

Using (1-2) this reduces to

$$\ddot{\mathbf{r}}_0 = -\frac{q^2}{m^2\omega^2}\left\langle \nabla_0 \frac{E^2}{2}\right\rangle - \frac{q^2}{m^2\omega^2}\left\langle \frac{\partial}{\partial t}(\mathbf{E}_0 \times \mathbf{B}_0)\right\rangle \quad (2\text{-}195)$$

As the energy flux $\mathbf{E} \times \mathbf{B}$ oscillates in a standing wave the last term vanishes, and we finally get

$$\ddot{\mathbf{r}}_0 = -\nabla \phi \quad (2\text{-}196)$$

where

$$\phi = \frac{q^2}{m^2\omega^2}\left\langle \frac{E_0{}^2}{2}\right\rangle \quad (2\text{-}197)$$

is the effective potential. The oscillation center drifts in the high-frequency field as if subjected to this potential. The force points in the direction of decreasing electric field. There is a formal similarity with the force formula (2-56), $E_0{}^2/2$ playing the role of $B_0{}^2/2$. Note, however, that the particle motion is different in the two cases, since for the guiding-center motion $\ddot{\mathbf{r}}$ is not proportional to the force. ϕ is also called the *ponderomotive potential*.

An energy equation can be derived by multiplying (2-196) by $\dot{\mathbf{r}}_0$:

$$\dot{\mathbf{r}}_0\ddot{\mathbf{r}}_0 + \frac{d\mathbf{r}_0}{dt}\frac{d\phi}{d\mathbf{r}_0} = 0 \quad (2\text{-}198)$$

which integrates to

$$[\tfrac{1}{2}\dot{\mathbf{r}}_0{}^2 + \phi] = \text{const} \quad (2\text{-}199)$$

Substituting the value of ϕ from (2-197) and that of $\mathbf{E}_0$ from (2-190), one obtains

$$\tfrac{1}{2}(\dot{r}_0{}^2 + \langle \dot{r}_1{}^2\rangle) = \text{const} \quad (2\text{-}200)$$

Thus in the time average there is no energy exchange between the electromagnetic field and the particle. It is simply an energy transformation between the oscillatory and translational kinetic energy. There is a tendency toward decreasing oscillatory energy as $\phi_{\min}$ corresponds to $\langle \dot{r}_1{}^2\rangle_{\min}$.

Using a standing wave, such as that of a cavity resonator, particles can be confined to certain field regions. Note that the forces depend on q^2; hence charges of both signs experience the same force direction and are therefore confined to the same region.

If a combination of harmonic oscillations is applied, with angular frequencies $\omega_1, \omega_2, ..., \omega_n$, the resulting effective potential is the combination

$$\phi = \frac{q^2}{m^2}\sum_k \frac{\langle E_k{}^2\rangle}{2\omega_k{}^2} \quad (2\text{-}201)$$

The average now has to be taken over a time containing many oscillation periods in each ω_k (see Exercise 2-31). With the help of this combination principle, a great variety of effective potential trough formations can be devised, in cavity resonators.

2-10. Summary

If the conditions for the guiding-center approximation are satisfied, the particle moves on a nearly circular path around its guiding center with the angular frequency

$$\omega = \frac{q}{m} B$$

while the guiding center drifts with the:

Electric drift: $\qquad \mathbf{w}^E = \dfrac{\mathbf{E} \times \mathbf{B}}{B^2}$

Gravitational drift: $\qquad \mathbf{w}^g = \dfrac{m}{q} \dfrac{\mathbf{g} \times \mathbf{B}}{B^2}$

Gradient drift: $\qquad \mathbf{w}^B = \dfrac{\mu_m}{qB^3} \mathbf{B} \times \nabla \dfrac{B^2}{2}$

Inertial drift: $\qquad \mathbf{w}^i = \dfrac{m}{qB^2} \left(\mathbf{B} \times \dfrac{d\mathbf{w}^0}{dt} \right)$

Two special cases of the inertial drift are worth mentioning:

Centrifugal drift: $\qquad \mathbf{w}^c = \dfrac{mv_\parallel^2}{qB^4} \mathbf{B} \times (\mathbf{B} \cdot \nabla)\mathbf{B}$

Polarization drift: $\qquad \mathbf{w}^P = \dfrac{m}{q} \dfrac{\dot{\mathbf{E}}_\perp}{B^2}$

Along the magnetic field lines the guiding center is accelerated:

$$\left(\frac{d\mathbf{w}}{dt} \right)_\parallel = \frac{q}{m} \mathbf{E}_\parallel - \frac{\mu_m}{mB} [(\mathbf{B} \cdot \nabla)\mathbf{B}]_\parallel$$

During the particle motion the magnetic moment

$$\mu_m = \frac{\frac{1}{2}mv_\perp^2}{B}$$

is an adiabatic invariant.

Another useful quantity is the longitudinal invariant:

$$J = \oint [2m\mu_m(B_{\text{turn}} - B)]^{1/2} \, ds$$

to be integrated along a magnetic field line. This invariant is a special case of the action integral

$$\oint P \, dQ$$

which is invariant for a quasi-periodic coordinate.

If there are symmetries in the fields such as to make one (or more) of the coordinates ignorable, the canonically conjugate momenta are constants of the motion. With an effective potential the number of degrees of freedom can be reduced.

When a particle moves through a region (or point) of vanishing magnetic field, μ_m is no longer conserved and the particle motion becomes nonadiabatic. The presence of ignorable coordinates often gives valuable insight into the characteristics of nonadiabatic particle motion. For instance, a particle originally at rest in an axially symmetric magnetic field at a distance R from the axis acquires the kinetic energy $E \geq q^2 R^2 B^2 / 2m$ when the direction of the field is reversed. It was assumed that B is uniform in the region $0 < r < R$.

A particle moving in a uniform static, magnetic, and time-varying electric field oscillates along the magnetic field line, while its motion in the perpendicular plane is a superposition of a rotation to the left and one to the right:

$$\mathbf{v}_L = \frac{q}{m} \frac{i}{\omega_c - \omega} \mathbf{E}_L \quad \text{and} \quad \mathbf{v}_R = -\frac{q}{m} \frac{i}{\omega + \omega_c} \mathbf{E}_R$$

Finally, a particle in a high-frequency electromagnetic field oscillates along the electric field line and drifts toward regions of small electric field as governed by the ponderomotive potential:

$$\phi = \frac{q^2}{m^2 \omega^2} \left\langle \frac{E_0^{\,2}}{2} \right\rangle$$

EXERCISES

2-1. Show that in the $v/c \ll 1$ limit, the electric field is transformed away in a coordinate system traveling with the velocity $\mathbf{v} = (\mathbf{E} \times \mathbf{B})/B^2$, where $\mathbf{E}$ and $\mathbf{B}$ are homogeneous and time-independent electric and magnetic fields, respectively. Prove that the same transformation does not change $\mathbf{B}$.

2-2. Use the special theory of relativity to investigate the motion of charged particles in crossed electric and magnetic fields when $E/B > c$. What coordinate transformation would you use?

2-3. Find the motion of a charged particle in a toroidal magnetic field using the guiding-center approximation. Show that the particles drift out of the field.

2-4. Which fraction of particles—having an isotropic velocity distribution at the symmetry plane of a mirror geometry—will be contained for a given mirror ratio $r = B_{max}/B_0$?

2-5. Describe the motion of the guiding center of a particle in the field of a straight current-carrying wire in the presence of a uniform electric field along the wire. Account for the energy balance of the gyrating particle in this process.

2-6. A particle moves in a slowly varying magnetic field $\mathbf{B} = B_z(x)\mathbf{e}_3$, and a uniform small electric field $\mathbf{E} = E_y\mathbf{e}_2$. Calculate the drifts and the rate of change of $\frac{1}{2}mv_\perp^2$. Show that the change in perpendicular kinetic energy can be accounted for by the drift along the electric field.

2-7. Prove that for a harmonic oscillator with slowly varying parameters W/ω is an adiabatic invariant.

2-8. There is a channel with reflecting walls and a slowly varying width $l(x)$ (Fig. 2-14). What fraction of the particles—having an isotropic velocity distribution at x_0, where $l = l_0$—will escape through the nozzle $l(x_1) < l_0$?

2-9. Prove that a particle moving in a magnetic mirror field, varying slowly in time, drifts in such a way (because of the induced electric field) that the flux encircled by the guiding center remains a constant.

2-10. A particle is confined along the axis of a magnetic mirror geometry between the turning points $\pm z_0$. The magnetic field along the axis can be described approximately by $B_z = \alpha z^2 + B_0$. Slowly varying the strength of the magnetic field in time, $[\alpha(t), B_0(t)]$,
(a) Describe the motion of the turning points z_0.
(b) Describe the variation of $v_\parallel$ and $v_\perp$ at $z = 0$.

2-11. Consider a charged particle in the earth's magnetic field (e.g., in the Van Allen belt). The guiding center oscillating between the turning points and drifting in the azimuthal direction sweeps out a surface. Although the magnetic field is not quite azimuthally symmetric, prove with the help of the first and second adiabatic invariants and the constancy of kinetic energy that this surface is closed. (*Hint:* Prove that the guiding center returns to the same field line with the same turning points.)

2-12. Prove that the bottom of the effective potential trough for a particle encircling the axis of a magnetic mirror has an ascending slope toward the mirrors.

2-13. Show, using the Hamiltonian method, that for an "electrified" mirror machine with a radial electric field, the minimum of the effective potential as a function of z for particles not encircling the axis can be expressed for small values of the electric field as $\psi_{min} = \frac{1}{2}m(E_r^2/B_z^2) + qV$. Give an interpretation for this result.

2-14. Describe the motion of charged particles in the earth's magnetic field far from the atmosphere (Van Allen belt):

(a) Using the guiding-center approximation.

(b) Using the Hamiltonian method. Show that there is no "strict" confinement for any one of the particles; there is an escape route to the poles.

Assume azimuthal symmetry.

2-15. Find the Hamiltonian, effective potential, and qualitative particle motion in a magnetic field of a straight current-carrying wire.

2-16. The *picket-fence geometry* consists of an infinite series of straight wires, carrying periodically alternating (spatially) currents (see Fig. 2-23). Describe the Hamiltonian and effective potential functions of this configuration. Describe qualitatively the motion of charged particles shot against the picket fence.

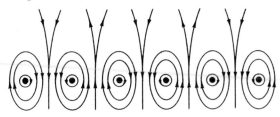

2-17. In the line-cusp configuration of Fig. 2-15 the third coordinate at right angles to the paper is ignorable. Write down the Hamiltonian of the particle motion and find an expression for the turning point. Use the action variable as an adiabatic invariant and assume $l \gg R$.

2-18. Consider Exercise 2-17 again, but this time take into account that particles are not reflected by the magnetic field as by a solid wall, but merge into it. How are $P_y(y)$ diagram and the adiabatic invariant affected by this consideration?

2-19. Prove, using the Hamiltonian formalism, that in a betatron the magnetic field at the particle orbit has to be exactly half of the average field inside the orbit, for accelerating particles in a circular orbit of constant radius.

2-20. Show that in a time-varying cylindrically symmetric magnetic field, the radial velocity of the zero effective potential surface is given by E_φ/B_z. This result does not depend on the "showness" of field variation, unlike the $E \times B$ drift of the guiding center.

2-21. A uniform magnetic field is switched on suddenly inside a cylindrical coil. Prove that a particle originally at rest, a distance r from the axis, moves on a circle of radius $r/2$ through the axis.

2-22. Investigate nonadiabatic particle motion in a magnetic field $\mathbf{B}(0,0,B_z)$, with the y and z coordinates ignorable. Find the flux conservation laws for the time variation of the field, and consider the effects of a reversal in the direction of the magnetic field.

2-23. An electron is situated in magnetic field $B = 1$ weber m^{-2}, $r = 10^{-1}$ m distant from the coil axis. What is the minimum energy acquired by the electron if the direction of current in the coil is reversed?

2-24. Describe qualitatively what happens to a particle traveling in the field of a straight current-carrying wire if the current is suddenly reversed. Construct graphically the effective potentials before and after field reversal.

2-25. Is there a flux conservation law which holds for the particles in the field of a current-carrying wire if the current changes in time?

2-26. What is the effective potential for a magnetic field of azimuthal symmetry, where neither B_φ nor B_z is zero? Plot $\psi(r)$ for the case where z is an ignorable coordinate. (The stellarator, to be discussed later, uses such a field.)

2-27. Prove that the energy picked up due to the polarization drift defined in (2-166), in the direction of the electric field, just suffices to cover the difference in zero-order electric drift energy, owing to the change in the electric field.

2-28. A particle is situated in a uniform magnetic field. At $t = 0$ an oscillating electric field is switched on. Investigate the particle motion if:
(a) At $t = 0$ the particle was at rest.
(b) At $t = 0$ the particle was gyrating in the magnetic field.
Prove that the solution is the superposition of the gyration and the motion obtained in a. What happens near the resonance $\omega \to \omega_c$?

2-29. A uniform magnetic field rotates around an axis perpendicular to the field.
(a) Describe the motion of the guiding center if the angular frequency $\omega \ll \omega_c$. Consider only lowest-order drifts.
(b) Use the oscillation-center approximation to describe the particle motion when $\omega \gg \omega_c$.

2-30. Use the Hamiltonian formalism for the description of the particle motion in a cylindrical cavity resonator, with $\mathbf{B}[0,0,B_z(r)]$, $\mathbf{E}[0,E_\varphi(r),0]$. Find the vector potential and cyclic coordinates and derive the equation of motion. Show that the latter agrees with (2-195) and (2-196). Prove directly that $(d/dt)\langle \mathcal{H} \rangle = 0$.

2-31. Prove the validity of the combination formula (2-201).

III

Plasma Equations: General Laws

3-1. The Boltzmann Equation

In Chapter II we studied the motion of a charged particle in a given electromagnetic field. Plasmas, however, consist of very large numbers of particles, and it would be very complicated indeed to follow the path of each of them. Actually we are not even interested in the fate of an individual particle, but prefer to look into the properties of the plasma as a macroscopic entity.

Studying the behavior of plasmas, a problem of a different nature presents itself: The motion of an individual particle is influenced by the presence of the others. The electromagnetic field created by the particles themselves can add up to a sizable fraction of the imposed field; therefore the electromagnetic field can no longer be considered as "given" but has to be found simultaneously with the equation of motion of particles, in a self-consistent way. In principle, of course, we could calculate and add up the contributions of each particle to the charge and current, but for a practical solution, other methods, better suited for the collective nature of plasma phemomena, are called for. A suitable way to treat the plasma as a collection of many charged particles moving in the self-consistent electromagnetic field is presented by the *Boltzmann equation.*

Let us represent each particle by a point in the six-dimensional configuration space-velocity space-hyperspace, with the "coordinates" (x,y,z,v_x,v_y,v_z) or, briefly, $(\mathbf{r},\mathbf{v})$. [Note that this representation is not identical with the representation in the familiar $(\mathbf{r},\mathbf{p})$ phase space, since in the presence of a vector potential the canonical momentum is not proportional to the velocity even in Cartesian coordinates.]

If a great number of particles are present, it suffices to know the density of points in this $(\mathbf{r},\mathbf{v})$ space, $f(\mathbf{r},\mathbf{v})$, or, since the points usually move in hyperspace and the density is not necessarily stationary, $f(\mathbf{r},\mathbf{v},t)$. More precisely, the number of particles with coordinates between x and $x + dx$, y and $y + dy$, z and $z + dz$, and velocity coordinates between v_x and $v_x + dv_x$, etc., is $f(\mathbf{r},\mathbf{v},t)\, dx\, dy\, dz\, dv_x\, dv_y\, dv_z$ or, briefly, $f(\mathbf{r},\mathbf{v},t)\, d^3r\, d^3v$.

If the velocity and acceleration of each particle is finite, the points in the $(\mathbf{r},\mathbf{v})$ space wander in continuous curves and the density function obeys an equation of continuity

$$\frac{\partial f}{\partial t} + \nabla_{\mathbf{r},\mathbf{v}} \cdot [f \cdot (\dot{\mathbf{r}},\dot{\mathbf{v}})] = 0 \qquad (3\text{-}1a)$$

where $\nabla_{\mathbf{r},\mathbf{v}}$ is the six-dimensional divergence operator and $(\dot{\mathbf{r}},\dot{\mathbf{v}})$ is the "velocity" vector in $(\mathbf{r},\mathbf{v})$ space. In Cartesian coordinates (3-1a) becomes

$$\frac{\partial f}{\partial t} + \frac{\partial}{\partial x_i} (f\dot{x}_i) + \frac{\partial}{\partial v_i} (f\dot{v}_i) = 0 \qquad (3\text{-}1b)$$

or, performing the differentiations,

$$\frac{\partial f}{\partial t} + f\left(\frac{\partial \dot{x}_i}{\partial x_i} + \frac{\partial \dot{v}_i}{\partial v_i}\right) + \frac{\partial f}{\partial x_i} \dot{x}_i + \frac{\partial f}{\partial v_i} \dot{v}_i = 0 \qquad (3\text{-}2)$$

We consider the six x_i, v_i coordinates as independent variables. Therefore the first term in the parentheses,

$$\frac{\partial \dot{x}_i}{\partial x_i} = \frac{\partial v_i}{\partial x_i} = 0 \qquad (3\text{-}3)$$

becomes zero. So does the second term,

$$\frac{\partial \dot{v}_i}{\partial v_i} = \frac{1}{m} \frac{\partial F_i}{\partial v_i} = 0 \qquad (3\text{-}4)$$

because the force F_i acting on a particle is velocity-independent for all but magnetic forces, and for those,

$$\frac{\partial F_i^m}{\partial v_i} = q \frac{\partial}{\partial v_i} (\mathbf{v} \times \mathbf{B})_i = 0 \qquad (3\text{-}5)$$

since the i component of the vector product contains all but the i component of $\mathbf{v}$.

Consequently,

$$\frac{\partial \dot{x}_i}{\partial x_i} + \frac{\partial \dot{v}_i}{\partial v_i} = \nabla_{\mathbf{r},\mathbf{v}} \cdot (\dot{\mathbf{r}},\dot{\mathbf{v}}) = 0 \qquad (3\text{-}6)$$

or the divergence of the six-dimensional velocity vector vanishes. Since in hydrodynamics fluids with this property are described as incompressible, we conclude that whatever the behavior of the particles in real space, their image points in the $(\mathbf{r},\mathbf{v})$ hyperspace behave as particles of an incompressible fluid

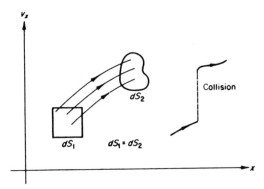

Fig. 3-1. Phase trajectories in the xv_x plane.

(Fig. 3-1). Using (3-6) in (3-2) one arrives at the so-called *collisionless Boltzmann equation*,

$$\frac{\partial f}{\partial t} + \frac{\partial f}{\partial x_i}\dot{x}_i + \frac{\partial f}{\partial v_i}\dot{v}_i = 0 \tag{3-7a}$$

or

$$\frac{\partial f}{\partial t} + \frac{\partial f}{\partial x_i}v_i + \frac{Fi}{m}\frac{\partial f}{\partial v_i} = 0 \tag{3-7b}$$

The motion of the points can also be described as a function of the time only, with $\mathbf{r} = \mathbf{r}(t)$, $\mathbf{v} = \mathbf{v}(t)$, and $f = f[\mathbf{r}(t),\mathbf{v}(t),t]$. With this interpretation (3-7a) becomes simply

$$df/dt = 0 \tag{3-8}$$

One can ascribe the following physical meaning to this equation: Following a point on its motion in $(\mathbf{r},\mathbf{v})$ space, one finds that the point density around it remains unchanged. This is again characteristic of an incompressible fluid. (Note, however, that the density need not be uniform.)

In deriving the Boltzmann equation we have used the equation of motion in the derivation of (3-6) but no other physical principle. Vice versa, the equation of motion can be derived from the Boltzmann equation. Hence the two principles are equivalent; neither one contains more information than the other. The latter is, however, better suited to handle collective phenomena, especially since simpler equations can be easily deduced from it by " destroying unnecessary information." Sometimes, for example, we do not care about the distribution in velocity space but wish to know the macroscopic particle density in configuration space. This can be deduced from the f function by " integrating out " over the velocity subspace:

$$n(\mathbf{r},t) = \int f(\mathbf{r},\mathbf{v},t)\, d^3v \tag{3-9}$$

Similarly, one might want to know—instead of the velocity of single particles—the mean plasma velocity, at a given point $\mathbf{r}$ as a function of time:

$$\mathbf{u}(\mathbf{r},t) = \langle \mathbf{v} \rangle = \frac{1}{n(\mathbf{r},t)} \int \mathbf{v} f(\mathbf{r},\mathbf{v},t) \, d^3 v \qquad (3\text{-}10)$$

where the angular brackets $\langle \; \rangle$ indicate "average over velocity space."
In general, any quantity $Q(\mathbf{r},\mathbf{v},t)$ possesses an average:

$$\langle Q(\mathbf{r},t) \rangle = \frac{1}{n(\mathbf{r},t)} \int Q f(\mathbf{r},\mathbf{v},t) \, d^3 v \qquad (3\text{-}11)$$

In the case of a plasma, one deals with charged particles and describes the
charge density as

$$\rho(\mathbf{r},t) = \sum_\sigma q^\sigma \int f^\sigma \, d^3 v = \sum_\sigma q^\sigma n^\sigma(\mathbf{r},t) \qquad (3\text{-}12)$$

and the current density as

$$\mathbf{J}(\mathbf{r},t) = \sum_\sigma q^\sigma \int \mathbf{v} f^\sigma \, d^3 v = \sum_\sigma q^\sigma n^\sigma(\mathbf{r},t) \mathbf{u}^\sigma(\mathbf{r},t) \qquad (3\text{-}13)$$

where the summation has to be carried over all particle species.

This formalism enables one to solve self-consistent problems. If only
electromagnetic forces are present, the Boltzmann equation for each particle
species becomes

$$\frac{\partial f^\sigma}{\partial t} + \mathbf{v} \cdot \frac{\partial f^\sigma}{\partial \mathbf{r}} + \frac{q^\sigma}{m^\sigma} (\mathbf{E} + \mathbf{v} \times \mathbf{B}) \frac{\partial f^\sigma}{\partial \mathbf{v}} = 0 \qquad (3\text{-}7\text{c})$$

often also called the *Vlasov equation.*

Equations (3-7c), (3-12), and (3-13) together with Maxwell's equations
yield a complete set of self-consistent equations. Knowing $\mathbf{E}(\mathbf{r},t)$ and $\mathbf{B}(\mathbf{r},t)$,
(3-7c) yields the $f^\sigma(r,t)$'s. These lead through (3-12) and (3-13) to the plasma
charges and currents; substituting them into Maxwell's equation we can solve
for $\mathbf{E}(\mathbf{r},t)$ and $\mathbf{B}(\mathbf{r},t)$. These equations should be solved simultaneously, of
course.

If sudden changes in the particle-velocity coordinates occur, because of
collisions, the phase paths of particles are no longer continuous. During a
collision a particle changes its velocity vector suddenly, which leads to the
disappearance of the representing point in one region of the $(\mathbf{r},\mathbf{v})$ space
and its simultaneous appearance somewhere else. The equation of continuity
(3-1) does not hold any longer but has to be corrected. It now reads

$$\frac{\partial f}{\partial t} + \nabla_{\mathbf{r},\mathbf{v}} \cdot [f \cdot (\mathbf{r},\mathbf{v})] = \left(\frac{\partial f}{\partial t} \right)_{\text{coll}} \qquad (3\text{-}14)$$

where $(\partial f/\partial t)_{\text{coll}}$ is the number of points per unit $(\mathbf{r,v})$ volume to appear, minus the number which disappears per unit time due to particle collision. Otherwise the derivation of (3-7) is unchanged, and we arrive at

$$\left(\frac{\partial f}{\partial t}\right) + v_i \frac{\partial f}{\partial x_i} + \frac{q}{m}\, (\mathbf{E} + \mathbf{v} \times \mathbf{B})_i\, \frac{\partial f}{\partial v_i} = \left(\frac{\partial f}{\partial t}\right)_{\text{coll}} \qquad (3\text{-}15)$$

The so-called *collision term* can be expressed in terms of the distribution functions and the interparticle forces. In this chapter, however, it will suffice to leave this term in symbolic form.

3-2. Moments of the Boltzmann Equation

As examples of how to gain general laws from the Boltzmann equation by destroying information, we develop moments of (3-15) by multiplying it with powers of the velocity vector and integrating over velocity space. To carry out the integrations we need to know something about the behavior of the distribution function at very large v values. Since no particle can have infinite velocity, it is safe to assume that f falls off very rapidly as $v \to \infty$. This assumption will assure that surface integrals, containing f in the numerator of the integrand, extended over a sphere of radius $v \to \infty$ in velocity space, vanish.

To evaluate the zero-order moment we multiply (3-15) by $v^0 = 1$ and integrate. The first term becomes

$$\int \frac{\partial f}{\partial t}\, d^3v = \frac{\partial}{\partial t}\, n \qquad (3\text{-}16)$$

the second

$$\int v_i \frac{\partial f}{\partial x_i}\, d^3v = \frac{\partial}{\partial x_i} \int f v_i\, d^3v = \frac{\partial}{\partial x_i}\, (nu_i) \qquad (3\text{-}17)$$

and the third

$$\int (\mathbf{E} + \mathbf{v} \times \mathbf{B})_i\, \frac{\partial f}{\partial v_i}\, d^3v = \int d\mathbf{S} - \int f \frac{\partial}{\partial v_i}\, (\mathbf{E} + \mathbf{v} \times \mathbf{B})_i d^3v = 0 \quad (3\text{-}18)$$

Here we integrated by parts, and $\int d\mathbf{S}$ is short for the surface integral extended over the large sphere $v \to \infty$ in velocity space, which vanished due to the aforementioned properties of f. The remaining integral vanishes for the same reasons as (3-5).

Finally,

$$\int (\partial f/\partial t)_{\text{coll}}\, d^3v = 0 \qquad (3\text{-}19)$$

since collisions simply displace points in velocity space but do not alter their density in configuration space. This leaves us with

$$\frac{\partial n}{\partial t} + \frac{\partial}{\partial x_i}(nu_i) = 0 \tag{3-20}$$

or the equation of continuity in configuration space. The average velocity u_i plays the role of the fluid velocity in hydrodynamics.

Now we multiply (3-15) by mv_k and integrate. The first term yields

$$m\frac{\partial}{\partial t}\int fv_k \, d^3v = \frac{\partial}{\partial t}(mnu_k) \tag{3-21}$$

The second becomes, with the help of (3-11),

$$m\int \frac{\partial}{\partial x_i} fv_iv_k \, d^3v = \frac{\partial}{\partial x_i}[mn\langle v_iv_k\rangle] \tag{3-22}$$

The third term is again integrated by parts:

$$\int F_i \frac{\partial f}{\partial v_i} v_k \, d^3v = -\int fF_i \frac{\partial v_k}{\partial v_i} \, d^3v = -\int fF_i \, \delta_{ik} \, d^3v = -n\langle F_k\rangle \tag{3-23}$$

where F_i is used for the force and δ_{ik} is the Kronecker symbol.

The collision term is again zero, since the momentum, summed over all the particles in velocity space, remains unchanged by collisions.

This yields for the first-moment equation,

$$\frac{\partial}{\partial t}(nmu_k) + \frac{\partial}{\partial x_i}(nm\langle v_iv_k\rangle) = n\langle F_k\rangle \tag{3-24}$$

This equation expresses the conservation of momentum. The first term is the rate of increase of momentum per unit volume. The second is the loss of momentum of the volume element through escaping particles (divergence of momentum flow). Hence the left side is the rate of change of momentum density, and the right side gives the force density.

The second moment of (3-15) yields the conservation of energy. We get it by multiplying by $\frac{1}{2}mv^2$ and integrating. The first term becomes

$$\frac{\partial}{\partial t}\int f\tfrac{1}{2}mv^2 \, d^3v = \frac{\partial}{\partial t}(n\langle \tfrac{1}{2}mv^2\rangle) \tag{3-25}$$

the second

$$\frac{\partial}{\partial x_i}\int fv_i\tfrac{1}{2}mv^2 \, d^3v = \frac{\partial}{\partial x_i}(n\langle \tfrac{1}{2}mv^2v_i\rangle) \tag{3-26}$$

and the third

$$\frac{1}{2}\int F_i \frac{\partial f}{\partial v_i} v^2 \, d^3v = -\frac{1}{2}\int fF_i \frac{\partial v^2}{\partial v_i} \, d^3v \tag{3-27}$$

Since

$$\frac{\partial}{\partial v_i} v_k v_k = 2 v_k \delta_{ik} \tag{3-28}$$

(3-27) becomes

$$\frac{1}{2} \int \frac{\partial f}{\partial x_i} F_i v^2 \, d^3v = - \int f F_i v_k \delta_{ik} \, d^3v = -nq \langle E_i v_i \rangle = -nq E_i u_i \tag{3-29}$$

Since kinetic energy is unaltered by collisions, the collision term vanishes again. This yields, finally, the energy conservation law

$$\frac{\partial}{\partial t} (n \langle \tfrac{1}{2} m v^2 \rangle) + \frac{\partial}{\partial x_i} (n \langle \tfrac{1}{2} m v^2 v_i \rangle) = nq E_i u_i \tag{3-30}$$

The first term is the rate of change of the energy density, the second the energy loss rate per volume element due to the energy flux (or heat transfer), and the right side gives the power fed into the system by the electric field (the work done by the magnetic field is, of course, zero).

We shall not evaluate higher moments, since they lack a simple physical meaning. Note that there are no quantities involving third or higher powers of v which are conserved by collisions; therefore the collision terms do not vanish in these equations either.

The equations of conservation of particle number, momentum, and energy are useful in making general statements about plasmas, but they cannot be considered as a closed system of plasma equations. The first of these equations contains two unknown functions, n and $\mathbf{u}$; the second these two plus $\langle v_i v_k \rangle$; the third $\langle v^2 v_i \rangle$ also. The number of unknowns always exceeds the number of equations. Neither can this chain be closed by introducing higher-order-moment equations.

3-3. Other Forms of the Conservation Laws

To carry out our program of finding self-consistent laws, we shall rewrite the first and second moments of the Boltzmann equation by expressing the self-fields through Maxwell's equations. A real plasma consists of at least two species of particles, each of which obeys (3-24) in the absence of collisions. If collisions are also present, it is easy to see that by adding the momentum equations for all particle species the collision terms cancel and one obtains

$$\frac{\partial}{\partial t} \sum_\sigma n^\sigma m^\sigma u_k^\sigma + \frac{\partial}{\partial x_i} \sum_\sigma n^\sigma m^\sigma \langle v_i v_k \rangle = \sum_\sigma n^\sigma \langle F_k \rangle^\sigma \tag{3-31}$$

Let us look first at the second term. It is the divergence of the tensor

$$P_{ik} = \sum_\sigma n^\sigma m^\sigma \langle v_i v_k \rangle^\sigma \tag{3-32}$$

which has the dimensions of pressure. For a gas with an isotropic velocity distribution,

$$f(\mathbf{r},\mathbf{v},t) = f(\mathbf{r},v^2,t) \tag{3-33}$$

For such a distribution (3-32) goes over into

$$P_{ik} = \sum_\sigma \tfrac{1}{3} n^\sigma m^\sigma \langle v^2 \rangle^\sigma \, \delta_{ik} \tag{3-34}$$

which is just the ordinary scalar pressure. We shall regard P_{ik} in (3-32) as the stress tensor of the plasma.

Now consider the right-hand term in (3-31). Although it is evident that it represents the electromagnetic force density acting on the particles, a derivation is instructive. Using the definition of the averages and (3-12) and (3-13),

$$\sum_\sigma n^\sigma \langle \mathbf{F} \rangle^\sigma = \sum_\sigma q^\sigma \int (\mathbf{E} + \mathbf{v} \times \mathbf{B}) f^\sigma \, d^3v$$

$$= \sum_\sigma q^\sigma n^\sigma \mathbf{E} + \sum_\sigma q^\sigma \left[\int f^\sigma \mathbf{v} \, d^3v \right] \times \mathbf{B} = \rho \mathbf{E} + \mathbf{J} \times \mathbf{B} \tag{3-35}$$

where ρ and $\mathbf{J}$ are charge and current densities produced by the plasma itself. It is well known that the electromagnetic force density can be represented—with the help of Maxwell's equations—as a divergence of a stress tensor, plus a radiation term. We transform the magnetic term, using Maxwell's equations and the vector identity (A-12):

$$\mathbf{J} \times \mathbf{B} = \frac{1}{\mu_0} (\nabla \times \mathbf{B}) \times \mathbf{B} - \dot{\mathbf{D}} \times \mathbf{B}$$

$$= \frac{1}{\mu_0} (\mathbf{B} \cdot \nabla)\mathbf{B} - \nabla \frac{B^2}{2\mu_0} - \dot{\mathbf{D}} \times \mathbf{B} \tag{3-36}$$

Since $\nabla \cdot \mathbf{B} = 0$, the first right-hand term can be written

$$\frac{B_i}{\mu_0} \frac{\partial B_k}{\partial x_i} = \frac{\partial}{\partial x_i} \frac{B_i B_k}{\mu_0} \tag{3-37}$$

which yields, finally,

$$(\mathbf{J} \times \mathbf{B})_k = \frac{\partial}{\partial x_i} \left(\frac{B_i B_k}{\mu_0} - \frac{B^2}{2\mu_0} \delta_{ik} \right) - (\dot{\mathbf{D}} \times \mathbf{B})_k \tag{3-38}$$

A similar transformation yields for the electric term (see Exercise 3-1),

$$\rho \mathbf{E}_k = \frac{\partial}{\partial x_i} \left(\varepsilon_0 E_i E_k - \frac{\varepsilon_0 E^2}{2} \delta_{ik} \right) - (\mathbf{D} \times \dot{\mathbf{B}})_k \tag{3-39}$$

The total force density becomes, adding (3-38) and (3-39),

$$(\rho \mathbf{E} + \mathbf{J} \times \mathbf{B})_k = -\frac{\partial T_{ik}}{\partial x_i} - \frac{\partial G_k}{\partial t} \tag{3-40}$$

where

$$T_{ik} = \left(\frac{\varepsilon_0 E^2}{2} + \frac{B^2}{2\mu_0}\right)\delta_{ik} - \left(\varepsilon_0 E_i E_k + \frac{B_i B_k}{\mu_0}\right) \tag{3-41}$$

is the electromagnetic stress tensor and

$$\mathbf{G} = \varepsilon_0 \mu_0 (\mathbf{E} \times \mathbf{H}) = \frac{\mathbf{E} \times \mathbf{H}}{c^2} \tag{3-42}$$

is the momentum density of the electromagnetic field.

This leads us to a new form of (3-31), in which the external forces are incorporated into the self-consistent fields:

$$\frac{\partial}{\partial t}(\Pi_k + G_k) + \frac{\partial}{\partial x_i}(P_{ik} + T_{ik}) = 0 \tag{3-43}$$

where

$$\Pi_k = \sum_\sigma n^\sigma m^\sigma u_k{}^\sigma \tag{3-44}$$

is the momentum density of the plasma. In a stationary state the partial time derivatives vanish and we arrive at the pressure-balance equation

$$\frac{\partial}{\partial x_i}(P_{ik} + T_{ik}) = 0 \tag{3-45a}$$

which says: In the stationary state the divergence of the total stress tensor (material + electromagnetic) vanishes. In integral form (3-45a) becomes

$$\int (P_{ik} + T_{ik})\, dS_i = 0 \tag{3-45b}$$

where the surface integral is extended over any closed surface inside the plasma. The symmetrical interdependence of the plasma and electromagnetic quantities is obvious in these equations. No longer is the electromagnetic field a given quantity and the plasma quantities calculated; the material pressure determines the electromagnetic pressure, and vice versa.

Let us take a simple case, where P_{ik} is a scalar, no electric fields are present, and the magnetic field lines are straight and parallel (but the field not necessarily uniform). In this case (3-45a) becomes, using (3-41) and noting that derivatives along the direction of magnetic field (say z) vanish,

$$\frac{\partial}{\partial x_i}\left(P + \frac{B^2}{2\mu_0}\right)\delta_{ik} = \frac{\partial}{\partial x_k}\left(P + \frac{B^2}{2\mu_0}\right) = 0 \tag{3-46}$$

The gradient of $P + (B^2/2\mu_0)$ is zero, consequently this quantity is uniform in space.

In this special case the pressure-balance equation has a particularly simple meaning: The sum of the magnetic and material pressures is a constant. Furthermore, outside the plasma, where $P = 0$, the magnetic field is necessarily larger than inside. This phenomenon is often referred to as *plasma diamagnetism*. These statements are, however, the consequence of our special assumptions; the general steady-state expression is (3-45a). We shall return later to have a closer look at this important equation.

The energy conservation law can also be rewritten in a self-consistent form. Summing (3-30) over all particle species and using (3-13), one finds

$$\frac{\partial}{\partial t} \sum_\sigma n^\sigma \langle \tfrac{1}{2} m v^2 \rangle^\sigma + \frac{\partial}{\partial x_i} \sum_\sigma n^\sigma \langle \tfrac{1}{2} m v^2 v_i \rangle^\sigma = \mathbf{J} \cdot \mathbf{E} \tag{3-47}$$

Multiplying (1-1) with $\mathbf{E}$ and (1-2) with $\mathbf{H}$ and subtracting, we obtain

$$\mathbf{E} \cdot (\nabla \times \mathbf{H}) - \mathbf{H} \cdot (\nabla \times \mathbf{E}) = \mathbf{J} \cdot \mathbf{E} + \varepsilon_0 \mathbf{E} \cdot \frac{\partial \mathbf{E}}{\partial t} + \mu_0 \mathbf{H} \cdot \frac{\partial \mathbf{H}}{\partial t} \tag{3-48}$$

The left side can be transformed with the help of a vector analytical identity (A-10), to find

$$\mathbf{J} \cdot \mathbf{E} = -\frac{\partial}{\partial t} \left(\frac{\varepsilon_0}{2} E^2 + \frac{B^2}{2\mu_0} \right) - \nabla \cdot (\mathbf{E} \times \mathbf{H}) \tag{3-49}$$

Substituting (3-49) in (3-47) one finds

$$\frac{\partial}{\partial t} \left(\sum n^\sigma \langle \tfrac{1}{2} m v^2 \rangle^\sigma + \frac{\varepsilon_0}{2} E^2 + \frac{B^2}{2\mu_0} \right) + \nabla \cdot \left(\sum_\sigma n^\sigma \langle \tfrac{1}{2} m v^2 \mathbf{v} \rangle^\sigma + \mathbf{E} \times \mathbf{H} \right) = 0 \tag{3-50}$$

The rate of change of the plasma kinetic, plus electromagnetic, field energy density equals the negative divergence of the material plus electromagnetic energy flow. This form of the energy conservation equation is again symmetrical in the plasma and electromagnetic quantities.

Finally we are going to express the momentum conservation equation in a form emphasizing the hydrodynamic features of plasma behavior. Let us consider the plasma as a fluid, each point of which moves with the average velocity

$$\mathbf{V} = \frac{1}{\sum\limits_\sigma m^\sigma n^\sigma} \sum_\sigma m^\sigma n^\sigma \mathbf{u}^\sigma \tag{3-51}$$

The fluid-pressure tensor should then be expressed in the local moving coordinate system, rather than in the rest system, as

$$p_{ik} = \sum_\sigma m^\sigma n^\sigma \langle (v_i - V_i)(v_k - V_k) \rangle^\sigma = \sum_\sigma (m^\sigma n^\sigma \langle v_i v_k \rangle^\sigma - m^\sigma n^\sigma V_i u_k^\sigma$$

$$- m^\sigma n^\sigma V_k u_i^\sigma + V_i V_k m^\sigma n^\sigma) = P_{ik} - V_i V_k \sum_\sigma m^\sigma n^\sigma \tag{3-52}$$

Summation of the momentum equations (3-24) over all particle species yields, with the help of (3-51) and (3-52),

$$\frac{\partial}{\partial t} \sum_\sigma m^\sigma n^\sigma V_k + \frac{\partial}{\partial x_i}\left(P_{ik} + V_i V_k \sum_\sigma m^\sigma n^\sigma \right) = \sum_\sigma n^\sigma \langle F_k \rangle^\sigma \tag{3-53}$$

This can be rewritten

$$V_k \frac{\partial}{\partial t} \sum_\sigma m^\sigma n^\sigma + \sum_\sigma m^\sigma n^\sigma \frac{\partial V_k}{\partial t} + \frac{\partial p_{ik}}{\partial x_i} + V_k \frac{\partial}{\partial x_i} \sum_\sigma m^\sigma n^\sigma V_i$$

$$+ \sum_\sigma m^\sigma n^\sigma V_i \frac{\partial V_k}{\partial x_i} = \sum_\sigma n^\sigma \langle F_k \rangle^\sigma \tag{3-54}$$

The multiplication of the equation of continuity (3-20) by m^σ and summation over all particle species leads to the new continuity equation

$$\frac{\partial}{\partial t} \sum_\sigma m^\sigma n^\sigma + \frac{\partial}{\partial x_i} \sum_\sigma m^\sigma n^\sigma V_i = 0 \tag{3-55}$$

This results in the cancellation of the first and fourth terms in (3-54), which becomes, finally,

$$\sum_\sigma m^\sigma n^\sigma \left(\frac{\partial V_k}{\partial t} + V_i \frac{\partial V_k}{\partial x_i} \right) + \frac{\partial p_{ik}}{\partial x_i} = \sum_\sigma n^\sigma \langle F_k \rangle^\sigma \tag{3-56a}$$

Since $V_k = V_k(\mathbf{r},t)$, the expression

$$\left(\frac{\partial V_k}{\partial t} + V_i \frac{\partial V_k}{\partial x_i} \right) = \left(\frac{\partial \mathbf{V}}{\partial t} + (\mathbf{V} \cdot \nabla)\mathbf{V} \right)_k = \left(\frac{d\mathbf{V}}{dt} \right)_k \tag{3-57}$$

is the total time derivative of $\mathbf{V}$. Therefore one may write (3-56a) in the alternative form

$$\sum_\sigma m^\sigma n^\sigma \frac{dV_k}{dt} + \frac{\partial p_{ik}}{\partial x_i} = \sum_\sigma n^\sigma \langle F_k \rangle^\sigma \tag{3-56b}$$

If one associates the local mean velocity of particles $\mathbf{V}$ with a "fluid velocity," (3-56b) can be considered an equation of motion of the fluid. For a scalar instead of a tensor pressure, this reduces to the familiar *Euler equation*.

The energy conservation law (3-30), after summation over the plasma constituents, may also be cast in a form which emphasizes the fluid-like characteristic of the plasma. Introducing the average "thermal" particle energy in the fluid frame,

$$\varepsilon = \tfrac{1}{2}m\langle(\mathbf{v} - \mathbf{V})^2\rangle \tag{3-58}$$

one may show (see Exercise 3-16) by a procedure similar to that used in deriving (3-56b) that

$$\sum_\sigma n^\sigma \frac{d\varepsilon^\sigma}{dt} + p_{ik} \frac{\partial V_k}{\partial x_i} = -\frac{\partial}{\partial x_i} \sum_\sigma n^\sigma \langle \tfrac{1}{2}m(\mathbf{v} - \mathbf{V})^2(v_i - V_i)\rangle^\sigma \tag{3-59}$$

The expression on the right is the divergence of the heat flux in the fluid frame.

3-4. Solution of the Vlasov Equation

In this section we restrict ourselves to investigation of the collisionless Boltzmann or Vlasov equation. It will be seen later that the neglect of the collision term is often well justified, especially in problems concerning very high temperatures.

The collisionless Boltzmann equation is a partial differential equation of the first order, and in principle it can be solved knowing the force $\mathbf{F}(\mathbf{r},\mathbf{v},t)$ and the initial and boundary conditions for the unknown function $f(\mathbf{r},\mathbf{v},t)$. If, however, there are constants of the (particle) motion at hand, particular solutions can be written down immediately. Any function

$$f(\mathbf{r},\mathbf{v},t) = f(c_1, c_2, \ldots) \tag{3-60}$$

which is a function of the constants of motion c_i only is a solution of the Boltzmann equation. Indeed,

$$\frac{df(c_1,\ldots)}{dt} = \frac{\partial f}{\partial c_i} \frac{dc_i}{dt} = 0 \tag{3-61}$$

by the definition of the constant of motion. This is in agreement with our previous statement that the Boltzmann equation contains no more and no less than the equation of motion of the particles. Each particle follows such a path in $(\mathbf{r},\mathbf{v})$ space so as to keep its constants of motion conserved. Hence an f function, which depends on the c_i's only, is in agreement with the equation of motion, and is therefore a solution of the Boltzmann equation.

The constants of motion—while constants for each particle—are, of course, functions of the $(\mathbf{r},\mathbf{v})$ space and of the time

$$c_i = c_i(\mathbf{r},\mathbf{v},t) \tag{3-62}$$

and this is how $f(c_1, c_2, \ldots)$ is a function of the same variables. Obviously if the constants of motion are time-independent (in this sense) we arrive at stationary solutions of the Boltzmann equation.

Let us consider, for example, one-dimensional motion with a time-independent constant of motion: $c(x,v)$. The equation $c(x,v) = \text{const}$ describes a line—or, prescribing different values for the constant, a family of lines—in the xv plane (Fig. 3-2). Each particle moves on such a line. If any strip between

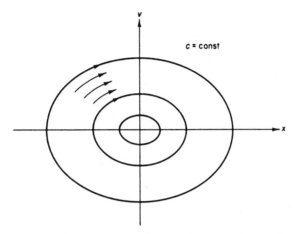

FIG. 3-2. Phase trajectories of one-dimensional particle dynamics are the $c(x,v) = \text{const}$ curves, where c is the constant of motion (e.g., energy for a time-independent potential).

two adjacent lines is populated by a uniform f point density, the points streaming around on the xv plane, like an incompressible fluid, correspond indeed to a stationary solution of the Boltzmann equation. This way of populating the xv plane with points, however, is equivalent to writing $f(x,v) = f(c)$.

Such a constant of motion is, for example, the total energy

$$W(v,x) = \tfrac{1}{2}mv^2 + qV(x) \tag{3-63}$$

in an electrostatic field. It is easy to prove that $f(W)$ is a solution of the stationary Boltzmann equation (see Exercise 3-3).

In the general case of three-dimensional motion, the $(\mathbf{r},\mathbf{v})$ space is six-dimensional. Knowledge of a single constant of motion c_1 restricts the motion of a phase point to the $c_1 = \text{const}$ five-dimensional subspace. Knowledge of a second constant c_2 restricts the motion of the phase points to the four-dimensional "intersection" of the two five-dimensional subspaces. Each additional constant of motion lowers the number of degrees of freedom by one. The more constants of motion are known, the more one knows of the actual particle motion. The motion of each particle is fully determined by the knowledge of six independent constants (usually obtained from the initial conditions).

Similarly, the more constants of motion one knows, the more general solutions of the Boltzmann equation one obtains. Not knowing any of the constants, we can immediately write down a trivial solution, namely,

$$f(\mathbf{r},\mathbf{v},t) = \text{const} \qquad (3\text{-}64)$$

where the distribution function is uniform in space as well as in velocity space, and does not vary in time either. The most general solution is

$$f(\mathbf{r},\mathbf{v},t) = f(c_1, c_2, ..., c_6) \qquad (3\text{-}65)$$

A possible set of constants are the initial coordinates of the particles: $c_1 = x_0$, $c_2 = y_0, ..., c_6 = v_{z0}$. This can obviously be fitted to any initial distribution:

$$f(\mathbf{r},\mathbf{v},0) = f(\mathbf{r}_0, \mathbf{v}_0) \qquad (3\text{-}66)$$

and yields, therefore, the solution of any initial-value problem. The trouble is that to make use of this solution we need to know each

$$c_i = c_i(\mathbf{r},\mathbf{v},t) \qquad (3\text{-}67)$$

function, which is equivalent to solving the equations of motion for all particles.

The most general stationary solution is obtained by knowing the five stationary constants of motion. The five five-dimensional subspaces cut out two-dimensional lines, the paths in $(\mathbf{r},\mathbf{v})$ space. (The sixth is the unimportant phase.)

It follows from these considerations that all the constants of motion obtained in Chapter II are of great use in the present collective treatment as well.

If we deal, for example, with a problem of particles moving in a stationary magnetic field, the particle energy W is a constant of motion. If the field is cylindrically symmetrical as the mirror field, the canonical momentum

$$P_\varphi = mr^2\dot{\varphi} + qA_\varphi r \qquad (3\text{-}68)$$

is also a constant. Since the field is time-independent, W and P_φ are stationary constants. A solution of the Boltzmann equation is, therefore,

$$f(\mathbf{r},\mathbf{v}) = f(W,P_\varphi) = f[\tfrac{1}{2}mv^2, m(x\dot{y} - y\dot{x}) + qA_\varphi(x,y,z) \cdot (x^2 + y^2)^{1/2}] \qquad (3\text{-}69)$$

If a special problem provides any more constants of motion, more general solutions can be built up that way, hence the larger the class of initial conditions to which they can be fitted.

Naturally what was said regarding constants of motion holds also for adiabatic invariants. The f functions, expressed as functions of adiabatic invariants, are solutions of the Boltzmann equation to the same approximation as the adiabatic invariants are good constants.

3-5. Thermodynamic Properties of Plasmas

We have seen that it is the direct consequence of the collisionless Boltzmann equation $df/dt = 0$ that the density of any bunch of points in $(\mathbf{r},\mathbf{v})$ space remains constant in the course of the motion. Since the *a priori* probability of finding a particle in the six-dimensional unit volume is uniform in phase space,† the thermodynamic probability of any system of points remains constant as well. This is equivalent to the conservation of entropy of any part of the system. This important theorem can be proved more formally, using the expression for the entropy

$$S = -k \iint f \log f \, d^3r \, d^3v + S_0 \tag{3-70}$$

and demonstrating that the time derivative of this expression vanishes (see Exercise 3-10).

In thermodynamics, entropy-conserving processes are called *adiabatic*. In this sense each process involving collisionless plasmas in any kind of electromagnetic field (including self-fields) is an adiabatic process.

Let us consider plasma particles occupying the volume V with an isotropic and uniform velocity distribution. If the plasma undergoes a (collisionless) process which leaves the velocity distribution isotropic, it follows from the constancy of phase volume that

$$4\pi v^2 \, dv \cdot V = 4\pi v_0^2 \, dv_0 \cdot V_0 \tag{3-71}$$

This equation can be satisfied by

$$v^3 = \frac{V_0}{V} v_0^{\,3} \tag{3-72}$$

Similarly, for a two-dimensional plasma,

$$2\pi v \, dv \, V = 2\pi v_0 \, dv_0 \cdot V_0 \tag{3-73}$$

where V is now a "two-dimensional volume," actually a surface area. The solution is

$$v^2 = \frac{V_0}{V} v_0^{\,2} \tag{3-74}$$

These results tally with the usual adiabatic equations of state. Defining a "kinetic temperature" T proportional to $\langle v^2 \rangle$ we obtain

$$\frac{T}{T_0} = \left(\frac{V_0}{V}\right)^{2/3} \tag{3-75}$$

---

† It might cause confusion that we are operating here in $(\mathbf{r},\mathbf{v})$ space instead of phase space. Note, however, that the transformation from one hyperspace to the other conserves the volume of each element while distorting its shape (see Exercise 3-9).

for the three-dimensional, and

$$\frac{T}{T_0} = \frac{V_0}{V}$$ (3-76)

for the two-dimensional case. For a Maxwellian distribution, the kinetic temperature goes over into the conventional temperature definition and (3-75) and (3-76) agree with the adiabatic equation of state, using $\gamma = \frac{5}{3}$ and $\gamma = 2$, respectively. Note that only the constancy of the phase volume (the collisionless Boltzmann equation) and the assumption of isotropic velocity distribution have been used. The same results can also be obtained from the energy conservation law (3-59) in the absence of heat flow (see Exercise 3-17).

All this is, of course, not true in the presence of collisions. We know from the general laws of thermodynamics that the entropy of a system tends to increase under the effect of collisions until it reaches thermal equilibrium. It is interesting to investigate what thermal equilibrium means for a plasma in an electromagnetic field. Quite generally, in thermal equilibrium, the distribution functions are of the form

$$F = \text{const} \exp[-\mathcal{H}/kT]$$ (3-77)

where $\mathcal{H}$ is the Hamiltonian of the system. Expressing $\mathcal{H}$ in terms of the x_i and v_i with the help of (2-132) and (2-131), one finds

$$F = \text{const} \exp\left[-\frac{\frac{1}{2}mv^2}{kT} - \frac{qV}{kT}\right]$$ (3-78)

It is interesting to note that the magnetic vector potential does not appear in this expression. In fact, (3-78) is completely independent of any imposed (or self-) magnetic fields. The significance of this result for the possibility of achieving magnetic confinement is the following:

We might force the plasma by magnetic forces into suitable artificial formations and produce confinement (e.g., in a mirror field). *This configuration, however, does not correspond to thermal equilibrium; hence in the course of time it will be destroyed by collisions. The plasma will finally assume an equilibrium configuration as though the magnetic field were not present.* Just how fast this is going to take place is an important question, and we shall investigate it later.

Note that the equilibrium velocity distribution is isotropic (F depends on v^2). This again supports our previous statement. A system with an isotropic velocity distribution is current-free, so it cannot carry magnetic forces.

We shall see later that even in the absence of collisions, many (in fact, most) artificially produced equilibrium distribution functions will run away from this stationary distribution when subjected to perturbations, however small these perturbations are. Such configurations are called *unstable*. Finding stable configurations is therefore of great importance. It is easy to prove that a spatially uniform field-free collisionless plasma with a monotone-decreasing velocity distribution $f(v^2)$ is always stable. Any change in the distribution function of a collisionless plasma can be thought of as a chain of exchanges of particles between equal phase-space volumes; any set of particles if transferred in phase space must occupy the same volume subsequently, pushing out those particles occupying that volume previously, and so on. This is the content of the statement $df/dt = 0$. If $f(v^2)$ is a monotone-decreasing function of v^2, two types of exchanges are possible: those on the same energy surface ($v^2 = \text{const}$) and those between different energy states. The first type does not change the distribution function, while the second transfers more low-energy particles (smaller v^2) into higher-energy positions. Consequently, energy is required for such a process which is unavailable in the field-free uniform plasma. Hence no change in the distribution function is possible and the theorem is proved.

Finally, we derive a virial theorem for plasmas. We rewrite the "hydrodynamic" equation (3-56b) in a self-consistent form using (3-40) for the force-density term

$$\sum_\sigma m^\sigma n^\sigma \frac{dV_k}{dt} + \frac{\partial}{\partial x_i}(p_{ik} + T_{ik}) + \frac{\partial}{\partial t} G_k = 0 \qquad (3\text{-}56c)$$

We multiply this equation by x_k and integrate over a volume in configuration space moving with the "fluid." The first term becomes

$$\sum_\sigma \int m^\sigma n^\sigma x_k \frac{dV_k}{dt} d^3r = \int x_k \frac{dV_k}{dt} d^3m \qquad (3\text{-}79)$$

where we changed from volume integration to integration over the mass. The meaning of the mass element d^3m is self-explanatory. Noting that $V_k = dx_k/dt$, (3-79) can be rewritten

$$\int x_k \frac{d^2 x_k}{dt^2} d^3m = \frac{1}{2}\frac{d^2}{dt^2}\int x_k x_k \, d^3m - \int \left(\frac{dx_k}{dt}\right)^2 d^3m = \frac{1}{2}\frac{d^2 I}{dt^2} - 2T \qquad (3\text{-}80)$$

where use has been made of the fact that moving with the fluid the mass elements do not vary in time. The first integral on the right side is the moment of inertia I, and the second is twice the kinetic energy T of the "fluid."

The second term can be integrated by parts:

$$\int_V x_k \frac{\partial}{\partial x_i}(p_{ik} + T_{ik}) \, d^3r = \int_S x_k(p_{ik} + T_{ik}) \, dS_i - \int_V \delta_{ik}(p_{ik} + T_{ik}) \, d^3r \qquad (3\text{-}81)$$

where S is the surface surrounding the volume of integration V, and we used $dx_k/dx_i = \delta_{ik}$. Using (3-52), the first term in the volume integral becomes

$$\sum_\sigma m^\sigma \int_V n^\sigma \langle (\mathbf{v} - \mathbf{V})^2 \rangle \, d^3r = 2U \qquad (3\text{-}82)$$

where U corresponds to the random "thermal" energy of the plasma. The electromagnetic term in the volume integral can be written, using (3-41),

$$\int_V \delta_{ik} T_{ik} \, d^3r = \int_V \left(\frac{\varepsilon_0 E^2}{2} + \frac{B^2}{2\mu_0} \right) d^3r = W^E + W^M \qquad (3\text{-}83)$$

where W^E and W^M are the electric and magnetic energy content of the volume considered. From (3-56c), (3-80), (3-82), and (3-83), the virial theorem becomes

$$\frac{1}{2} \frac{d^2}{dt^2} I + \int_V x_k \frac{\partial G_k}{\partial t} \, d^3r = 2(T + U) + W^E + W^M - \int x_k (p_{ik} + T_{ik}) \, dS_i \qquad (3\text{-}84)$$

As an application we consider the possibility of self-confinement of a finite plasma (sometimes called a *plasmoid*). *Can a plasmoid generate electromagnetic fields which act on the self-charges and currents in such a way as to provide for a stationary configuration?* We extend the volume to cover the entire plasma and field. This renders the surface integral zero. For a stationary configuration the time derivatives vanish and the remainder of (3-84) becomes

$$2(T + U) + W^E + W^M = 0 \qquad (3\text{-}85)$$

This obviously leads to contradiction, since each energy term on the left side is positive. *Consequently no self-confinement is possible.* The volume-integral term, including the Poynting vector in (3-84), is usually very small compared to the others. Therefore a plasmoid (left alone) is going to expand, since

$$\frac{d^2 I}{dt^2} > 0 \qquad (3\text{-}86)$$

The more energy (electromagnetic, kinetic, thermal) the plasmoid contains, the larger is this term and the faster the expansion.

3-6. Fluids and Plasmas

Let us look now at the physical picture presented by a collisionless plasma. Such a plasma consists of charged particles moving freely through space, influenced only by electromagnetic fields if they happen to be present. In general, particles close in space with completely different velocity vectors go off in different directions to remote parts of the plasma. The motion of the local center of mass is a fiction that does not correspond to any real mass motion. It is rather fortunate, therefore, to have found that these fictitious quantities obey some quasi-hydrodynamic equations: an equation of continuity (3-55) and an equation of motion (3-56b). This is about as far as the analogy can be stretched. Unfortunately these equations do not suffice to provide solutions for the great number of unknown quantities. In addition to the n^σ's and V, the six independent components of the symmetric tensor p_{ik} are all unknown. Furthermore, these equations should be coupled to Maxwell's equations to provide the electromagnetic force-density term in (3-56b).

With a fluid or a dense gas—or for that matter a plasma with a high-enough collision rate—the situation is quite different. The particles are restricted in their motion by collisions with neighbors. As a result they tend to stick together, and their motion may be represented by the superposition of the local mass motion, plus an isotropic distribution of velocities. These properties provide V with a physical meaning as the fluid velocity, and reduce p_{ik} to a scalar. For a one-component gas, with known fields, the equation of continuity and the equation of motion contain only the unknowns n, V, and p, and an additional equation of state connecting n and p renders the system solvable.

Fortunately one can get a step closer to real fluid-type behavior in collisionless plasmas by sacrificing some of the generality of treatment. We have seen that if the magnetic field is sufficiently strong, and its spatial and time variation slow, particles are forced into nearly circular paths around a drifting guiding center. The principal motion of the guiding center, perpendicular to the magnetic field direction, is the electric drift, common to all particles. We might suspect that this drift could play the role of local mass motion, while the gyrating motion of many particles with different phases exhibits the character of an isotropic velocity distribution. Thus the effect of the magnetic field might to a certain extent replace the role of collisions. We see, however, that this works only perpendicular to the magnetic field lines, while the motion along field lines remains chaotic. To force a closer analogy to fluid dynamics we shall have to make assumptions concerning longitudinal motion.

We have investigated in Secs. 2-2 and 2-3 the domain of validity of the guiding-center approximation for a particle. For a plasma we should add that the space should be densely populated by particles and the particle distribution function f should vary only slowly in a range of the order of the cyclotron radii, and during a gyration period.

We proceed now as follows. Starting out from the precise "hydrodynamic" equations (3-55) and (3-56b), we consider spatial and time derivatives as "small." Since the only term not containing such derivatives is the electromagnetic force density, this approximation is equivalent to the recognition of the leading role of this term. The approximate equations will consequently be composed of the zero-order mass velocity and pressure tensor, while retaining higher-order terms in computing the force density.

The velocity of a single particle is in the zero-order approximation

$$\mathbf{v} = \mathbf{u}_\perp + \mathbf{w}_\perp{}^0 + \mathbf{w}_\parallel{}^0 \tag{3-87}$$

where $\mathbf{u}_\perp$ is again the gyration velocity about the guiding center and $\mathbf{w}_\perp{}^0$ and $\mathbf{w}_\parallel{}^0$ are the zero-order drifts. To express $\mathbf{V}$ we average over particles of the same kind and over particle species σ. To simplify the notation, all this averaging will be included in angular brackets $\langle \ \rangle$. This leads to

$$\mathbf{V} = \langle \mathbf{u}_\perp \rangle + \langle \mathbf{w}_\perp{}^0 \rangle + \langle \mathbf{w}_\parallel{}^0 \rangle = \langle \mathbf{w}_\perp{}^0 \rangle + \langle \mathbf{w}_\parallel{}^0 \rangle \tag{3-88}$$

since as a consequence of our approximation the phases of particle velocities as well as the guiding centers are uniformly distributed (to the zero order), hence $\langle \mathbf{u}_\perp \rangle = 0$. We recognize furthermore that $\mathbf{w}_\perp{}^0 = \mathbf{w}^E$ is equal for all particles.

In the same approximation the pressure tensor becomes

$$p_{ik} = \sum_\sigma m^\sigma n^\sigma \langle (\mathbf{u}_\perp + \mathbf{w}_\parallel{}^0 - \langle \mathbf{w}_\parallel{}^0 \rangle)_i (\mathbf{u}_\perp + \mathbf{w}_\parallel{}^0 - \langle \mathbf{w}_\parallel{}^0 \rangle)_k \rangle \tag{3-89}$$

After introducing $\mathbf{u}_\parallel = \mathbf{w}_\parallel{}^0 - \langle \mathbf{w}_\parallel{}^0 \rangle$ one obtains

$$p_{ik} = \sum_\sigma m^\sigma n^\sigma (\langle u_{\perp i} u_{\perp k} \rangle + \langle u_{\perp i} u_{\parallel k} \rangle + \langle u_{\parallel i} u_{\perp k} \rangle + \langle u_{\parallel i} u_{\parallel k} \rangle) \tag{3-90}$$

Since $\mathbf{u}_\perp$ and $\mathbf{u}_\parallel$ are uncorrelated and both quantities average out to zero, the mixed terms vanish and we are left with

$$p_{ik} = \sum_\sigma m^\sigma n^\sigma (\langle u_{\perp i} u_{\perp k} \rangle + \langle u_{\parallel i} u_{\parallel k} \rangle) \tag{3-91}$$

We see that in the lowest order the pressure tensor splits into a longitudinal and a transverse component. It follows again from the isotropic $\mathbf{u}_\perp$ distribution that in a local "magnetic" coordinate system with the third axis along $\mathbf{B}$, $\langle u_i \cdot u_k \rangle$ vanishes for $i \neq k$ and the two perpendicular (to $\mathbf{B}$) components agree. Consequently,

$$p_{ik} = \begin{pmatrix} p_\perp & 0 & 0 \\ 0 & p_\perp & 0 \\ 0 & 0 & p_\parallel \end{pmatrix} \tag{3-92}$$

in this local system. In the laboratory system this tensor is given by

$$p_{ik} = p_\perp(\delta_{ik} - t_i t_k) + p_\parallel t_i t_k \tag{3-93}$$

where $\mathbf{t} = \mathbf{B}/B$. Since for $\mathbf{t}(0,0,1)$ (the local system) (3-93) reduces to (3-92), considering p_{ik}'s tensor-transformation properties, its choice must be the correct one.

Because of the importance of this so-called "hydromagnetic" description of plasmas, we shall approach it now from another, more formal point of view, where the approximations made are more precisely defined. We start out from the collisionless Boltzmann equation

$$\frac{\partial f}{\partial t} + \mathbf{v} \cdot \frac{\partial f}{\partial \mathbf{r}} + \frac{q}{m}(\mathbf{E} + \mathbf{v} \times \mathbf{B}) \cdot \frac{\partial f}{\partial \mathbf{v}} = 0 \qquad (3\text{-}94)$$

for a particle species. The guiding-center approximation suggests that the cyclotron frequency $(q/m)B$ is to be taken much larger than other "frequencies." These are $\partial/\partial t$ and $\mathbf{v} \cdot \partial/\partial \mathbf{r}$, both of the dimension 1 sec^{-1}. It means, of course, that these operators applied to f should result in "small" values, requiring a correspondingly slow space-time variation of the distribution function. Thus in the zero order one has

$$\frac{q}{m}(\mathbf{E} + \mathbf{v} \times \mathbf{B}) \cdot \frac{\partial f^0}{\partial \mathbf{v}} = 0 \qquad (3\text{-}95)$$

We introduce now the new variable $\mathbf{u}$ through $\mathbf{u} = \mathbf{v} - (\mathbf{E} \times \mathbf{B}/B^2)$. It follows that

$$\mathbf{u} \times \mathbf{B} = (\mathbf{E} + \mathbf{v} \times \mathbf{B}) - \mathbf{E}_\| \qquad (3\text{-}96)$$

We assume now that $\mathbf{E}_\| = 0$. This reduces (3-95) to

$$(\mathbf{u} \times \mathbf{B}) \cdot \frac{\partial f^0}{\partial \mathbf{v}} = 0 \qquad (3\text{-}97)$$

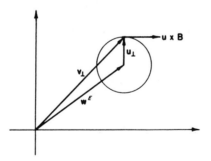

FIG. 3-3. Velocity-space trajectory of a particle in a strong magnetic field in the $\mathbf{v} \cdot \mathbf{B} = 0$ plane.

To understand the meaning of this equation we take a look at the plane of velocity space perpendicular to the magnetic field, where $\mathbf{v} \cdot \mathbf{B} = 0$ (Fig. 3-3). The gradient of f^0 in velocity space contains no component in the $\mathbf{u} \times \mathbf{B}$ direction, according to (3-97). Since this is true for any $\mathbf{u}$, f must be constant on $u_\perp = $ const circles, hence its distribution isotropic (in the plane perpendicular to $\mathbf{B}$) around the center $\mathbf{w}^E$. This is equivalent to saying that

$$f^0(\mathbf{r},\mathbf{v},t) = f^0(u_\perp, v_\parallel, \mathbf{r}, t) \tag{3-98}$$

This leads immediately to $\mathbf{V} = \mathbf{w}^E$ and to the form of the pressure tensor as given by (3-92) (see Exercises 3-13 and 3-14). The first-order approximation f^1 can be obtained by using f^0 in the next term of a series expansion in terms of q/m ($q/m \gg 1$). The first-order equation reads

$$\frac{\partial f^0}{\partial t} + \mathbf{v}\,\frac{\partial f^0}{\partial \mathbf{r}} + \frac{q}{m}\,(\mathbf{E}^0 + \mathbf{v} \times \mathbf{B}^0)\,\frac{\partial f^1}{\partial \mathbf{v}} = 0 \tag{3-99}$$

Instead of pursuing this line further we use (3-88) and (3-93) in the hydrodynamic equations (3-55) and (3-56b). The equation of continuity becomes

$$\frac{\partial v}{\partial t} + \nabla \cdot (v\mathbf{V}) = 0 \tag{3-100}$$

where v has been introduced for the mass density and $\mathbf{V} = \mathbf{w}^E + \langle \mathbf{w}_\parallel{}^0 \rangle$.

To transcribe (3-56b) we need to evaluate the divergence of p_{ik}. This goes as follows:

$$\frac{\partial}{\partial x_i}\,p_{ik} = (\delta_{ik} - t_i t_k)\,\frac{\partial}{\partial x_i}\,p_\perp + (p_\parallel - p_\perp)t_i\,\frac{\partial t_k}{\partial x_i}$$
$$+ t_k\!\left[t_i\,\frac{\partial p_\parallel}{\partial x_i} + (p_\parallel - p_\perp)\,\frac{\partial t_i}{\partial x_i} \right] \tag{3-101a}$$

The divergence vector splits up easily into a parallel and a perpendicular component. Transforming

$$(\delta_{ik} - t_i t_k)\,\frac{\partial}{\partial x_i} = \nabla_{\perp k} \tag{3-102}$$

into a $\mathbf{t}(0,0,1)$ system, its perpendicular character can be immediately recognized. We have already met the perpendicular vector

$$t_i\,\frac{\partial t_k}{\partial x_i} = (\mathbf{t} \cdot \nabla)\mathbf{t} \tag{3-103}$$

in (2-63), and the scalar term multiplied by $\mathbf{t}$ yields the parallel component. Thus (3-101a) breaks up into

$$\left[\frac{\partial}{\partial x_i}\,p_{ik} \right]_\perp = \nabla_\perp p_\perp + (p_\parallel - p_\perp)(\mathbf{t} \cdot \nabla)\mathbf{t} \tag{3-101b}$$

and

$$\left[\frac{\partial}{\partial x_i}\,p_{ik} \right]_\parallel = \mathbf{t} \cdot \nabla p_\parallel + (p_\parallel - p_\perp)\,\nabla \cdot \mathbf{t} \tag{3-101c}$$

As a result, we write down the perpendicular component of (3-56b). This reads, in our approximation,

$$v(d\mathbf{V}/dt)_\perp + \nabla_\perp p_\perp + (p_\| - p_\perp)(\mathbf{t} \cdot \nabla)\mathbf{t} = \mathbf{f}_\perp \qquad (3\text{-}104)$$

where we introduced $\mathbf{f}$ for the electromagnetic force density. The parallel equation reads

$$v(d\mathbf{V}/dt)_\| + \mathbf{t} \cdot \nabla p_\| + (p_\| - p_\perp)\nabla \cdot \mathbf{t} = 0 \qquad (3\text{-}105a)$$

since by assumption $\mathbf{E}_\| = 0$. From $\mathbf{t} = \mathbf{B}/B$, using $\nabla \cdot \mathbf{B} = 0$, it follows that $\nabla \cdot \mathbf{t} = -(\mathbf{B}/B^2) \cdot (\nabla B)$ and (3-105a) becomes

$$v(d\mathbf{V}/dt)_\| + \nabla_\| p_\| + (p_\perp - p_\|)(\nabla B/B)_\| = 0 \qquad (3\text{-}105b)$$

where $\nabla_\| = \mathbf{t} \cdot \nabla$.

Equations (3-100), (3-104), (3-105b), and Maxwell's equations for the electromagnetic field, with $\mathbf{w}^E = (\mathbf{E} \times \mathbf{B}/B^2)$ providing a connection between the two sets, present an almost complete system of equations for the unknowns $\mathbf{E}$, $\mathbf{B}$, v, $p_\|$, $p_\perp$, and $\mathbf{V}$. We miss two *equations of state* coupling v to $p_\|$ and $p_\perp$. Since in the zero order the velocity distribution is isotropic in the perpendicular plane and in the longitudinal direction, the two- and one-dimensional adiabatic expressions can be used, in the zero-order approximation, as was shown in Sec. 3-5, provided there is no energy transport from one region of the plasma to another. It is easy to show (Exercise 3-15) that this expression is

$$\frac{p_\perp^2 p_\|}{v^5} = \text{const} \qquad (3\text{-}106)$$

The last missing equation of state can be obtained by recognizing the connection between the magnetic field strength and the perpendicular component of the particle energy. Owing to the constancy of magnetic moments,

$$\frac{\langle \tfrac{1}{2}mu_\perp^2 \rangle}{B} = \frac{p_\perp}{vB} = \text{const} \qquad (3\text{-}107)$$

Usually further simplifications are made regarding the electromagnetic properties of plasmas. Charge neutrality and hence the absence of electrostatic forces, on the one hand, and the neglect of the displacement current in comparison with plasma currents, on the other, are well-justified assumptions for an important class of plasmas. This leads to†

$$\mathbf{f} = \mathbf{J} \times \mathbf{B} = \frac{1}{\mu_0}(\mathbf{B} \cdot \nabla)\mathbf{B} - \nabla\frac{B^2}{2\mu_0} \qquad (3\text{-}108)$$

† Note that the zero-order (isotropic) distribution function leads to $\mathbf{f} = 0$. However, as mentioned earlier, in the evaluation of the force density we proceed to a higher-order approximation.

Substituting this in (3-104) one obtains

$$v \left[\frac{d}{dt} \mathbf{V} \right]_\perp + \nabla_\perp \left(p_\perp + \frac{B^2}{2\mu_0} \right) - \left[\frac{(\mathbf{B} \cdot \nabla)\mathbf{B}}{\mu_0} \right]_\perp \left(\frac{p_\perp - p_\parallel}{B^2/\mu_0} + 1 \right) = 0 \quad (3\text{-}109)$$

From the definition of $\mathbf{w}^E$ and our assumption $E_\parallel = 0$, one finds

$$\mathbf{E} = -\mathbf{w}^E \times \mathbf{B} = -\mathbf{V} \times \mathbf{B} \qquad (3\text{-}110)$$

Combining this with Maxwell's $\nabla \times \mathbf{E} = -\dot{\mathbf{B}}$ results in

$$\nabla \times (\mathbf{V} \times \mathbf{B}) = \frac{\partial \mathbf{B}}{\partial t} \qquad (3\text{-}111)$$

We have eliminated the electric-field variable from our equations completely. We now have (3-100), (3-105b), (3-109), (3-111), and the equations of state (3-106) and (3-107) for determining $\mathbf{V}, p_\perp, p_\parallel, v, \mathbf{B}$.

This set of equations is referred to as the Chew-Goldberger-Low (CGL) set of equations, or the *double-adiabatic approximation* as a reference to the equations of state (3-106) and (3-107). They are rather complicated nonlinear equations. Under certain conditions they can be further simplified to approach the equations which describe the physics of conductive fluids. The branch of physics which deals with conductive fluids is called *magnetohydrodynamics or hydromagnetics*.

3-7. Summary

The motion of a large number of particles in electromagnetic fields can be described by the distribution function which obeys the Boltzmann equation,

$$\frac{df}{dt} = \frac{\partial f}{\partial t} + \mathbf{v} \cdot \frac{\partial f}{\partial \mathbf{r}} + \frac{q}{m} (\mathbf{E} + \mathbf{v} \times \mathbf{B}) \cdot \frac{\partial f}{\partial \mathbf{v}} = \left(\frac{\partial f}{\partial t} \right)_c$$

where the right side signifies the time rate of change of the distribution function due to collisions alone. Often collisions can be neglected and one arrives at the collisionless Boltzmann or Vlasov equation.

Moments of the Boltzmann equation in velocity space (multiplication by powers of $\mathbf{v}$ and integration over d^3v) yield equations in which the variables depend only on space and time. The first three of these yield the equation of continuity (conservation of particles), equation of motion (conservation of momentum), and equation of state (conservation of energy). These "fluid-like" equations contain a tensor pressure p_{ik} and a heat-flow term in the equation of state. They do not form a closed set of equations, since the number of unknowns exceeds the number of equations, nor is this difficulty alleviated by the inclusion of higher-order moments.

One can approach a more fluid-like behavior by imposing a sufficiently strong magnetic field and assuming that there is no heat flow along the lines of force. The pressure tensor now becomes diagonal, containing only $p_\parallel$ and $p_\perp$. The absence of heat flow across the field (the particles gyrate about the lines), and following our assumption, along the field, results in two adiabatic equations of state, one in the parallel and the other in the perpendicular direction. The resulting equations form a closed set describing a fluid pressure (and temperature) different along and perpendicular to the field in a plasma that moves with the electric drift velocity $\mathbf{w}^E$ across the field.

This set of equations is known as the CGL equations or the double-adiabatic approximation. They consist of the equation of continuity

$$\frac{\partial v}{\partial t} + \nabla \cdot (v\mathbf{V}) = 0$$

two equations of motion

$$v\left(\frac{d\mathbf{V}}{dt}\right)_\parallel + \nabla_\parallel p_\parallel + (p_\perp - p_\parallel)\left(\frac{\nabla B}{B}\right)_\parallel = 0$$

$$v\left(\frac{d\mathbf{V}}{dt}\right)_\perp + \nabla_\perp\left(p_\perp + \frac{B^2}{2\mu_0}\right) - \left[\frac{(\mathbf{B}\cdot\nabla)\mathbf{B}}{\mu_0}\right]_\perp\left(\frac{p_\perp - p_\parallel}{B^2/2\mu_0} + 1\right) = 0$$

an equation describing the relationship between plasma velocity and the magnetic field

$$\nabla \times (\mathbf{V} \times \mathbf{B}) = \dot{\mathbf{B}}$$

and the two equations of state

$$\frac{d}{dt}\left(\frac{p_\perp^2 p_\parallel}{v^5}\right) = 0 \quad \text{and} \quad \frac{d}{dt}\left(\frac{p_\perp}{vB}\right) = 0$$

One often seeks solutions to the collisionless Boltzmann equation. If the constants of particle motion $c_i(\mathbf{r},\mathbf{v},t)$ are known, the most general solutions can be easily constructed, in the form $f(c_1 \cdots c_6)$, where the arbitrary function f is to be determined from the initial conditions. If some (but not all) of the constants are known, solutions can be constructed in the same way, but now they are not general and hence they cannot be fitted to arbitrary initial conditions.

Some general laws follow from the structure of the Boltzmann equation: The virial theorem, for instance, shows that no self-confined stationary plasma can exist confined by electromagnetic fields generated by the plasma alone. A plasma confined by external fields cannot be in thermal equilibrium, while a collisionless plasma cannot change its entropy; hence it seems impossible to pass from a confined state to (unconfined) thermal equilibrium by collisionless processes alone. Later we shall see that the plasma is usually able to overcome this obstacle and destroy confinement through collisionless instabilities. A uniform Maxwellian (thermal) plasma is, of course, stable (it possesses the highest entropy it can have). Another (rather trivial) example of a spatially uniform, in velocity space monotonically decreasing, distribution was shown to be stable against collisionless perturbations.

EXERCISES

3-1. Derive (3-39) in analogy with the derivation of (3-38).

3-2. Find the expression of the pressure-balance equation (3-45a) in cylindrical coordinates for an azimuthally symmetric system $(\partial/\partial t = \partial/\partial\varphi = 0)$.

3-3. Prove by direct substitution that for a one-dimensional motion of charged particles in the potential field $V(x)$, $f(W) = f(\frac{1}{2}mv^2 + qV)$ is a stationary solution of the Boltzmann equation.

3-4. Prove that the Vlasov equation in cylindrical coordinates assumes the following form:

$$\frac{\partial f}{\partial t} + \dot{r}\frac{\partial f}{\partial r} + \dot{\varphi}\frac{\partial f}{\partial\varphi} + r\dot{\varphi}^2\frac{\partial f}{\partial\dot{r}} - \frac{2\dot{r}\dot{\varphi}}{r}\frac{\partial f}{\partial\dot{\varphi}} + \dot{z}\frac{\partial f}{\partial z}$$

$$+ \frac{1}{m}\left(F_r\frac{\partial f}{\partial\dot{r}} + \frac{F_\varphi}{r}\frac{\partial f}{\partial\dot{\varphi}} + F_z\frac{\partial f}{\partial\dot{z}}\right) = 0$$

(*Hint*: $df/dt = 0$ is independent of the coordinate system.)

3-5. By substitution into the Boltzmann equation in cylindrical coordinates (see Exercise 3-4) show that

$$f = f(\tfrac{1}{2}mv^2, mr^2\dot{\varphi} + qrA_\varphi)$$

is a stationary solution for an azimuthally symmetric magnetic field.

3-6. Noninteracting charged particles (charge q, mass m) move in the uniform magnetic field $\mathbf{B} = B_0\mathbf{e}_3$. The distribution function is stationary and independent of y and z. What is the general form of the distribution function $f(x, v_x, v_y, v_z)$?

3-7. Consider a collisionless plasma confined by an azimuthally symmetric $B_z(r)$ field. Show that the kinetic angular momentum L obeys the equation

$$\partial L/\partial t = \int \rho r E_\varphi \, dV - \int J_r B_z r \, dV$$

(*Hint:* Use the conservation of P_φ for individual particles.)

3-8. Derive the general form of the second moment of the Boltzmann equation in the absence of collisions, multiplying by $mv_i v_k$ and integrating over velocity space.

3-9. Prove that the Jacobian of the transformation from $(\mathbf{r},\mathbf{v})$ to phase space is a constant, hence the transformation is volume-preserving.

3-10. Prove that the total time derivative of the entropy (3-70) vanishes for a system obeying the collisionless Boltzmann equation.

3-11. Derive the adiabatic equation of state for a one-dimensional collisionless gas.

3-12. Many people claim to have seen "ball lightning," a glowing spherical object (presumably a plasma), 10^{-1} to 1 m in diameter, drifting through the air for seconds. Owing to the virial theorem, these balls must be confined by external forces, probably by atmospheric pressure. Give the upper limit of the energy content of ball lightning, assuming a 1-m diameter.

3-13. Show that the average (or "hydrodynamic") velocity, corresponding to the distribution function given by (3-98), is $\mathbf{w}^E$.

3-14. Show that the pressure tensor of a distribution function (3-98) reduces to that given in (3-92).

3-15. Consider a collisionless gas where the motion of particles in the $z(\|)$ direction is independent of the motion in the $xy(\perp)$ plane. Prove that the adiabatic equation of state for such a gas is $p_\perp{}^2 p_\| v^{-5} = \text{const.}$

3-16. Prove that the energy conservation law (3-30) summed over the particle species may be cast in the form (3-59). [Make use of the equation of continuity and the momentum conservation law (3-76b).]

3-17. Use (3-59) in the absence of a heat-flow term to derive the adiabatic equations of state (3-75) and (3-76).

IV

Magnetohydrodynamics of
Conductive Fluids

___

4-1. Fundamental Equations

We saw in Chapter III that plasma equations, under certain limiting conditions, can be simplified and brought to a quasi-hydrodynamic form. In addition to its hydrodynamic properties, a plasma possesses equally important electromagnetic properties, a fact reflected in its basic equations. The question can be asked, therefore: "What kind of fluids do plasmas resemble in the limiting cases under consideration?" As we shall see presently, the answer is: "Fluids that possess an infinite conductivity." Therefore in this chapter we are going to forget about plasmas, at least for a while, and investigate the properties of conductive fluids. Later on we shall compare our results with the plasma equations studied in the last chapter and examine the limits of applicability of this hydromagnetic treatment to plasmas.

Real fluids are subject to the equation of continuity

$$\frac{\partial v}{\partial t} + \nabla \cdot (v\mathbf{v}) = 0 \tag{4-1a}$$

and the equation of motion

$$v\frac{d\mathbf{v}}{dt} = -\nabla p + \mathbf{f} \tag{4-2}$$

where $d/dt = (\partial/\partial t) + (\mathbf{v} \cdot \nabla)$, $\mathbf{f}$ is the force density acting on the unit fluid volume, and the pressure p is now a scalar. The electromagnetic field is, of course, governed by Maxwell's equations. Since fluids of high conductivity will be considered, it is only natural to neglect the displacement currents compared to conduction currents inside the fluid. By the same token, accumulation of space charges inside the conductor will be neglected. For convenience we write Maxwell's equations in these reduced forms:

$$\nabla \times \mathbf{H} = \mathbf{J} \tag{4-3}$$

$$\nabla \times \mathbf{E} = -\partial \mathbf{B}/\partial t \tag{4-4}$$

$$\nabla \cdot \mathbf{D} = 0 \tag{4-5}$$

$$\nabla \cdot \mathbf{B} = 0 \tag{4-6}$$

The fluid equation and the equations describing the electromagnetic field are coupled by the electromagnetic forces acting on the fluid, namely,

$$\mathbf{f} = \mathbf{J} \times \mathbf{B} \tag{4-7}$$

and by Ohm's law relating the current density to field quantities:

$$\mathbf{J} = \sigma(\mathbf{E} + \mathbf{v} \times \mathbf{B}) = \sigma\mathbf{E}' \tag{4-8a}$$

where $\mathbf{E}' = \mathbf{E} + \mathbf{v} \times \mathbf{B}$ is the electric field felt by the observer (or an electron) moving with the fluid element.

The description of the fluid behavior is not complete without an equation of state. This can assume various forms, depending on the nature of the fluid and the processes involved. For an incompressible fluid it is

$$\nabla \cdot \mathbf{v} = 0 \tag{4-9a}$$

for an isothermal process

$$\frac{d}{dt}\left(\frac{p}{v}\right) = 0 \tag{4-9b}$$

or for an adiabatic process

$$\frac{d}{dt}(pv^{-\gamma}) = 0 \tag{4-9c}$$

While (4-9a) simplifies the equations considerably, we usually take (4-9c), which allows for closer analogy with plasmas. Equations (4-1) to (4-9) form a closed system to be solved for any one of the fluid or electromagnetic variables. To reduce the number of equations one eliminates the electric field variable. We substitute (4-7) in (4-2) with the help of (4-3):

$$v\frac{d\mathbf{v}}{dt} = -\nabla p + \frac{1}{\mu_0}(\nabla \times \mathbf{B}) \times \mathbf{B} = -\nabla\left(p + \frac{B^2}{2\mu_0}\right) + \frac{(\mathbf{B} \cdot \nabla)\mathbf{B}}{\mu_0} \tag{4-10}$$

and (4-8a) in (4-4) using (4-3) and (4-6):

$$\frac{\partial \mathbf{B}}{\partial t} = \nabla \times (\mathbf{v} \times \mathbf{B}) + \frac{1}{\mu_0\sigma}\nabla^2\mathbf{B} \tag{4-11a}$$

Equations (4-1), (4-6), (4-10), (4-11a), and one of equations (4-9) contain all the information needed to solve for the four variables v, $\mathbf{v}$, p, and $\mathbf{B}$. For infinite conductivity, $\sigma \to \infty$, (4-11a) becomes

$$\frac{\partial \mathbf{B}}{\partial t} = \nabla \times (\mathbf{v} \times \mathbf{B}) \tag{4-11b}$$

The analogy with (3-100), (3-105b), (3-109), and (3-111), except for the isotropic pressure, is obvious. Note that for infinite conductivity (4-8a) reduces to

$$\mathbf{E} + \mathbf{v} \times \mathbf{B} = 0 \tag{4-8b}$$

the same as (3-110), where the electric drift $\mathbf{w}^E$ plays the role of fluid velocity. The electric drift transforms the electric field away and so does the fluid velocity in a fluid of infinite conductivity. In such a fluid any finite electric field in the co-moving fluid system would draw infinite currents, which is, of course, impossible.

To solve our basic equations, knowledge of the boundary conditions is required also. Consider a fluid-vacuum interface. If we denote by $\mathbf{n}$ the unit vector normal to the interface, and by square brackets [ ] the increment in any quantity across the boundary, we have from (4-6)

$$\mathbf{n} \cdot [\mathbf{B}] = 0 \tag{4-12a}$$

We consider a fluid of infinite conductivity, with $\mathbf{E}$ and $\mathbf{B}$ coupled through (4-8b). We multiply this equation vectorially with $\mathbf{n}$, on both sides of the interface, subtract the two, apply (4-12a), and readily obtain

$$\mathbf{n} \times [\mathbf{E}] = \mathbf{n} \cdot \mathbf{v}[\mathbf{B}] \tag{4-13}$$

which says in effect that the tangential component of the electric field felt by the co-moving observer, $\mathbf{E}'_t$, is continuous. If surface currents are present with the density $\mathbf{J}^*$ (4-3) yields at the surface

$$\mathbf{n} \times [\mathbf{B}] = \mu_0 \mathbf{J}^* \tag{4-14}$$

Denoting an infinitesimal displacement from one side of the boundary to the other by $\delta\mathbf{x}$, the equation of motion (4-10) becomes

$$\delta\mathbf{x} \cdot v \frac{d\mathbf{v}}{dt} = -\delta\left(p + \frac{B^2}{2\mu_0}\right) + \frac{(\mathbf{B} \cdot \nabla)\mathbf{B}}{\mu_0} \cdot \delta\mathbf{x} \tag{4-15}$$

In the limit $\delta x \to 0$, (4-15) yields

$$\delta\left(p + \frac{B^2}{2\mu_0}\right) = \left[p + \frac{B^2}{2\mu_0}\right] = 0 \tag{4-16}$$

In the presence of a surface current density $\mathbf{J}^*$, one can avoid infinite force densities tangential to the surface, only by requiring that

$$\mathbf{n} \cdot \mathbf{B} = 0 \tag{4-12b}$$

Equations (4-12b), (4-13), (4-14), and (4-16) are the boundary conditions for a fluid-vacuum interface. For an interface between two fluids,

$$\mathbf{n} \cdot [\mathbf{v}] = 0 \tag{4-17}$$

should be added. The fluid is sometimes in contact with a rigid, perfectly conductive wall. For this case (4-17) and (4-13) reduce to

$$\mathbf{n} \cdot \mathbf{v} = 0 \tag{4-18}$$

and

$$\mathbf{n} \times [\mathbf{E}] = \mathbf{n} \times \mathbf{E} = 0 \tag{4-19}$$

since $\mathbf{E}$ vanishes in the conductor. Equations (4-4) and (4-19) yield the condition on the surface of the rigid conductor,

$$\mathbf{n} \cdot \frac{\partial \mathbf{B}}{\partial t} = 0 \tag{4-20}$$

We conclude this section by proving that the system of hydromagnetic equations possesses an energy integral. Considering again infinite conductivity and the adiabatic equation of state (4-9c), we multiply (4-10) by $\mathbf{v}$:

$$v\mathbf{v} \cdot \frac{d\mathbf{v}}{dt} = -\mathbf{v} \cdot \nabla p + \frac{1}{\mu_0} \mathbf{v} \cdot (\nabla \times \mathbf{B}) \times \mathbf{B} \tag{4-21}$$

The term on the left side can be transformed:

$$v\mathbf{v} \cdot \left[\frac{\partial \mathbf{v}}{\partial t} + (\mathbf{v} \cdot \nabla)\mathbf{v} \right] = \tfrac{1}{2} v \left(\frac{\partial v^2}{\partial t} + \mathbf{v} \cdot \nabla v^2 \right)$$

$$= \frac{\partial}{\partial t} (\tfrac{1}{2} v v^2) - \frac{v^2}{2} \frac{\partial v}{\partial t} + \tfrac{1}{2} v\mathbf{v} \cdot \nabla v^2$$

$$= \frac{\partial}{\partial t} (\tfrac{1}{2} v v^2) + \frac{v^2}{2} \nabla \cdot (v\mathbf{v}) + \frac{v\mathbf{v}}{2} \cdot \nabla v^2$$

$$= \frac{\partial}{\partial t} (\tfrac{1}{2} v v^2) + \nabla \cdot \left(\frac{v^2}{2} v\mathbf{v} \right) \tag{4-22}$$

where (A-12) and (4-1a) have been used. For transforming the next term we write (4-1a) and (4-9c) in the form

$$\frac{dv}{dt} = -v\nabla \cdot \mathbf{v} \tag{4-1b}$$

and

$$\frac{dp}{dt} v^{-\gamma} - \gamma v^{-(\gamma+1)} p \frac{dv}{dt} = 0 \tag{4-9c'}$$

Combining the last two equations one obtains

$$\frac{\partial p}{\partial t} + \mathbf{v} \cdot \nabla p + \gamma p \nabla \cdot \mathbf{v} = 0 \tag{4-23}$$

or

$$\frac{\partial p}{\partial t} + (1 - \gamma)\mathbf{v} \cdot \nabla p + \gamma \nabla \cdot (p\mathbf{v}) = 0 \tag{4-24}$$

which leads to

$$\mathbf{v} \cdot \nabla p = \frac{1}{\gamma - 1} \frac{\partial p}{\partial t} + \frac{\gamma}{\gamma - 1} \nabla \cdot (p\mathbf{v}) \tag{4-25}$$

Finally the last term of (4-21) may be written

$$\mathbf{v} \cdot (\nabla \times \mathbf{B}) \times \mathbf{B} = -(\mathbf{v} \times \mathbf{B}) \cdot (\nabla \times \mathbf{B})$$

$$= \mathbf{E} \cdot (\nabla \times \mathbf{B}) = \mathbf{B} \cdot \nabla \times \mathbf{E} - \nabla \cdot (\mathbf{E} \times \mathbf{B}) \tag{4-26}$$

where use has been made of (4-8b) and (A-10). Using (4-4) we arrive at

$$\frac{1}{\mu_0} \mathbf{v} \cdot (\nabla \times \mathbf{B}) \times \mathbf{B} = -\frac{\partial}{\partial t} \frac{B^2}{2\mu_0} - \nabla \cdot (\mathbf{E} \times \mathbf{B}) \tag{4-27}$$

Using (4-22), (4-25) and (4-27) in (4-21) yields the energy conservation law

$$\frac{\partial}{\partial t}\left(\tfrac{1}{2}vv^2 + \frac{p}{\gamma - 1} + \frac{B^2}{2\mu_0}\right) + \nabla \cdot \left[\frac{v^2}{2} v\mathbf{v} + \frac{\gamma}{\gamma - 1} p\mathbf{v} + \mathbf{E} \times \mathbf{H}\right] = 0 \tag{4-28}$$

Integrating over the entire fluid-plus-vacuum volume, the divergence term yields a surface integral with the first two terms vanishing, since v, p and $\mathbf{v}$ are zero outside the fluid. The remaining surface term is the surface integral of the Poynting vector. For an isolated system this term vanishes and one obtains the energy conservation law

$$\int \left(\tfrac{1}{2}vv^2 + \frac{p}{\gamma - 1} + \frac{B^2}{2\mu_0}\right) d\tau = U = \text{const} \tag{4-29}$$

The first integral is the kinetic energy of the fluid, the second the thermal free energy, and the last one the total energy of the magnetic field. It is often useful to split the energy into a kinetic-energy part

$$K = \int \tfrac{1}{2}vv^2 \, d\tau \tag{4-30}$$

leaving

$$W = \int \left(\frac{p}{\gamma - 1} + \frac{B^2}{2\mu_0}\right) d\tau \tag{4-31}$$

to play the role of potential energy.

4-2. Magnetic Field Lines

Equation (4-11b) describes the coupling between the magnetic field and fluid motion for a perfectly conductive fluid. To examine the meaning of this relationship we integrate over an arbitrary surface inside the fluid and apply

Stokes' theorem

$$\frac{\partial}{\partial t}\int \mathbf{B} \cdot d\mathbf{S} - \oint (\mathbf{v} \times \mathbf{B}) \cdot d\mathbf{s} = 0 \qquad (4\text{-}32)$$

where the line integral is extended over the periphery of surface **S**. This expression can be rewritten as

$$\frac{\partial \phi}{\partial t} + \oint \mathbf{B} \cdot (\mathbf{v} \times d\mathbf{s}) = 0 \qquad (4\text{-}33)$$

The first term represents the rate of change of flux through the fixed surface **S**, while the second is the additional increment of flux, swept out per unit time by the periphery moving with the local fluid velocity **v** (Fig. 4-1). The left side of

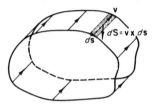

FIG. 4-1. Illustration of the flux conservation law for perfectly conductive fluids.

(4-33) therefore yields the total rate of change of flux through a "material" surface fixed to and moving with the fluid. Hence (4-33) expresses the constancy of flux through any material surface in a perfectly conducting fluid

$$\boxed{d\phi/dt = 0} \qquad (4\text{-}34)$$

The concept of the magnetic line of force is an abstraction. In general no identity can be attached to these lines (they cannot be labeled in a varying field) nor can we speak of a "motion" of field lines. In a perfect conductor, however, owing to the theorem of flux conservation, the concept of individual field lines becomes meaningful with measurable consequences.

Consider a material line in the fluid (e.g., a chain of labeled droplets), defined by intersecting material surfaces. Let us choose these surfaces everywhere tangential to the magnetic field, at $t = 0$. Consequently, the flux through both surfaces is zero and their intersection defines a field line at this particular moment. Following the theorem of flux conservation, the surfaces remain flux-free in the course of the motion. Since this is true for any two surfaces intersecting in the same line at $t = 0$, their intersection (the marked material line) always remains a field line. There is no contradiction in the assumption that it is the "same" field line. Labeling of the material line serves as a label to the field line as well, and the local fluid velocity $\mathbf{v}(\mathbf{r},t)$ is the velocity of the local section of a field line, in brief, *the field line is attached to or "frozen in" the fluid.*

In a fluid of finite conductivity this is no longer true. Surface integration of (4-11a) yields

$$\frac{d\phi}{dt} = \frac{1}{\mu\sigma} \int \nabla^2 \mathbf{B} \cdot d\mathbf{S} \tag{4-35}$$

the right side giving rise to a "slippage" of magnetic flux through a closed material line.

Note that (3-111) yields the same results for collisionless plasmas in the guiding-center approximation as for a perfect conductor, provided the electric drift is substituted for fluid velocity. It can be said therefore that *magnetic field lines are attached to the particle guiding centers in magnetoplasmas in the zero-order approximation.*

Magnetic fields can carry forces in a medium. It is useful at this point to review the various forms in which these forces can be recognized.

We are well acquainted with the formula

$$\mathbf{f} = \mathbf{J} \times \mathbf{B} \tag{4-36}$$

for the force density (Fig. 4-2a). Since current density and magnetic field are

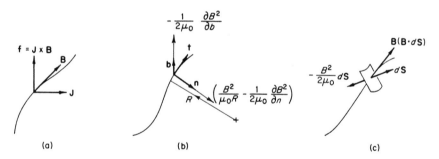

FIG. 4-2. Forces in magnetic fields.

connected through (4-3), knowledge of the former is not necessary for expressing **f**, if in addition to the local magnetic field, the field at adjacent points (or the partial derivatives) is given also. This way one arrives at the expression

$$\mathbf{f} = (\nabla \times \mathbf{H}) \times \mathbf{B} = \frac{1}{\mu_0} (\mathbf{B} \cdot \nabla)\mathbf{B} - \nabla \frac{B^2}{2\mu_0} \tag{4-37}$$

We get a clearer picture of the meaning of this expression by expanding it in components in a local coordinate system fitted to the magnetic field (Fig. 4-2b). Let us choose **t**, **n**, and **b** as the local unit vectors tangential, normal, and binormal to the field line, respectively (Frenet coordinate system). Further-more, we denote the differentiation along the field line, in the normal direc-tion, and in the binormal direction by $\partial/\partial s$, $\partial/\partial n$, and $\partial/\partial b$, respectively. In this system

$$(\mathbf{B} \cdot \nabla)\mathbf{B} = B \frac{\partial}{\partial s}(B\mathbf{t}) = \mathbf{t} \frac{\partial}{\partial s}\frac{B^2}{2} + \mathbf{n}\frac{B^2}{R} \tag{4-38}$$

since

$$\frac{\partial \mathbf{t}}{\partial s} = \frac{\mathbf{n}}{R} \tag{4-39}$$

where R is the local radius of curvature. As a result (4-37) becomes in these coordinates,

$$\mathbf{f} = 0 \cdot \mathbf{t} + \left(\frac{B^2}{\mu_0 R} - \frac{\partial}{\partial n}\frac{B^2}{2\mu_0}\right)\mathbf{n} - \left(\frac{\partial}{\partial b}\frac{B^2}{2\mu_0}\right)\mathbf{b} \tag{4-40}$$

There is, of course, no component along the field. In the plane normal to the field the force can be thought of as a superposition of a pressure action—as if the magnetic field were a fluid with the "magnetic pressure" $B^2/2\mu_0$—and a tension—as if the field lines were elastic cords each one with the tension B/μ_0—pressing against the fluid. The force density is the negative gradient of the pressure, while each "magnetic cord" presses with the force $B/\mu_0 R$ per unit length in the $\mathbf{n}$ direction, giving rise to the force density $B(B/\mu_0 R)$.

Finally, observing that $\nabla \cdot \mathbf{B} = \partial B_k/\partial x_k = 0$, we write (4-37) as

$$f_k = -\frac{\partial}{\partial x_k} T_{ik} = \frac{\partial}{\partial x_k}\left(\frac{B_i B_k}{\mu_0} - \frac{B^2}{2\mu_0}\delta_{ik}\right) \tag{4-41}$$

Through Gauss' theorem the force acting on a fluid volume can be expressed as a surface integral:

$$\int_v \mathbf{f} \, d\tau = -\int \mathbf{T} \cdot d\mathbf{S}$$

$$= \int \left[\frac{\mathbf{B}(\mathbf{B} \cdot d\mathbf{S})}{\mu_0} - \frac{B^2}{2\mu_0}d\mathbf{S}\right] \tag{4-42}$$

The geometric meaning of this expression is pictured in Fig. 4-2c. Note that since $\mathbf{B} = B\mathbf{t}$, the magnetic pressure tensor can be written

$$T_{ik} = \frac{B^2}{2\mu_0}(\delta_{ik} - t_i t_k) - \frac{B^2}{2\mu_0}t_i t_k$$

$$= T_\perp(\delta_{ik} - t_i t_k) + T_\parallel t_i t_k \tag{4-43}$$

with

$$T_\perp = B^2/2\mu_0 \tag{4-44}$$

and

$$T_\parallel = -B^2/2\mu_0 \tag{4-45}$$

Comparing (4-43) with (3-93) we see that in a local "magnetic" coordinate system, off-diagonal components of the magnetic pressure tensor vanish and it becomes

$$T_{ik} = \begin{pmatrix} B^2/2\mu_0 & 0 & 0 \\ 0 & B^2/2\mu_0 & 0 \\ 0 & 0 & -B^2/2\mu_0 \end{pmatrix} \tag{4-46}$$

Both the material pressure of a plasma in the guiding-center approximation and the magnetic pressure tensor can be diagonalized in the same coordinate system, where both yield identical components in any direction perpendicular to **B**.

4-3. Magnetohydrostatics

In static equilibrium the velocity **v** and the time derivatives are zero, and the set of equations of magnetohydrodynamics reduce to the magnetohydrostatic equations

$$\mathbf{J} \times \mathbf{B} = \nabla p \tag{4-47}$$

$$\nabla \times \mathbf{H} = \mathbf{J} \tag{4-48}$$

$$\nabla \cdot \mathbf{B} = 0 \tag{4-49}$$

or, eliminating **J**, to the equivalent set

$$\frac{1}{\mu_0} (\nabla \times \mathbf{B}) \times \mathbf{B} = \nabla p \tag{4-50a}$$

and

$$\nabla \cdot \mathbf{B} = 0 \tag{4-49}$$

It is useful to rewrite (4-50a) in the form of a pressure-balance equation,

$$\nabla\left(p + \frac{B^2}{2\mu_0}\right) = \frac{(\mathbf{B} \cdot \nabla)\mathbf{B}}{\mu_0} \tag{4-50b}$$

with the magnetic pressure $B^2/2\mu_0$ playing the same role as the fluid pressure.

A special case is the force-free field, with

$$(\nabla \times \mathbf{B}) \times \mathbf{B} = 0 \tag{4-51}$$

and

$$\nabla \cdot \mathbf{B} = 0 \tag{4-49}$$

Equation (4-51) can be satisfied with the nontrivial solution

$$\nabla \times \mathbf{B} = \alpha(\mathbf{r})\mathbf{B} \tag{4-52}$$

Taking the divergence of this equation and using (4-49) one obtains

$$\mathbf{B} \cdot \nabla \alpha = 0 \tag{4-53}$$

The α = const surfaces are "magnetic surfaces" in the sense that they are made up of magnetic field lines.

In the general case $\nabla p \neq 0$, (4-47) to (4-50a) have some simple consequences. From (4-47),

$$\mathbf{J} \cdot \nabla p = 0 \tag{4-54}$$

and

$$\mathbf{B} \cdot \nabla p = 0 \tag{4-55}$$

The p = const surfaces are both "magnetic surfaces" and "current surfaces." It follows from (4-48) that

$$\nabla \cdot \mathbf{J} = 0 \tag{4-56}$$

and from (4-50b) that

$$\nabla \times (\mathbf{B} \cdot \nabla)\mathbf{B} = 0 \tag{4-57}$$

Since both $\mathbf{B}$ and $\mathbf{J}$ are divergence-free, the magnetic field as well as the current lines either extend to infinity or are closed (or perhaps ergodic) in the finite. The p = const magnetic and current surfaces are either tubes extending to infinity or assume the form of toroids. This is true at least for "smooth" variation of the variables with p = const surfaces containing no edges.

First we shall investigate configurations extending to infinity. Consider a system of straight and parallel magnetic field lines, created, for example, by a solenoid of infinite length (Fig. 4-3). This is the θ pinch configuration. In the absence of a conductive fluid the field is uniform. Introducing a fluid—being careful not to disturb symmetry along the lines—(4-50b) becomes

$$\nabla\left(p + \frac{B^2}{2\mu_0}\right) = 0 \tag{4-58}$$

while (4-49) is automatically satisfied by our prescription for field lines. According to (4-58) the magnetic field is reduced in the region occupied by the fluid. This is just a special case of the pressure-balance equation discussed in Sec. 3-3. If the external field (the field outside the fluid, which agrees here with the field before introducing the fluid) is B_0, then the maximum fluid pressure to be confined is

$$p_{\max} = B_0^2/2\mu_0 \tag{4-59}$$

If B_0 is known (boundary condition), and the distribution of fluid pressure $p(x, y)$ is given, the magnetic pressure can be immediately "filled in" in (4-58) and the magnetic field computed. In a special case a region can be filled with a fluid of pressure $p_{\max}$, putting fluid nowhere else. This creates a sharp boundary between a region of no fluid with magnetic field strength B_0 and a field-free fluid region.

Another example of interest is the so-called *pinch configuration*. Consider an infinite cylindrical fluid column, carrying currents in the z direction, thereby creating azimuthal magnetic fields (Fig. 4-4). The $\mathbf{J} \times \mathbf{B}$ force points

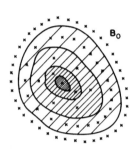

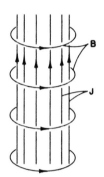

FIG. 4-3. Magnetofluid confined by straight parallel field lines (θ pinch).

FIG. 4-4. Magnetofluid column confined by fields generated by axial currents in the column (pinch geometry).

radially inward and counteracts the pressure gradient in the fluid. The $p =$ const surfaces are concentric cylinders. The radial component of the pressure-balance equation becomes for this case [using, e.g., (4-40)]

$$\frac{\partial}{\partial r}\left(p + \frac{B^2}{2\mu_0}\right) + \frac{B^2}{\mu_0 r} = 0 \tag{4-60}$$

while the φ- and z-component equations are trivial. In integral form,

$$\left[p + \frac{B^2}{2\mu_0}\right]_{r_1}^{r_2} + \frac{1}{\mu_0}\int_{r_1}^{r_2}\frac{B^2}{r}\, dr = 0 \tag{4-61}$$

Here again the divergence equation is automatically satisfied, and with a given pressure distribution $p(r)$, (4-60) can be solved for $B(r)$. There is no upper limit to p, but it is evident from (4-61) that $p(0) - p(r) > 0$; hence the pressure reaches its maximum value at $r = 0$. Note that the pinch configuration creates the magnetic field needed for its confinement itself, and therefore requires no external coils.

An interesting relationship exists between the pinch current and the amount of fluid confined at a given temperature T. If we choose $r_2 = R$, the radius of the pinch, and $r_1 = r$, (4-61) becomes

$$p(r) = \frac{B^2(R)}{2\mu_0} - \frac{B^2(r)}{2\mu_0} + \frac{1}{\mu_0}\int_r^R\frac{B^2(\xi)}{\xi}\, d\xi \tag{4-62}$$

because $p(R) = 0$. If one multiplies by $2\pi r$ and integrates from 0 to R, the last term,

$$\frac{2\pi}{\mu_0} \int\limits_0^R \left[r \int\limits_r^R \frac{B^2}{\xi} d\xi \right] dr = \frac{2\pi}{\mu_0} \left[\frac{r^2}{2} \int\limits_r^R \frac{B^2}{\xi} d\xi \right]_0^R + \frac{2\pi}{\mu_0} \int\limits_0^R \frac{r^2}{2} \frac{B^2}{r} dr = \frac{2\pi}{\mu_0} \int\limits_0^R \frac{rB^2}{2} dr$$

(4-63)

just cancels the second term on the right side of (4-62). Since for an isothermal system $p = nkT$, where n is the particle density, (4-62) becomes

$$NkT = \frac{\pi R^2 B^2(R)}{2\mu_0} = \frac{\mu_0 I^2}{8\pi}$$

(4-64)

where $N = \int_0^R n 2\pi r \, dr$ is the number of particles per unit length and I is the pinch current.

It might be thought that finite toroidal arrangements can be obtained simply by bending the configurations examined above into toruses. It is easy to see, however, that this does not work, because the pressure-balance equation cannot be satisfied. Let us examine, for example, the arrangements sketched in Fig. 4-5.

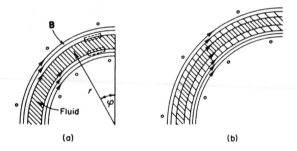

(a)   (b)

FIG. 4-5. Illustration of the impossibility of confinement by a toroidal magnetic field: (a) sharp fluid boundary, (b) diffuse boundary.

The magnetic field for the configurations of Fig. 4-5 is produced by a toroidal coil. This field falls off as $1/r$ in the absence of the fluid. Consider first the case of sharp boundaries. Here inside the field-free fluid $\nabla p = 0$ the fluid is isobaric. Consider now a closed surface cutting through the fluid interface with small linear dimensions normal to the interface. According to the integral form of the pressure-balance equation the surface integral of the total pressure over any closed surface is zero in equilibrium. On the inner surface we get $p \, dS$, while outside, according to (4-42), the result is $(B^2/2\mu_0) \, dS$, $(\mathbf{B} \cdot d\mathbf{S} = 0)$. In contrast with the uniformity of p everywhere, the magnetic field decreases as $1/r$. As a consequence, no static equilibrium can exist. By the same token we are led to the conclusion that the condition of static equilibrium for a fluid with sharp boundaries is the constancy of the magnitude of B on the entire surface. The nonexistence of equilibrium for the configuration of Fig. 4-5b with diffuse boundaries can also be proved without difficulty (see Exercise 4-5).

Similar considerations yield the same answer for the toroidal pinch. Instead of resorting to calculations based on the pressure-balance equation one can use the virial theorem as obtained in Sec. 3-5. The general plasma equations used there as a starting point include, of course, as a special case, conducting fluids with scalar pressure. The impossibility of self-confinement of finite plasmoids applies to the case of the toroidal pinch as well. Notice that the same argument could not be used for the previous arrangement containing a coil supported by external (not fluid) forces.

Fortunately not all the toroidal geometries suffer from this serious shortcoming. Consider a combination of the arrangements of Figs. 4-3 and 4-4 carrying longitudinal as well as azimuthal currents and magnetic field components. The resulting magnetic field lines form a family of nested helices (Fig. 4-6a). Bending these configurations into a torus we arrive at a family of

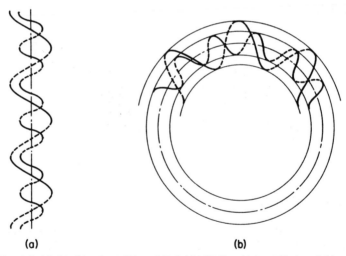

(a)   (b)

FIG. 4-6. (a) Combination of B_z and B_θ fields. (b) Toroidal equilibrium field.

nested toroids formed by the magnetic field lines, with the innermost toroid degenerating into a single closed curve—the magnetic axis. Such configurations can be produced by external coils (*stellerators*), or by a combination of coils, which generate the toroidal field, and toroidal plasma currents, which give rise to the poloidal field component (*tokomaks*) (Fig. 4-6b). It can be shown that various field constructions of this kind satisfy the pressure-balance equation and hence yield static configurations.

Another class of static configurations of finite size can be obtained if one allows sharp cusps in the plasma magnetic-field boundary. Such configurations are sketched in Fig. 4-7. Figure 4-7a shows the sharp boundary hydromagnetic equivalent of the mirror geometry. Figure 4-7b shows the spindle-shaped monocusp, which contains one line cusp and two point cusps. The external field is produced by two circular coils carrying opposing currents.

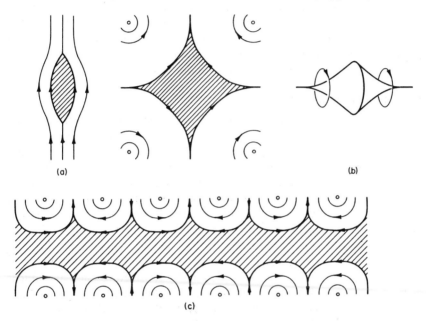

FIG. 4-7. Cusp and picket-fence field geometries.

Lining up more coils, as in Fig. 4-7c, and alternating the current direction, one arrives at cusp geometries of higher order. Figure 4-7c with diffuse boundaries is referred to as a *picket fence*. Edges and cusps are characteristic of these geometries.

We turn now to the quasi-stationary processes. In these processes changes are slow enough so that the inertial term in (4-10) can be neglected. This leaves us again with the static pressure balance and $\nabla \cdot \mathbf{B} = 0$ equations, plus the equation of continuity, equation of state, and the flux conservation law (4-11b) or, for finite conductivity, (4-11a). At any instant during a process the static equations yield the familiar pressure and field distributions. These distributions, however, change in time in such a way as to leave the mass of fluid elements constant and the magnetic field lines frozen into them (for $\sigma \to \infty$). Consider any one of the configurations with a sharp boundary between fluid and field. Slowly increasing the external field (e.g., by driving more current through the coils or the pinched fluid itself) compresses the fluid to pressures just high enough to balance the increasing magnetic pressures. Because of the conservation of flux, the fluid when originally field-free always remains so. If the adiabatic equation of state holds, this can be a means of heating the fluid, by adiabatic compression. In this process the magnetic field acts like a piston, but the fluid need not be in contact with material objects. In a similar fashion, decreasing the magnetic field leads to expansion. If the fluid also contains magnetic fields, compression and expansion can be obtained the same way, taking the conservation of flux law into account. In case of finite conductivity, flux conservation has to be replaced by (4-35), and the solution becomes considerably more involved.

4-4. Hydromagnetic Waves

We are now going to examine time-dependent phenomena described by the complete set of hydromagnetic equations. As in electromagnetic theory we are looking for periodic wave solutions. In electromagnetic (and any other linear) theory, these solutions are the most general ones, in the sense that any solution can be described as a suitable linear composition of these. Unfortunately, this cannot be said of hydromagnetic systems described by the nonlinear set (4-1a), (4-6), (4-9), (4-10), and (4-11a). The superposition of two solutions is not necessarily a solution any longer, and the possibilities are not exhausted by finding all the periodic solutions. Nevertheless we shall examine the possibility of obtaining wave equations. Since these equations are nonlinear we shall linearize the hydromagnetic equations by considering small amplitudes.

Looking at the hydromagnetic equations, the existence of some wave solutions is immediately apparent. With $\mathbf{B} = 0$ these equations reduce to the regular fluid equations, and yield after linearization the well-known sound waves. Apart from these trivial solutions one suspects that the coupling of fluid and electromagnetic phenomena might result in new types of waves. To simplify our investigations we restrict ourselves first to an incompressible fluid with infinite conductivity.

Consider an unbounded conductive fluid embedded in a homogeneous magnetic field of magnitude $\mathbf{B}_0$. We disturb this static equilibrium by shifting a pillar in the y direction with velocity $\mathbf{v}$. If the fluid were nonconductive, an electric field $\mathbf{E}' = \mathbf{v} \times \mathbf{B}_0$ could be measured in the co-moving frame (Fig. 4-8a). In a moving conductor, however, currents arise (Fig. 4-8b) in order to set up the charge distribution necessary to cancel out this electric field. These currents interact with the magnetic field and produce forces impeding the motion of the pillar and accelerating the adjacent layers. In this way the motion is transferred and a wave motion arises. Hydromagnetic waves of this kind are the *Alfven waves*, named after their discoverer.

The next step is the linearization of the equations by assuming that the

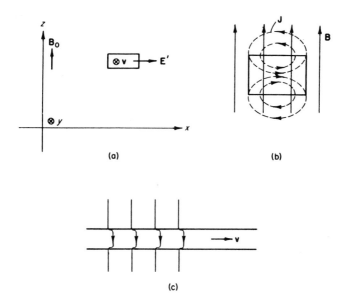

FIG. 4-8. Illustration of the physical mechanism of Alfven waves. [After H. Alfven and C.-G. Fälthammar, "Cosmical Electrodynamics," 2nd ed. © 1963, Oxford Univ. Press, Oxford.]

wave constitutes only a small perturbation compared to the static situation. One writes the magnetic field as $\mathbf{B}(\mathbf{r},t) = \mathbf{B}_0 + \mathbf{b}(\mathbf{r},t)$, with $b \ll B$, and since the velocity is zero in the static case, $\mathbf{v}$ is also a small first-order quantity. Products of small quantities will be neglected. Equation (4-11b) becomes, in the first order,

$$\frac{\partial \mathbf{b}}{\partial t} = \nabla \times (\mathbf{v} \times \mathbf{B}_0) \tag{4-65}$$

while (4-10) yields

$$v \frac{\partial \mathbf{v}}{\partial t} = -\nabla\left(p + \frac{B^2}{2\mu_0}\right) + \frac{1}{\mu_0} (\mathbf{B} \cdot \nabla)\mathbf{b} \tag{4-66}$$

The equation of state for incompressible fluids,

$$\nabla \cdot \mathbf{v} = 0 \tag{4-9a}$$

when substituted into the equation of continuity (4-1a) yields

$$dv/dt = 0; \qquad v_0 = \text{const} \tag{4-67}$$

without linearization.

Equation (4-65) can be transformed with the help of the identity (A-11),

$$\nabla \times (\mathbf{v} \times \mathbf{B}_0) = (\mathbf{B}_0 \cdot \nabla)\mathbf{v} - (\mathbf{v} \cdot \nabla)\mathbf{B}_0 + \mathbf{v}(\nabla \cdot \mathbf{B}_0) - \mathbf{B}_0(\nabla \cdot \mathbf{v}) \tag{4-68}$$

Owing to (4-6) and (4-9a) the last two terms vanish. So does the previous one, since $\mathbf{B}_0$ is a constant. This reduces (4-65) to

$$\partial\mathbf{b}/\partial t = (\mathbf{B}_0 \cdot \nabla)\mathbf{v} \tag{4-69}$$

We set up our coordinate axes in such a way that $\mathbf{B}_0$ points in the z direction, and look for plane waves with

$$\frac{\partial}{\partial x} = \frac{\partial}{\partial y} = 0; \qquad \mathbf{B}_0(0,0,B_z) \tag{4-70}$$

Equations (4-69) and (4-66) now become

$$\frac{\partial\mathbf{b}}{\partial t} = B_0 \frac{\partial\mathbf{v}}{\partial z} \tag{4-71}$$

and

$$v_0 \frac{\partial\mathbf{v}}{\partial t} = \frac{1}{\mu_0} B_0 \frac{\partial\mathbf{b}}{\partial z} - \frac{\partial}{\partial z}\left(p + \frac{B^2}{2\mu_0}\right)\mathbf{e}_3 \tag{4-72}$$

where $\mathbf{e}_3$ is the unit vector in the z direction.

From (4-6) and (4-9a) one obtains

$$\partial v_z/\partial z = 0 \tag{4-73}$$

and

$$\partial b_z/\partial z = 0 \tag{4-74}$$

Consequently v_z and b_z are uniform in space. One neglects only trivial (not wave) solutions by taking both constants zero. Looking now at (4-72) one may see that the first and second terms have no z components; consequently,

$$\frac{\partial}{\partial z}\left(p + \frac{B^2}{2\mu_0}\right) = 0 \tag{4-75}$$

The linearized equations (4-71) and (4-72) appear now in symmetrical form:

$$\frac{\partial\mathbf{b}}{\partial t} = B_0 \frac{\partial\mathbf{v}}{\partial z} \tag{4-71}$$

$$\frac{\partial\mathbf{v}}{\partial t} = \frac{B_0}{\mu_0 v_0} \frac{\partial\mathbf{b}}{\partial z} \tag{4-76}$$

Combining these equations after differentiation leads to

$$\frac{\partial^2\mathbf{b}}{\partial t^2} = \frac{B_0^2}{\mu_0 v_0} \frac{\partial^2\mathbf{b}}{\partial z^2} \tag{4-77}$$

and

$$\frac{\partial^2\mathbf{v}}{\partial t^2} = \frac{B_0^2}{\mu_0 v_0} \frac{\partial^2\mathbf{v}}{\partial z^2} \tag{4-78}$$

These are wave equations representing plane waves traveling in the z direction with the phase velocity

$$V = \pm \frac{B_0}{(\mu_0 v_0)^{1/2}} \tag{4-79}$$

There is no dispersion, since V is independent of the frequency. These waves are transverse, since $\mathbf{b}$ and $\mathbf{v}$ are free of z components and it follows from (4-71) or (4-76) that both are polarized in the same direction. With solutions of the form $\mathbf{A} \exp[-i(\omega t - \mathbf{k} \cdot \mathbf{r})]$ (linear polarization), the operator ∇ can be replaced by $i\mathbf{k}$ and (4-3) yields

$$i\mathbf{k} \times \mathbf{H} = \mathbf{J} \tag{4-80}$$

Thus the current is also transverse to the direction of propagation, as expected. Furthermore, from the equation of motion,

$$v_0 \frac{\partial \mathbf{v}}{\partial t} = \mathbf{J} \times \mathbf{B} \tag{4-81}$$

(remembering that ∇p has no transverse components) we see that $\mathbf{J}$ is normal to $\partial \mathbf{v}/\partial t$, hence to $\mathbf{v}$ as well. Substituting the exponential forms of $\mathbf{v}$ and $\mathbf{b}$ into (4-71) one obtains

$$-\omega \mathbf{b} = B_0 k \mathbf{v} \tag{4-82}$$

Since $V = \omega/k = B_0/(\mu_0 v_0)^{1/2}$

$$\mathbf{v} = -\frac{\mathbf{b}}{(\mu_0 v_0)^{1/2}} \tag{4-83}$$

Another way to look at these waves is also illuminating. We know that the magnetic field is frozen into a perfect conductor. Shifting the fluid column of Fig. 4-8 distorts the field lines (see Fig. 4-8c). Owing to the bending of field lines, forces arise tending to straighten them out again. These forces and the inertia of the fluid give rise to wave propagation. This argument can be made somewhat more quantitative. The transverse vibration of a string with mass per unit length m and tension S is expressed by

$$m \frac{\partial^2 y}{\partial t^2} = S \frac{\partial^2 y}{\partial z^2} \tag{4-84}$$

in the first-order approximation. The phase velocity is

$$V = \pm (S/m)^{1/2} \tag{4-85}$$

We have seen in Sec. 2 that magnetic forces can be expressed as a sum of a scalar pressure transverse to the field and a tension B/μ_0 per field line. Here $\nabla(p + B^2/2\mu_0) = 0$, the field-pressure forces, are neutralized by the fluid pressure. The fluid mass per unit length and per field line is v/B. Therefore, substituting B_0/μ_0 for S and v_0/B_0 for m, (4-84) becomes

$$\frac{\partial^2 y}{\partial t^2} = \frac{B_0^2}{\mu_0 v_0} \frac{\partial^2 y}{\partial z^2} \tag{4-86}$$

and the phase velocity

$$V = \pm \frac{B_0}{(\mu_0 v_0)^{1/2}} \tag{4-87}$$

as in (4-78) and (4-79).

Although we found these plane waves as solutions of the linearized equations only, we will show by direct substitution that they satisfy the original nonlinear equations as well. This is true even in a more general form with

$$\mathbf{B} = \mathbf{B}_0 + \mathbf{b}(x,y,z \pm Vt) \tag{4-88}$$

and

$$\mathbf{v} = \mp \mathbf{b}/(\mu_0 v_0)^{1/2} \tag{4-89}$$

where $\mathbf{b}$ and $\mathbf{v}$ need not lie in the xy plane and can be functions of x and y. We again set

$$\nabla\left(\frac{B^2}{2\mu_0} + p\right) = 0 \tag{4-90}$$

Equations (4-10) and (4-11b) now become

$$v_0\left[\frac{\partial \mathbf{v}}{\partial t} + (\mathbf{v} \cdot \nabla)\mathbf{v}\right] = \frac{1}{\mu_0}(\mathbf{B} \cdot \nabla)\mathbf{B} \tag{4-91}$$

and

$$\partial\mathbf{B}/\partial t = (\mathbf{B} \cdot \nabla)\mathbf{v} - (\mathbf{v} \cdot \nabla)\mathbf{B} + \mathbf{v}(\nabla \cdot \mathbf{B}) - \mathbf{B}(\nabla \cdot \mathbf{v}) \tag{4-92}$$

where the last two terms vanish again. Denoting the derivative of $\mathbf{b}$ with respect to the argument $z \pm Vt$ by $\mathbf{b}'$ and using (4-89), one obtains from (4-91),

$$v_0\left[\frac{\pm V}{(\mu_0 v_0)^{1/2}} \mathbf{b}' + \frac{1}{\mu_0 v_0}(\mathbf{b} \cdot \nabla)\mathbf{b}\right] = \frac{1}{\mu_0}(\mathbf{B}_0 \cdot \nabla)\mathbf{b} + \frac{1}{\mu_0}(\mathbf{b} \cdot \nabla)\mathbf{b} \tag{4-93}$$

Recognizing that two terms cancel and B_0 points in the z direction,

$$\pm V\mathbf{b}' = \frac{B_0}{(\mu_0 v_0)^{1/2}} \mathbf{b}' \tag{4-94}$$

Equation (4-92) becomes, in a similar fashion,

$$\pm V\mathbf{b}' = \frac{B_0}{(\mu_0 v_0)^{1/2}} \mathbf{b}' + \frac{(\mathbf{b} \cdot \nabla)\mathbf{b}}{(\mu_0 v_0)^{1/2}} - \frac{(\mathbf{b} \cdot \nabla)\mathbf{b}}{(\mu_0 v_0)^{1/2}} \tag{4-95}$$

It is clear from both (4-94) and (4-95) that (4-88) to (4-90) are solutions of the nonlinear equations with $V = \pm B_0/(\mu_0 v_0)^{1/2}$.

As long as wave amplitudes are small, the linearized equations present a fairly good approximation, and the superpositions of various solutions can be regarded as solutions in the same approximation. If the fluid is bounded, solutions should be added to match boundary conditions. Since there is no dispersion, the group velocity agrees with the phase velocity V for waves traveling along the field lines. It is clear from inspection of (4-69) that Alfven waves traveling normal to the main field ($\partial/\partial z = 0$) do not exist.

The inclusion of compressibility complicates matters slightly. Considering an adiabatic law we have instead of (4-9a) and (4-67) the linearized equations

$$\frac{\partial v}{\partial t} + v_0 \nabla \cdot \mathbf{v} = 0 \tag{4-96}$$

and

$$\frac{\partial v}{\partial t} = \frac{v_0}{p_0 \gamma} \frac{\partial p}{\partial t} \tag{4-97}$$

or, eliminating v by combining these equations,

$$\nabla \cdot \mathbf{v} + \frac{1}{p_0 \gamma} \frac{\partial p}{\partial t} = 0 \tag{4-98}$$

Equations (4-65), (4-66), and (4-98) plus $\nabla \cdot \mathbf{B} = 0$ constitute our new set of equations. Looking again for plane waves traveling along the field, one finds easily that these waves split up into transverse waves, which coincide with the Alfven waves found before, and longitudinal waves, which are unaffected by the magnetic field. These longitudinal waves turn out to be ordinary sound waves (see Exercise 4-11).

There are now also waves propagating normal to the field. These waves are longitudinal in $\mathbf{v}$, transverse in $\mathbf{b}$, and are called *magnetoacoustic waves*, since they are sound waves modified by the presence of magnetic fields. The compression waves, in addition to compressing and expanding the fluid, must act against the magnetic pressure. The phase-velocity formula (see Exercise 4-12)

$$V = \left[\frac{p_0 \gamma + (B^2/\mu_0)}{v_0} \right]^{1/2} \tag{4-99}$$

is a reminder that the magnetic pressure $B^2/2\mu_0$ plays the same role as the fluid pressure p_0, noting that the magnetic pressure is only two-dimensional (perpendicular to the field), hence $\gamma_{magn} = 2$. With vanishing magnetic field (4-99) reduces to the familiar formula of sound velocity, while if the magnetic pressure largely exceeds p_0, the velocity formula for Alfven waves is regained.

The inclusion of finite, but very large conductivity leads to damped waves (see Exercise 4-13). Small conductivity results in uninteresting aperiodic solutions, while vanishing conductivity uncouples electromagnetic and fluid phenomena and we are back at the source-free Maxwell's equations and fluid equations.

Consider an electromagnetic plane wave impinging on the surface of a hydromagnetic fluid. If the conductor were rigid, the wave would be attenuated exponentially, the penetration being measured by the skin depth

$$\delta = \frac{1}{(\mu_0 \sigma \omega)^{1/2}} \tag{4-100}$$

For infinite conductivity there is no penetration at all. However, the situation is quite different in a conductive fluid. When the fluid-vacuum interface is perpendicular to $\mathbf{B}_0$, electromagnetic waves impinging on the fluid ($\mathbf{k} \parallel \mathbf{B}_0$) penetrate as Alfven waves, the penetration depth *increasing* with increasing conductivity, contrary to the case of a rigid conductor. This problem is considered in more detail in Exercise 4-14.

When the interface is parallel to $\mathbf{B}_0$, the component of the incoming wave, with the magnetic field vector polarized parallel to $\mathbf{B}_0$, can propagate as a magnetoacoustic wave inside the fluid. If the conductivity is large but finite, two types of magnetoacoustic waves emerge (see Exercise 4-17); one is unattenuated and propagates with the magnetoacoustic velocity (4-99), the other one decays exponentially, with a penetration depth V_S/V_{MA} times smaller than the skin depth (4-100) in a rigid conductor, where V_S and V_{MA} are the sound and magnetoacoustic velocities respectively.

For bounded fluids another class of interesting solutions exists. One finds that the dispersion relation often yields solutions where ω is not real. For infinite conductivity, dissipation does not exist, and the frequencies are either real or imaginary. While the first ones lead to waves, the latter ones lead to instabilities. Since they are of great practical importance, Chapter V will be devoted to these instabilities.

4-5. Domain of Validity of the Hydromagnetic Equations

A satisfactory description of a plasma is given by the combined Boltzmann equation (collision term included) and Maxwell's equations. We have seen that the moments in velocity space reduce the Boltzmann equation to continuum equations in three-dimensional space and time, but these equations do not constitute a closed chain of equations. There are several limiting cases, however, where these moments reduce to a solvable set of macroscopic equations.

A conventional limiting case is the collision-dominated domain of ordinary gas dynamics. If the collision time is short compared to other characteristic times, or $(\partial f/\partial t)_{coll}$ dominates over other terms in the Boltzmann equation, a series expansion can be carried out. In the zero-order approximation the

Boltzmann equation reduces to

$$(\partial f^0/\partial t)_{\text{coll}} = 0 \qquad (4\text{-}101)$$

The resulting solution is a Maxwellian distribution superimposed on an arbitrary transport velocity. This velocity as well as the temperature and particle density are undetermined functions of space and time. The pressure is a scalar, and each fluid element obeys the adiabatic equation of state in three dimensions. One can now insert f^0 into the left side of the Boltzmann equation, while $(\partial f^1/\partial t)_{\text{coll}}$ is put for the right side. This equation serves to determine the first-order distribution function, which leads to the determination of the coefficients of thermal and electrical conductivity, and viscosity. This procedure results in a closed set of fluid equations which can be coupled with Maxwell's equations to yield the system of hydromagnetic equations.

If the viscosity is small, and the thermal conductivity either small enough so that the adiabatic equation of state can be used or so large that the isothermal law applies, these equations reduce to those discussed in this chapter. The adjectives "large" and "small" are, of course, only meaningful when related to the characteristic quantities of the phenomena under investigation. The same considerations hold for the applicability of the infinite-conductivity approximation. From (4-11a), the resistive term can be dropped if

$$L \gg (\tau/\mu_0\sigma)^{1/2} \qquad (4\text{-}102)$$

where L is the characteristic length and τ a characteristic time. For periodic time dependence (e.g., a hydromagnetic wave) $\tau \sim \omega^{-1}$ and (4-103) reduces to

$$L \gg \delta \qquad (4\text{-}103)$$

where δ is the skin depth. When comparing the two terms on the right side of (4-11a), one finds that the resistive term can be neglected if

$$R_M = \sigma v \mu_0 L \gg 1 \qquad (4\text{-}104)$$

R_M is the *magnetic Reynolds number*. While the infinite-conductivity hydromagnetic model fits many astrophysical applications, it rarely applies to laboratory plasmas.

In Sec. 3-6 we investigated in detail the opposite limiting case of no collisions and small cyclotron radii. The equations we arrived at were slightly more complicated than the hydromagnetic ones, because of the anisotropy along and perpendicular to field lines. If for some reason $p_\parallel = p_\perp$, the equations of motion (3-105b) and (3-109) reduce to the hydromagnetic equation (4-10) and the equations of state reduce to the three-dimensional adiabatic equation (4-9c). Note that the other equations—the equation of continuity, $\nabla \cdot \mathbf{B} = 0$, and the equation of flux conservation—agree exactly with the corresponding equations for the fluid. Therefore, for example, everything said in Sec. 4-3 is valid for plasmas with small cyclotron radii and scalar pressure.

More frequently one deals with two-dimensional geometries in which field quantities vary only perpendicular to the straight and parallel field lines. Now (3-105b) becomes meaningless and (3-109) reduces to

$$v \frac{d\mathbf{v}}{dt} + \nabla \left(p + \frac{B^2}{2\mu_0} \right) = 0 \qquad (4\text{-}105)$$

where the subscript $\perp$ has been dropped. This agrees with the hydromagnetic equation of motion for the same geometry. Among the equations of state, the parallel one is meaningless, while the perpendicular one is the adiabatic equation with $\gamma = 2$ for two dimensions. This applies, for instance, to the θ pinch geometry.

If however, $p_{\parallel}$ does not equal $p_{\perp}$ nor is the configuration two-dimensional, the original double-adiabatic equations should be applied. We get then, for example, for the static pressure-balance equations,

$$\nabla_{\perp} \left(p_{\perp} + \frac{B^2}{2\mu_0} \right) - \left[\frac{(\mathbf{B} \cdot \nabla)\mathbf{B}}{\mu_0} \right]_{\perp} - \frac{\mathbf{n}}{R} (p_{\perp} - p_{\parallel}) = 0 \qquad (4\text{-}106)$$

and

$$\nabla_{\parallel} p_{\parallel} + (p_{\perp} - p_{\parallel}) \left(\frac{\nabla B}{B} \right)_{\parallel} = 0 \qquad (4\text{-}107)$$

instead of (4-50b). These in conjunction with $\nabla \cdot \mathbf{B} = 0$ can now be used to solve for static configurations (see Exercise 4-19).

Wave phenomena are also affected by the anisotropic pressure, e.g., the propagation velocity of Alfven waves depends on $p_{\perp} - p_{\parallel}$ (see Exercise 4-20). Nevertheless, hydromagnetic waves, magnetoacoustic waves, etc., also exist in the $p_{\parallel} \neq p_{\perp}$ case but in a modified form (see Exercise 4-18). It is reasonable, therefore, to approach many of the complicated plasma phenomena through the simpler hydromagnetic model. One should keep in mind, however, that the plasma equations are considerably richer than the macroscopic hydromagnetic equations and therefore contain phenomena nonexistent in the hydromagnetic approximation. Therefore, whenever possible one should try to reach back to more and more exact equations, possibly to the Boltzmann equation.

4-6. Summary

The behavior of conductive fluids is described by the set of hydromagnetic equations

$$\frac{\partial v}{\partial t} + \nabla \cdot (v\mathbf{v}) = 0$$

$$v \frac{d\mathbf{v}}{dt} = -\nabla p + \frac{1}{\mu_0} (\nabla \times \mathbf{B}) \times \mathbf{B}$$

$$\frac{\partial \mathbf{B}}{\partial t} = \nabla \times (\mathbf{v} \times \mathbf{B}) + \frac{1}{\mu_0 \sigma} \nabla^2 \mathbf{B}$$

$$\nabla \cdot \mathbf{B} = 0$$

and an equation of state, usually the adiabatic

$$\frac{d}{dt} (pv^{-\gamma}) = 0$$

The conductivity was assumed to be large enough so that the displacement current can be neglected compared to the conduction current. On the fluid-vacuum boundary, the boundary conditions

$$\mathbf{n} \cdot [\mathbf{B}] = 0$$

$$\mathbf{n} \times [\mathbf{E}] = \mathbf{n} \cdot \mathbf{v}[\mathbf{B}]$$

$$\mathbf{n} \times [\mathbf{B}] = \mu_0 \mathbf{J}^*$$

and

$$\left[p + \frac{B^2}{2\mu_0} \right] = 0$$

apply. For the special case $\sigma \to \infty$, the fluid-vacuum system conserves the sum of kinetic and "potential" energy:

$$K + W = \text{const}$$

where

$$K = \int \tfrac{1}{2} v v^2 \, d\tau$$

and

$$W = \int \left(\frac{p}{\gamma - 1} + \frac{B^2}{2\mu_0} \right) d\tau$$

where integration extends over the entire fluid-plus-vacuum volume.

The perfectly conductive fluid also has the property of conserving magnetic flux. As a result, magnetic field lines can be considered frozen into the fluid, and identity can be ascribed to individual lines by marking the fluid elements they pass through.

Magnetohydrostatic equilibria are easy to find. They contain nested $p = \text{const}$ surfaces made up of $\mathbf{B}$ and $\mathbf{J}$ lines. From $\nabla \cdot \mathbf{B} = \nabla \cdot \mathbf{J} = 0$ it follows that these surfaces either form topological toroids, or else extend to infinity. The pinch, θ pinch, stellarator, mirror, and cusp geometries are some of the better-known equilibria. Slow time variations of the confining magnetic field lead to quasi-static situations. The fluid passes through a series of static configurations, while the pressure, temperature, and density vary in accordance with the adiabatic law, and the (internal) magnetic field follows the law of flux conservation. Adiabatic heating of the fluid can be achieved this way.

Fluid elements lying along a magnetic field line can be thought of as being mechanically connected by the (elastic) field line. An oscillation of fluid elements transverse to the magnetic field propagates along the field lines, with the Alfven velocity

$$V_A = B/(\mu_0 v)^{1/2}$$

Compressional acoustic waves (except when propagating along the field), couple to the magnetic field, giving rise to magnetoacoustic waves. Such waves, when propagating perpendicular to the field with the phase velocity

$$V_{MA} = \left[\frac{p_0 \gamma + (B^2/\mu_0)}{v_0}\right]^{\frac{1}{2}}$$

can be easily interpreted; compression (and expansion) of a fluid element is linked with compression (and expansion) of the flux frozen therein, adding to $p_0 \gamma$ the quantity $(B^2/2\mu_0) \times 2$, in accordance with the notion of magnetic pressure and $\gamma = 2$ for its two-dimensional equation of state.

EXERCISES

4-1. Prove that the following two equations follow from the basic equations of magnetohydrostatics: $(\mathbf{B} \cdot \nabla)\mathbf{J} = (\mathbf{J} \cdot \nabla)\mathbf{B}$ and $\nabla \cdot (\mathbf{B} \times \nabla p) = 0$.

4-2. Analyze the magnetic forces acting on a pinch using various forms of the expression for magnetic forces and pressures.

4-3. What is the current necessary to confine 10^{16} particles in a pinch discharge of 0.1 m length at $T = 10^7 \,°\text{K}$?

4-4. What is the magnetic pressure acting on the interior of a solenoid carrying a magnetic field $B = 20$ webers m^{-2}?

4-5. Prove that no static equilibrium is possible with the field configuration of Fig. 4-5b. [*Hint*: Use (4-57) to show that no variation of the magnetic field in the z direction is compatible with the pressure-balance equation.]

4-6. In the case of finite conductivity, field lines can penetrate the fluid. Use (4-11a) to estimate the time constant of penetration for mercury with linear dimensions $L \approx 10^{-1}$ m.

4-7. Show that α as defined in (4-52) for the force-free field can be expressed as $\alpha = \mathbf{t} \cdot \nabla \times \mathbf{t}$.

4-8. Prove that ∇B lies in the osculating plane (**tn** plane) for a force-free field.

4-9. Find the velocity of an Alfven wave in mercury in a magnetic field $B_0 = 100$ gauss $= 10^{-2}$ webers m^{-2}.

4-10. Find the equations and phase velocity of an Alfven wave in an incompressible fluid with a propagation vector oblique to $\mathbf{B}_0$.

4-11. Derive the equations for a hydromagnetic plane wave traveling along $\mathbf{B}_0$ in a case where the fluid obeys the adiabatic law. Prove that the equations split into parts parallel and perpendicular to $\mathbf{B}_0$, the former designating a longitudinal sound wave, the latter a transversal Alfven wave.

4-12. Find the equations for magnetoacoustic waves propagating normal to $\mathbf{B}_0$. Find the polarization of the vectors **v** and **b** and show that the phase velocity is $V = [(1/v_0)(p_0\gamma + B^2/\mu_0)]^{1/2}$.

4-13. Include finite conductivity in the derivation of plane Alfven waves. Prove that solutions of the form $\exp(i\omega t + \alpha z)$ satisfy the linearized equations and find α.

4-14. A plane electromagnetic wave impinges on the plane boundary of a hydromagnetic fluid with a uniform background magnetic field. $\mathbf{B}_0 \| \mathbf{k}$, **k** is perpendicular to the interface, σ is large but finite. Use the result of exercise 4-13 to find:
(a) How deeply the wave penetrates the fluid. Analyze this result for variation as a function of σ and B_0.
(b) What fraction of the incoming wave energy penetrates the fluid.

4-15. Prove the equipartition of kinetic and magnetic (wave) energy in a small-amplitude Alfven wave.

4-16. Consider two adjacent material points in a perfectly conductive fluid situated on the same magnetic line of force. Prove that the quantity $B/v\Delta l$ remains constant in the course of the motion if Δl is the distance of the material points.

4-17. A plane electromagnetic wave impinges on a conductive fluid embedded in a uniform magnetic field $\mathbf{B}_0$. Take $\mathbf{k} \perp \mathbf{B}_0$ and $\mathbf{k}\|\mathbf{n}$ where **n** is the surface normal. Assume that σ is large but finite and that the magnetic field vector of the incoming wave is parallel to $\mathbf{B}_0$.
Prove that there are two wave modes penetrating the fluid, where one mode is an unattenuated magnetoacoustic wave, and the other has an effective skin depth reduced by the ratio V_S/V_{MA} compared to the skin depth in a rigid conductor (V_S and V_{MA} are the sound and magnetoacoustic velocities respectively).

4-18. Derive the equations of magnetoacoustic waves in a plasma propagating perpendicular to the uniform magnetic field. Prove that the phase velocity is $V = [(2p_\perp + B_0{}^2/\mu_0)/v_0]^{1/2}$. Compare this result with (4-99).

4-19. Use (4-106) and (4-107) to find the pressure-balance equation for a pinch configuration with a tensor pressure.

4-20. Find the propagation velocity for Alfven waves in a plasma with $p_\perp \neq p_\parallel$. Use the CGL approximation.

4-21. Prove that in a perfect conductor electric energy density/magnetic energy density $= (v/c)^2$, where v is the fluid velocity and c is the speed of light.

V

Hydromagnetic Stability

5-1. The Problem of Stability

As we have seen, there is no difficulty in constructing systems in magneto-hydrostatic equilibrium. The usefulness of such an equilibrium configuration depends, however, on its stability. Before investigating the complex problem of hydromagnetic stability it is useful to review briefly the well-known question of stability of simple mechanical systems.

Consider a simple one-dimensional system of a mass point in a potential field, as in Fig. 5-1. (One can think, for example, of a ball in a vertical gravitational field on a surface represented in the figure.) The positions A and B are

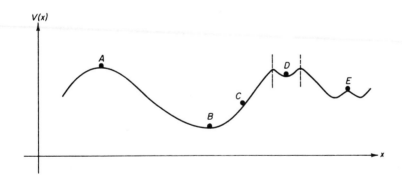

Fig. 5-1. Illustration of the problem of stability in one dimension.

both equilibria; the force (proportional to the slope) is zero. C is not an equilibrium position; the ball if placed there is acted on by a force accelerating it toward B.

The equilibria A and B are, however, very different. The slightest misplacement of the ball in the former position results in a force which accelerates it *away* from the equilibrium position, while in the latter position it is pushed *toward* the equilibrium to execute oscillations around the position *B*. Hence *A* is called an *unstable* and *B* a *stable equilibrium*. Since a small misplacement can never be completely avoided, it is clear that for practical purposes the unstable equilibrium *A* is no better than the nonequilibrium *C*.

Another, equally good, test of stability leaves the ball initially at its equilibrium position but gives it a small push. At *A* the ball with a small initial velocity will run away, while at *B* it will oscillate. This "dynamical" test of stability makes use of a displacement in velocity space instead of a displacement in coordinate space.

It should be noted that stability depends on the behavior of the system if subjected to small (in fact, infinitesimal) displacements. Position *D* is therefore considered stable irrespective of the fact that a larger displacement may result in the ball rolling away from *D*, and *E* is unstable although the ball is soon returned by an uphill slope.

One way to find out about the stability of an equilibrium position is by finding the solution of the equation of motion in the immediate neighborhood of the equilibrium. If the coordinate of the equilibrium is x_0 and the force $F(x)$, one finds

$$m \frac{d^2x}{dt^2} = F(x) = F(x_0) + F'(x_0) \cdot (x - x_0) + \cdots \tag{5-1}$$

Since x_0 corresponds to an equilibrium position, $F(x_0) = 0$. For small displacements the higher-order terms can be neglected, and it is convenient to introduce for the displacement $x - x_0 = \xi$. The resulting equation of motion,

$$m \frac{d^2\xi}{dt^2} = F'(x_0)\xi \tag{5-2}$$

yields the solution

$$\xi = \xi_0 \exp[F'(x_0)/m]^{1/2}t = \xi_0 \exp(i\omega t) \tag{5-3}$$

where $\omega^2 = -F'(x_0)/m$. At *A*, $F'(x_0) > 0$ (which is just another way of saying that an infinitesimal displacement in the positive direction results in a positive force, one in the negative direction in a negative force) and the displacement grows exponentially in time. At *B*, $F'(x_0) < 0$, hence ω real and the solution oscillatory, as expected. Hence stability can be characterized by the sign of ω^2, $\omega^2 > 0$ stability, $\omega^2 < 0$ instability, $\omega^2 = 0$ neutral.

Another way of expressing the criterion for stability is by using $F(x) = -V'(x)$. If $V''(x_0)$ is positive so is ω^2, and stability follows, while $V''(x_0) < 0$ indicates instability. This can also be seen directly from the energy principle. The system is conservative, the sum of kinetic and potential energy constant. If V'' is positive at an equilibrium (bottom of a well) it follows that a displacement in either direction increases the potential energy, hence decreases the kinetic energy. Therefore, the ball in static equilibrium (zero kinetic energy) cannot climb out of the well without external help. On the top of a hill, however ($V'' < 0$), the farther the ball rolls the more kinetic energy it acquires and the faster it runs away.

It is clear for our simple system that ω^2 is always real, hence ω is real or purely imaginary. In the presence of friction ω^2 and ω become complex and yield a damped solution. One can also construct systems with a positive feedback. Here an external energy source feeds energy into the oscillating system at a rate proportional to the amplitude. As a result the amplitude of the oscillation increases exponentially, which is described by a complex ω. The latter case is called "overstability," but since it leads to runaway solutions as an instability does, in plasma physics the two are usually not distinguished.

Both damping and overstability are characterized by external sinks or sources of energy. It is easy to show in general that in a conservative system, where the sum of the potential energy (function of position) and the kinetic energy (function of velocity) is conserved, ω^2 is always real. Look at any point the system reaches in the course of its oscillatory motion, for example, the turning point. At this point the kinetic energy vanishes, and the potential energy equals the total energy $V_{turn} = E$. If ω^2 is complex, the turning point shifts at subsequent oscillations, in case of damping closer, in case of overstability farther away from the equilibrium point. Since V is a function of position only, V_{turn} and E vary in the course of the oscillation in contradiction with our assumption of a conservative system. Hence in a conservative system ω^2 is always real.

Consider now a two-dimensional conservative system (Fig. 5-2). Among the equilibria one now finds hilltops (a), wells (b), and saddle points (c) and (d).

In (a) both $\partial^2 V/\partial x^2$ and $\partial^2 V/\partial y^2$ are negative; in (b) both are positive; in (c) the former is positive and the latter negative and in (d) it is the other way around. Clearly only (b) is stable, while in the other three configurations at least some displacements lead to runaway solutions. Now we have two equations of motion, one for $\xi = x - x_0$ and one for $\eta = y - y_0$. If the co-ordinate systems are chosen properly, so that the axes point in the direction of steepest descent (or ascent), the two equations are uncoupled—one depends on ξ, the other only on η. This choice of coordinate systems results in normal coordinates. Now one obtains two ω's for an equilibrium. If both are real the equilibrium is stable; in any other case it is unstable. Since a real ω is again associated with a positive second derivative of V, the system is stable if and only if both $\partial^2 V/\partial x^2 > 0$ and $\partial^2 V/\partial y^2 > 0$, as in case (b). Further generalization to more than two degrees of freedom follows the same lines: *The sufficient and necessary condition of stability is that all ω's are real, or that any one of $\partial^2 V/\partial x_i^2 > 0$.*

Inspection of Fig. 5-1 shows that in the one-dimensional case the number

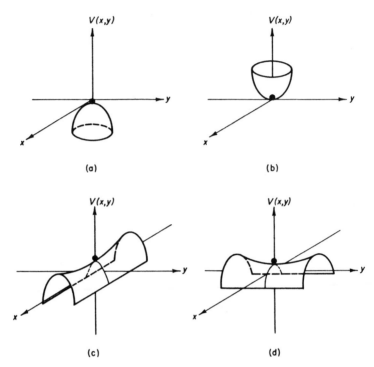

FIG. 5-2. Equilibria in a two-dimensional potential: (a), (c), and (d) unstable; (b) stable.

of stable and unstable equilibria are approximately equal. This means that after finding an equilibrium [if nothing is known of the sign of $V''(x)$], chances are about 50 per cent that it will turn out to be stable. On a two-dimensional "potential landscape," cases (a), (b), (c), and (d) can also be found with equal probability. Hence not more than 25 per cent of the equilibria are stable. In the case of n degrees of freedom, the chance that an equilibrium is stable reduces to 2^{-n}. Hence in a confined plasma, or in a magnetofluid in which $n \to \infty$, there is little cause for optimism if one finds an equilibrium configuration without any knowledge of its stability; chances are overwhelming that it will turn out to be unstable. Hence the need for general stability criteria—so one can immediately choose the stable equilibria.

As an example of a system with an infinite number of degrees of freedom, consider two incompressible fluids in a gravitational field in hydrostatic equilibrium (Fig. 5-3). If the two-fluid interface is a horizontal plane, the forces are in equilibrium at every point. The question of stability can be settled if one considers the behavior of the system under displacements of

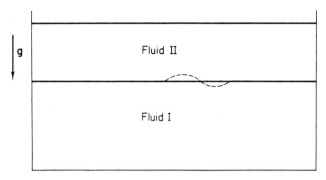

FIG. 5-3. Hydrostatic equilibrium of two fluids in a gravitational field.

droplets from this equilibrium. It is easy to see that rearrangement of droplets inside fluid I or fluid II does not change the equilibrium. The only interesting displacements are those which perturb the boundaries, either the interface or the upper surface of fluid II. Let us investigate the former, from the point of view of the energy principle.

Consider a ripple in the two-fluid interface. If the specific weights of the fluids (say water and mercury) differ, this ripple will change the potential energy of the system; some of fluid II is now lower, the same volume of fluid I higher than before. If the specific weight of fluid I exceeds that of fluid II ($\rho_I > \rho_{II}$) the potential energy has increased, while if the upper fluid is heavier ($\rho_{II} > \rho_I$) the potential energy has decreased. In the former case the displacement leads to stable oscillations around the equilibrium, in the latter to an exponentially growing ripple which leads finally to the upper fluid breaking through and the two fluids changing place to establish a stable configuration. It is intuitively clear that any small perturbation of the interface leads to the same result. The instability for $\rho_{II} > \rho_I$ is the Rayleigh-Taylor instability.

It is worth noting that the application of the energy principle permitted a quick answer to the question of stability. The process of setting up and solving the equations of motion is lengthier. First one has to search for the normal coordinates to describe the perturbation (they turn out to be the Fourier-analyzed amplitudes), and then look for their change in time. The results are, of course, identical with ours, and in addition one obtains the values of the frequencies for the stable case and growth rates for the unstable case.

In addition to the above-mentioned two cases (water up, mercury down; mercury up, water down), there are an infinite number of equilibria for the two-fluid system [layers of water and mercury with various thickness (Fig. 5-4)]. Each one is unstable except for the water up, mercury down version, but despite the infinite number of possibilities there is no difficulty in finding the stable one.

We have already seen (Sec. 4-1) that a hydromagnetic fluid possesses an energy integral. This integral was split up into two parts, one to be called the *kinetic energy* and the other the *potential energy*. It is required that the kinetic energy be a function of the time derivative of coordinates and the potential of the coordinates only.

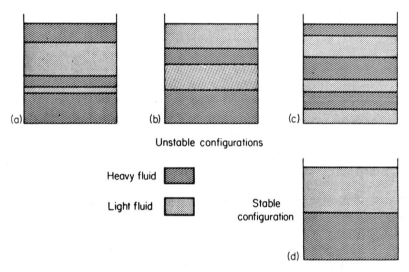

Unstable configurations

Heavy fluid

Light fluid

Stable configuration

FIG. 5-4. Hydrostatic equilibria of two fluids: (a), (b), and (c) unstable; (d) stable.

Consider a perfectly conductive fluid with given initial conditions at $t = 0$. At a later time t, a fluid element initially at $\mathbf{r}_0$ will have moved to $\mathbf{r}$. One may introduce the displacements $\boldsymbol{\xi}(\mathbf{r}_0) = \mathbf{r} - \mathbf{r}_0$ as "coordinates" of the system. Given the initial pressure distribution and the displacements of all fluid elements, the new pressure distribution is uniquely determined. Since the magnetic lines of force are frozen into the fluid, the same holds for the magnetic field. Consequently the potential energy

$$W = \int \left(\frac{p}{\gamma - 1} + \frac{B^2}{2\mu_0} \right) d\tau \qquad (4\text{-}31)$$

is a function of the coordinates only. Since the velocity is the time derivative of the displacement, clearly

$$K = \tfrac{1}{2} \int v v^2 \, d\tau \qquad (4\text{-}30)$$

is a function of $\dot{\boldsymbol{\xi}}$ only. Hence a hydromagnetic system is conservative in the same sense as a mechanical system, and the laws outlined above are applicable for the stability analysis of hydromagnetic systems as well.

5–2. The Problem of Hydromagnetic Stability

Consider a perfectly conducting fluid in magnetohydrostatic equilibrium. The pressure and magnetic field satisfy the equilibrium equation

$$\nabla p_0 = \frac{1}{\mu_0}(\nabla \times \mathbf{B}_0) \times \mathbf{B}_0 \qquad (4\text{-}50a)$$

If a small, space-dependent perturbation is applied, the second-order $(\mathbf{v} \cdot \nabla)\mathbf{v}$ term can be dropped and v replaced by its equilibrium value in the equation of motion of the fluid (4-10), which then becomes

$$v_0\dot{\mathbf{v}} = -\nabla p + \frac{1}{\mu_0}(\nabla \times \mathbf{B}) \times \mathbf{B} \qquad (5\text{-}4)$$

where the dot stands for $\partial/\partial t$. Differentiating with respect to time one obtains

$$v_0\ddot{\mathbf{v}} = -\nabla \dot{p} + \frac{1}{\mu_0}(\nabla \times \dot{\mathbf{B}}) \times \mathbf{B} + \frac{1}{\mu_0}(\nabla \times \mathbf{B}) \times \dot{\mathbf{B}} \qquad (5\text{-}5)$$

The expressions for $\dot{p}$ and $\dot{\mathbf{B}}$ as functions of $\mathbf{v}$ are available from Sec. 4-1:

$$\dot{p} = -\mathbf{v} \cdot \nabla p - \gamma p \nabla \cdot \mathbf{v} \qquad (4\text{-}23)$$

and

$$\dot{\mathbf{B}} = \nabla \times (\mathbf{v} \times \mathbf{B}) \qquad (4\text{-}11b)$$

These can be inserted into (5-5), which becomes

$$v_0\ddot{\mathbf{v}} = \nabla(\mathbf{v} \cdot \nabla p_0 + \gamma p_0 \nabla \cdot \mathbf{v}) + \frac{1}{\mu_0}\{\nabla \times [\nabla \times (\mathbf{v} \times \mathbf{B}_0)]\} \times \mathbf{B}_0 +$$

$$+ \frac{1}{\mu_0}(\nabla \times \mathbf{B}_0) \times \{\nabla \times (\mathbf{v} \times \mathbf{B}_0)\} \qquad (5\text{-}6)$$

where p and $\mathbf{B}$ are replaced by their equilibrium value to keep the equation linear in the small quantities (all quantities deviate but little from their equilibrium values).

Equation (5-6) is the linearized equation of motion, which yields the velocity distribution $\mathbf{v}(\mathbf{r},t)$ in the fluid which was subjected to the "dynamical" initial perturbation $\mathbf{v}(\mathbf{r},0)$ at $t = 0$. A runaway $\mathbf{v}(\mathbf{r},t)$ indicates instability, while if $\mathbf{v}(\mathbf{r},t)$ is bounded for *any* initial perturbation, the stability of the configuration is proved.

It is often convenient to describe the equation of motion in terms of the Lagrangian variable ξ. This is simply the displacement of any fluid element from its equilibrium position and is expressed as a function of the equilibrium position $\mathbf{r}_0$ as $\xi(\mathbf{r}_0, t)$ (Fig. 5-5). The time derivative $\dot{\xi}(\mathbf{r}_0, t)$ describes (at fixed $\mathbf{r}_0$) the velocity of the same fluid element as it varies in time. Equation (5-6), on the other hand, uses a Eulerian description, where $\mathbf{v}(\mathbf{r},t)$ signifies (at fixed $\mathbf{r}$)

the velocity history at a given point in space, as various fluid elements pass through it. In the first order, however, the two velocities do not differ:

$$\mathbf{v}(\mathbf{r},t) = \mathbf{v}(\mathbf{r}_0, t) + (\xi \cdot \nabla)\mathbf{v} + \cdots \approx \mathbf{v}(\mathbf{r}_0, t) \qquad (5\text{-}7)$$

[**v** being small and a smooth function of position, $(\xi \cdot \nabla)\mathbf{v}$ is of second order].

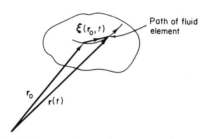

Path of fluid element

$\xi(r_0, t)$

r_0

$r(t)$

FIG. 5-5. Perturbation of a fluid equilibrium.

Consequently one puts $\dot{\xi} = \mathbf{v}$ in (5-6) and integrates with respect to time. Since in our dynamical stability experiment $\xi(\mathbf{r}_0, 0) = \dot{\xi}(\mathbf{r}_0, 0) = 0$, one finds

$$v_0\ddot{\xi} = \nabla(\xi \cdot \nabla p_0 + \gamma p_0 \nabla \cdot \xi) + \frac{1}{\mu_0} \{\nabla \times [\nabla \times (\xi \times \mathbf{B}_0)]\} \times \mathbf{B}_0$$
$$+ \frac{1}{\mu_0} (\nabla \times \mathbf{B}_0) \times \{\nabla \times (\xi \times \mathbf{B}_0)\} \qquad (5\text{-}8)$$

We see that the displacement from equilibrium obeys the same differential equation as the velocity. Since the left side is the mass density times the acceleration, the right side must be the linearized force density. Equation (5-8) can be abbreviated

$$v_0\ddot{\xi} = \mathbf{F}(\xi) \qquad (5\text{-}9)$$

where **F** is the time-independent linear operator in (5-8). One can separate $\xi(\mathbf{r}_0, t)$ into a product of a space- and a time-dependent part $\xi(\mathbf{r}_0, t) = T(t) \cdot \xi(\mathbf{r}_0)$, which leads at once to the separation of (5-9) into

$$\ddot{T} = -\omega_k^2 T \qquad (5\text{-}10)$$

and

$$-v_0\xi_k(\mathbf{r}_0)\omega_k^2 = \mathbf{F}(\xi_k) \qquad (5\text{-}11)$$

where the values of the separation constant ω_k^2 are determined from the eigenvalues of the operator equation (5-11). Equation (5-10) results, of course, in periodic time dependence, and the general solution of (5-8) can be written

$$\xi(\mathbf{r}_0, t) = \sum_k a_k \exp(i\omega_k t)\xi_k(\mathbf{r}_0) \qquad (5\text{-}12)$$

Negative values of ω_k^2 lead again to runaway solutions. Hence the criterion for stability can be formulated as follows: *A hydromagnetic equilibrium is stable if and only if all eigenvalues of the operator* $\mathbf{F}/v_0$ *are negative.*

Since our system is conservative, neither damping nor overstability is possible, and ω_k^2 must be real. From this fact one suspects that the operator $\mathbf{F}$ is Hermitian. This can indeed be proved by considering the conservation of the system's energy. The change in potential energy density due to a small perturbation ξ is

$$-\int_0^{\xi} \mathbf{F}(\eta) \cdot d\eta = -\tfrac{1}{2}\xi \cdot \mathbf{F}(\xi) \tag{5-13}$$

since $\mathbf{F}$ is a linear function of its argument. Since the rate of change of the kinetic plus potential energy must be zero,

$$\frac{\partial}{\partial t} \int [\tfrac{1}{2} v_0 \dot{\xi}^2 - \tfrac{1}{2}\xi \cdot \mathbf{F}(\xi)] \, d\tau_0$$
$$= \int v_0 \dot{\xi} \cdot \ddot{\xi} \, d\tau_0 - \tfrac{1}{2} \int \dot{\xi} \cdot \mathbf{F}(\xi) \, d\tau_0 - \tfrac{1}{2} \int \xi \cdot \mathbf{F}(\dot{\xi}) \, d\tau_0 = 0 \tag{5-14}$$

where the integration is extended over the equilibrium fluid volume. Using (5-9) in (5-14) one obtains

$$\int \dot{\xi} \cdot \mathbf{F}(\xi) \, d\tau_0 = \int \xi \cdot \mathbf{F}(\dot{\xi}) \, d\tau_0 \tag{5-15}$$

Since ξ and $\dot{\xi}$ are arbitrary independent functions satisfying the same boundary conditions, it follows that $\mathbf{F}$ is indeed Hermitian.

The eigenfunctions of $\mathbf{F}$ are the normal modes; they represent displacement fields, each one of which preserves its shape and only changes its amplitude in time, according to the harmonic differential equation (5-10). The amplitudes a_k of the normal modes are the normal coordinates, in accordance with our previous definition.

It follows from the Hermitian character of $\mathbf{F}$ that eigenfunctions corresponding to different eigenvalues are orthogonal. The degenerate ones can be orthogonalized in the usual manner. It is convenient to put the orthonormality condition in the form

$$\tfrac{1}{2} \int v_0 \xi_k \cdot \xi_l \, d\tau_0 = \delta_{kl} \tag{5-16}$$

In analogy with our considerations in the previous section, one expects an energy condition to hold: The system is stable if and only if all possible small perturbations ξ make the change in the potential energy

$$\delta W = -\tfrac{1}{2} \int \xi \cdot \mathbf{F}(\xi) \, d\tau \tag{5-17}$$

positive. In fact, inserting (5-12) and using (5-11) and (5-16) in (5-17).

$$\delta W = \sum_k \sum_l a_k a_l \, \mathrm{Re} \, \exp(i\omega_k t) \, \mathrm{Re} \, \exp(i\omega_l t)\omega_l{}^2 \int \tfrac{1}{2}\nu_0 \boldsymbol{\xi}_k \cdot \boldsymbol{\xi}_l \, d\tau_0$$

$$= \sum_k a_k{}^2 \cos{}^2 \omega_k t \, \omega_k{}^2 \tag{5-18}$$

δW can only be made negative if at least one eigenvalue $\omega_k{}^2$ is negative. Similarly if one $\omega_k{}^2$ is negative δW can be made negative by choosing $a_l{}^2 = 0$ for $l \neq k$. The stability criterion based on the energy principle is therefore equivalent to the one based on the equation of motion.

So far we have only discussed what goes on inside the fluid. Actually we are interested mainly in finding out about the stability of magnetically confined configurations, where the fluid occupies only a finite region in space, confined by a vacuum magnetic field. In investigating the stability of such a configuration, surface displacements affecting the confining field must also be considered. To this end we shall transform the expression for the change in potential energy (5-17), to separate the surface contribution and vacuum energy from the fluid part, and derive the linearized boundary conditions. First we write (5-17) explicitly

$$\delta W = -\tfrac{1}{2} \int \boldsymbol{\xi} \cdot \left[\nabla(\boldsymbol{\xi} \cdot \nabla p + \gamma p \nabla \cdot \boldsymbol{\xi}) + \frac{1}{\mu_0} (\nabla \times \mathbf{Q}) \times \mathbf{B} \right.$$

$$\left. + \frac{1}{\mu_0} (\nabla \times \mathbf{B}) \times \mathbf{Q} \right] d\tau \tag{5-19}$$

where the index 0 on the unperturbed quantities has been dropped and we abbreviated

$$\nabla \times (\boldsymbol{\xi} \times \mathbf{B}) = \mathbf{Q} \tag{5-20}$$

The integral is again extended over the whole volume τ_0 occupied by the unperturbed fluid. $\mathbf{Q}$ has a simple physical meaning: $\mathbf{Q} = \delta\mathbf{B}$ is the change in the local magnetic field intensity due to the perturbation (see Exercise 5-3).

Using the identities

$$\nabla \cdot [\boldsymbol{\xi}(\boldsymbol{\xi} \cdot \nabla p + \gamma p \nabla \cdot \boldsymbol{\xi})] = \boldsymbol{\xi} \cdot \nabla(\boldsymbol{\xi} \cdot \nabla p + \gamma p \nabla \cdot \boldsymbol{\xi})$$

$$+ \nabla \cdot \boldsymbol{\xi}(\boldsymbol{\xi} \cdot \nabla p + \gamma p \nabla \cdot \boldsymbol{\xi}) \tag{5-21}$$

$$\nabla \cdot [(\boldsymbol{\xi} \times \mathbf{B}) \times \mathbf{Q}] = \mathbf{Q} \cdot \nabla \times (\boldsymbol{\xi} \times \mathbf{B}) - (\boldsymbol{\xi} \times \mathbf{B}) \cdot \nabla \times \mathbf{Q}$$

$$= \mathbf{Q}^2 + \boldsymbol{\xi} \cdot [(\nabla \times \mathbf{Q}) \times \mathbf{B}] \tag{5-22}$$

$$\boldsymbol{\xi} \cdot [(\nabla \times \mathbf{B}) \times \mathbf{Q}] = -(\nabla \times \mathbf{B}) \cdot (\boldsymbol{\xi} \times \mathbf{Q}) \tag{5-23}$$

and Gauss' theorem in (5-19) one finds

$$\delta W = \frac{1}{2} \int \left[\frac{Q^2}{\mu_0} + \frac{1}{\mu_0} (\nabla \times \mathbf{B}) \cdot (\xi \times \mathbf{Q}) + (\nabla \cdot \xi)\xi \cdot \nabla p + \gamma p (\nabla \cdot \xi)^2 \right] d\tau$$

$$- \frac{1}{2} \int \left[\frac{1}{\mu_0} (\xi \times \mathbf{B}) \times \mathbf{Q} + \xi(\xi \cdot \nabla p + \gamma p \nabla \cdot \xi) \right] \cdot d\mathbf{S} \qquad (5\text{-}24)$$

where the surface integral is extended over the fluid-vacuum boundary.

The perturbed field quantities have to satisfy the boundary conditions discussed in Sec. 4-1. In the vacuum, the displacement of the fluid boundary changes the magnetic field by $\delta \mathbf{B}_v$, and the small electric field $\delta \mathbf{E}_v$ is generated by induction. A boundary condition requires that the tangential component of the effective electric field $\mathbf{E}'$ be continuous, (4-13):

$$\mathbf{n} \times \delta \mathbf{E}_v + \mathbf{n} \times (\mathbf{v} \times \mathbf{B}) = (\mathbf{n} \cdot \mathbf{v})\mathbf{B}_v - (\mathbf{n} \cdot \mathbf{v})\mathbf{B} \qquad (5\text{-}25)$$

since on the fluid side $\mathbf{E} = -\mathbf{v} \times \mathbf{B}$. $\mathbf{B}_v$ is the unperturbed vacuum magnetic field and $\mathbf{n} = \mathbf{n}_0$ the unperturbed surface normal, to keep the equation linear in small quantities. Expanding the triple vector product and using the boundary condition $\mathbf{n} \cdot \mathbf{B} = 0$, one finds

$$\mathbf{n} \times \delta \mathbf{E}_v = (\mathbf{n} \cdot \mathbf{v})\mathbf{B}_v \qquad (5\text{-}26)$$

To find the linearized form of (4-16), which describes the equality of pressures on both sides of the boundary, we need to determine the magnetic field on the inside of the displaced boundary. Using

$$\frac{d\mathbf{B}}{dt} = \frac{\partial \mathbf{B}}{\partial t} + (\mathbf{v} \cdot \nabla)\mathbf{B} \qquad (5\text{-}27)$$

and integrating with respect to time from $t = 0$ one obtains

$$\mathbf{B}(\mathbf{r},t) - \mathbf{B}(\mathbf{r}_s, 0) = \mathbf{Q}(\mathbf{r}_s, t) + (\xi \cdot \nabla)\mathbf{B}(\mathbf{r}_s, t) \qquad (5\text{-}28)$$

where $\mathbf{r}_s$ designates the position of a fluid element on the unperturbed fluid surface and $\mathbf{r}$ is its present location on the perturbed boundary. On the vacuum side a similar equation holds, where $\mathbf{B}$ is replaced by $\mathbf{B}_v$ and $\mathbf{Q}$ by $\delta \mathbf{B}_v$.

The pressure at the displaced boundary follows from (4-23), which can be written

$$dp/dt = -\gamma p \nabla \cdot \mathbf{v} \qquad (5\text{-}29)$$

to yield after integration

$$p(\mathbf{r},t) - p(\mathbf{r}_s, 0) = -\gamma p(\mathbf{r}_s, 0)\nabla \cdot \xi \qquad (5\text{-}30)$$

Inserting (5-28) and (5-30) in

$$\left(p + \frac{B^2}{2\mu_0} \right)_{\text{fluid}} = \frac{B_v{}^2}{2\mu_0} \qquad (5\text{-}31)$$

at the displaced boundary and making use of the same condition for the equilibrium state, one obtains

$$-\gamma p \nabla \cdot \boldsymbol{\xi} + \frac{\mathbf{B}}{\mu_0} \cdot [\mathbf{Q} + (\boldsymbol{\xi} \cdot \nabla)\mathbf{B}] = \frac{\mathbf{B}_v}{\mu_0} \cdot [\delta \mathbf{B}_v + (\boldsymbol{\xi} \cdot \nabla)\mathbf{B}_v] \qquad (5\text{-}32)$$

where all quantities are to be taken at the equilibrium boundary.

Finally we use the boundary conditions to transform the surface integral in (5.24). If one expands $(\boldsymbol{\xi} \times \mathbf{B}) \times \mathbf{Q}$ and uses $(\mathbf{B} \cdot d\mathbf{S}) = 0$ and (5-32), the integral becomes

$$\int_S \left(-\frac{\mathbf{B} \cdot \mathbf{Q}}{\mu_0} + \boldsymbol{\xi} \cdot \nabla p + \gamma p \nabla \cdot \boldsymbol{\xi} \right) \boldsymbol{\xi} \cdot d\mathbf{S}$$

$$= \int_S \left\{ \frac{\mathbf{B}}{\mu_0} \cdot (\boldsymbol{\xi} \cdot \nabla)\mathbf{B} - \frac{1}{\mu_0} \mathbf{B}_v \cdot \delta\mathbf{B}_v - \frac{1}{\mu_0} \mathbf{B}_v \cdot (\boldsymbol{\xi} \cdot \nabla)\mathbf{B}_v + \boldsymbol{\xi} \cdot \nabla p \right\} \boldsymbol{\xi} \cdot d\mathbf{S}$$

$$(5\text{-}33)$$

With the help of the identity

$$\mathbf{B} \cdot (\boldsymbol{\xi} \cdot \nabla)\mathbf{B} = B_i \xi_k \frac{\partial}{\partial x_k} B_i = \xi_k \frac{\partial}{\partial x_k} \frac{B_i B_i}{2} = (\boldsymbol{\xi} \cdot \nabla) \frac{B^2}{2} \qquad (5\text{-}34)$$

the surface integral reduces to

$$\int \left\{ \boldsymbol{\xi} \cdot \nabla \left(p + \frac{B^2}{2\mu_0} \right) - \boldsymbol{\xi} \cdot \nabla \frac{B_v^2}{2\mu_0} - \frac{\mathbf{B}_v \cdot \delta\mathbf{B}_v}{\mu_0} \right\} \boldsymbol{\xi} \cdot d\mathbf{S} \qquad (5\text{-}35)$$

Since (5-31) holds all over the boundary, the tangential component of $\nabla(p + (B^2/2\mu_0))$ must also be continuous. Hence only the normal components survive and (5-35) can be written

$$-\int_S (\boldsymbol{\xi} \cdot \mathbf{n})^2 \left[\nabla \left(p + \frac{B^2}{2\mu_0} \right) \right] \cdot d\mathbf{S} - \int_S \frac{\mathbf{B}_v \cdot \delta\mathbf{B}_v}{\mu_0} \boldsymbol{\xi} \cdot d\mathbf{S} \qquad (5\text{-}36)$$

where the boldface square brackets [ ] again denote the increment across the boundary. It remains only to transform the last integral. To this end we introduce the perturbation vector potential $\delta\mathbf{A}$, to generate the first-order vacuum fields by

$$\nabla \times \delta\mathbf{A} = \delta\mathbf{B}_v \qquad (5\text{-}37)$$

and

$$-\delta\dot{\mathbf{A}} = \delta\mathbf{E}_v \qquad (5\text{-}38)$$

where the Coulomb gauge has been adopted so that the scalar potential does not appear. Integrating (5-26) with respect to time yields

$$-\mathbf{n} \times \delta\mathbf{A} = (\mathbf{n} \cdot \boldsymbol{\xi})\mathbf{B}_v \qquad (5\text{-}39)$$

which can be inserted into the last integral in (5-36) to give

$$\frac{1}{\mu_0} \int (\nabla \times \delta\mathbf{A}) \cdot (d\mathbf{S} \times \delta\mathbf{A}) = -\frac{1}{\mu_0} \int_{\tau_v} \nabla \cdot \{\delta\mathbf{A} \times (\nabla \times \delta\mathbf{A})\} \, d\tau_v \quad (5\text{-}40)$$

where the volume integral is extended over the vacuum. The sign changes because $d\mathbf{S}$ points away from the fluid into the vacuum. It is assumed that either the vacuum field falls off at infinity or the vacuum is terminated by a metallic wall where the tangential component of the electric field and $\delta\mathbf{A}$ vanish. Since in the vacuum there is no current,

$$\nabla \times (\nabla \times \delta\mathbf{A}) = 0 \quad (5\text{-}41)$$

and the integral becomes

$$-\frac{1}{\mu_0} \int (\nabla \times \delta\mathbf{A})^2 \, d\tau_v = -\int \frac{\delta B_v^{\,2}}{\mu_0} \, d\tau_v \quad (5\text{-}42)$$

We found that the change in the potential energy due to the perturbation splits into three parts. The fluid-energy part can be expressed as

$$\delta W_F = \frac{1}{2} \int \left\{ \frac{Q^2}{\mu_0} + \frac{1}{\mu_0} (\nabla \times \mathbf{B}) \cdot (\boldsymbol{\xi} \times \mathbf{Q}) + (\nabla \cdot \boldsymbol{\xi}) \boldsymbol{\xi} \cdot \nabla p + \gamma p (\nabla \cdot \boldsymbol{\xi})^2 \right\} d\tau_F$$

$$(5\text{-}43)$$

where the first term gives the change in magnetic energy, the second the work done against the unbalanced magnetic forces in the fluid, and the last two the change in internal energy of the fluid due to nonmagnetic forces.

The surface energy

$$\delta W_s = \frac{1}{2} \int (\boldsymbol{\xi} \cdot \mathbf{n})^2 \left[\nabla \left(p + \frac{B^2}{2\mu_0} \right) \right] \cdot d\mathbf{S} \quad (5\text{-}44)$$

signifies the work done against the surface current by displacing the boundary by $\boldsymbol{\xi}$ (see Exercise 5-4). The vacuum energy

$$\delta W_v = \int \frac{\delta B_v^{\,2}}{2\mu_0} \, d\tau_v \quad (5\text{-}45)$$

is the change in magnetic energy in the vacuum region.

Note that each potential energy term is of second order in the small quantities. Since the initial system was in equilibrium, the first-order terms in the perturbed potential energy have automatically vanished.

5-3. Some Applications of the Equation of Motion

The linearized equation of motion can be used to investigate motion around equilibrium, both oscillatory or runaway. This method, also called *normal-mode analysis*, has the advantage over the energy method that it yields the *value* of ω^2 immediately; hence the frequency of oscillations, or the growth rate of the instability, is obtained.

The case of an infinite uniform fluid in a homogeneous background magnetic field is stable, and small deviations from equilibrium result in hydromagnetic waves. One need only solve the eigenvalue equation

$$-\nu\xi\omega^2 = F(\xi) \tag{5-11}$$

with the operator

$$F(\xi) = \gamma p \nabla\nabla \cdot \xi + \frac{1}{\mu_0}(\nabla \times Q) \times B \tag{5-46}$$

The perturbed magnetic field can be expanded:

$$Q = \nabla \times (\xi \times B) = (B \cdot \nabla)\xi - B(\nabla \cdot \xi) \tag{5-47}$$

since $\nabla \cdot B = 0$.

Looking for perturbations with $\nabla \cdot \xi = 0$, the equation of motion becomes

$$\mu_0 \nu \omega^2 \xi = B \times \left(\nabla \times B \frac{\partial \xi}{\partial z}\right) = B^2 \frac{\partial}{\partial z}[e_3 \times (\nabla \times \xi)] \tag{5-48}$$

if one takes the z axis along the field. Since the right side is perpendicular to B, $\xi_z = 0$. Expanding,

$$[e_3 \times (\nabla \times \xi)] = e_3 \times \left(-e_1 \frac{\partial \xi_y}{\partial z} + e_2 \frac{\partial \xi_x}{\partial z}\right) = -\frac{\partial \xi}{\partial z} \tag{5-49}$$

The solution of the equation

$$\mu_0 \nu \omega^2 \xi = -B^2 \frac{\partial^2 \xi}{\partial z^2} \tag{5-50}$$

is

$$\xi = \xi_0 e^{ikz} \tag{5-51}$$

with

$$\omega^2 = \frac{B^2 k^2}{\mu_0 \nu} \tag{5-52}$$

always positive. These are, of course, the Alfven waves, propagating along the field ($k = e_3 k$), with displacements transverse to the field ($\xi_z = 0$), and the phase velocity given by the Alfven speed

$$V_A = \frac{\omega}{k} = \frac{B}{(\mu_0 \nu)^{1/2}} \tag{5-53}$$

If the fluid is compressible one has to use (5-46) in its complete form. Since F and the ∇ operator commute, the eigenfunctions of F are those of the del operator, namely,

$$\xi = \xi_0 e^{ik \cdot x} \tag{5-54}$$

This is also seen by direct substitution, which yields

$$v\omega^2 \xi_0 = \gamma p k(k \cdot \xi_0) + \frac{1}{\mu_0} [k \times (B \cdot k)\xi_0] \times B - \frac{1}{\mu_0} [k \times B(k \cdot \xi_0)] \times B$$

$$(5\text{-}55)$$

This is a system of algebraic equations for the determination of the unknown ξ_0, to be written

$$\xi_0 = \overset{\leftrightarrow}{\kappa} \cdot \xi_0 = \kappa_{il}\xi_{0l} \tag{5-56}$$

with $\overset{\leftrightarrow}{\kappa}$ a 3×3 matrix. A solution can only be found if the determinant $|\kappa_{il} - \delta_{il}| = 0$. This establishes a relationship between the components of k and ω^2, a dispersion relation. The finding of the explicit form of $\overset{\leftrightarrow}{\kappa}$ and the dispersion relation is left as an exercise to the reader (Exercise 5-5).

Waves propagating in the principal directions, $k \parallel B$ and $k \perp B$, can be obtained at once from (5-55). If $k \parallel B$,

$$v\omega^2 \xi_0 = \gamma p(k \cdot \xi_0)k + \frac{(B \cdot k)^2}{\mu_0} \xi_0 - \frac{\xi_0 \cdot B}{\mu_0} (B \cdot k)k \tag{5-57}$$

Two of the vectors in this equation point in the ξ_0, two others in the k, direction. Therefore either $k \parallel \xi$ and

$$v\omega^2 \xi = \gamma p k^2 \xi \tag{5-58}$$

which leads to acoustic waves with $V = \omega/k = (\gamma p/v)^{1/2}$, or the coefficient of the ξ_0 vectors vanish, which leads at once to the Alfven waves.

If $k \perp B$,

$$v\omega^2 \xi_0 = \gamma p(k \cdot \xi_0)k + \frac{B^2}{\mu_0} (k \cdot \xi_0)k \tag{5-59}$$

and since ξ_0 is clearly parallel to k, there is only one mode with

$$\frac{\omega}{k} = \left[\frac{\gamma p + (B^2/\mu_0)}{v} \right]^{1/2} \tag{5-60}$$

as we have already seen in (4-99). In a general direction there are three waves: a transverse Alfven wave and two magnetoacoustic waves (see Exercise 5-6).

As an example of an unstable configuration, consider the hydromagnetic version of the Rayleigh-Taylor arrangement. Here a hydromagnetic fluid is supported against the gravitational field by a magnetic field (in Fig. 5-3 the lower fluid is replaced by a uniform horizontal magnetic field).

To treat this case we have to include in the equation of motion the effect of a gravitational force density

$$f_g = -v \nabla \phi \tag{5-61}$$

where ϕ is the gravitational potential. This can be done without difficulty (see Exercise 5-7) to obtain the first-order gravitational force density

$$\mathbf{F}_g = \nabla \cdot (v\xi) \, \nabla \phi \qquad (5\text{-}62)$$

For a uniform gravitational field acting in the $-y$ direction the equilibrium equation reads

$$\frac{\partial}{\partial y} \left(p + \frac{B^2}{2\mu_0} \right) + vg = 0 \qquad (5\text{-}63)$$

and the first-order equation becomes

$$-v\omega^2\xi = -\nabla \, \delta p + \frac{1}{\mu_0} \{ (\nabla \times \mathbf{Q}) \times \mathbf{B} + (\nabla \times \mathbf{B}) \times \mathbf{Q} \} + \mathbf{e}_2 g \, \nabla \cdot (v\xi)$$

$$(5\text{-}64)$$

We are going to investigate the stability of an incompressible fluid instead of an adiabatic one; hence the first-order pressure change was left in its general form. Furthermore, we take $\mathbf{B} = B\mathbf{e}_3$, assume that equilibrium quantities depend on y only, and consider only perturbations which do not vary along the magnetic field (see Fig. 5-6),

$$(\mathbf{e}_3 \cdot \nabla)\xi = 0 \qquad (5\text{-}65)$$

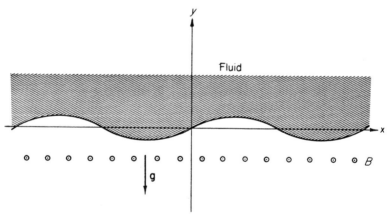

FIG. 5-6. Perturbation of a hydromagnetic fluid suspended by a magnetic field.

With the help of the identities (A-11) and (A-12) one may write

$$\mathbf{Q} = \nabla \times (\xi \times \mathbf{B}) = -\xi_y \frac{dB}{dy} \mathbf{e}_3 \qquad (5\text{-}66)$$

and

$$(\nabla \times \mathbf{Q}) \times \mathbf{B} + (\nabla \times \mathbf{B}) \times \mathbf{Q} = -\nabla(\mathbf{B} \cdot \mathbf{Q}) \qquad (5\text{-}67)$$

Equation (5-64) now becomes

$$v\omega^2 \xi = \nabla\left(\delta p + \frac{\mathbf{B} \cdot \mathbf{Q}}{\mu_0}\right) - \mathbf{e}_2 g \xi \cdot \nabla v \qquad (5\text{-}68)$$

where use has been made of the "equation of state"

$$\nabla \cdot \xi = 0 \qquad (5\text{-}69)$$

The translational symmetry of the system in the x direction permits solutions of the form e^{ikx} and the component equations of (5-68) become

$$v\omega^2 \xi_x = ik\left(\delta p + \frac{1}{\mu_0} \mathbf{B} \cdot \mathbf{Q}\right) \qquad (5\text{-}70)$$

and

$$v\omega^2 \xi_y = \frac{\partial}{\partial y}\left(\delta p + \frac{1}{\mu_0} \mathbf{B} \cdot \mathbf{Q}\right) - g\xi_y \frac{dv}{dy} \qquad (5\text{-}71)$$

Expressing the sum of the first-order fluid and magnetic pressures from (5-70) and inserting in (5-71) the latter becomes

$$v\omega^2 \xi_y = \frac{\omega^2}{ik} \frac{\partial}{\partial y}(v\xi_x) - g\xi_y \frac{dv}{dy} \qquad (5\text{-}72)$$

ξ_x can be eliminated by differentiation of (5-72) with respect to x and using (5-69) to obtain

$$ikv\omega^2 \xi_y = -\frac{\omega^2}{ik} \frac{\partial}{\partial y}\left(v \frac{\partial \xi_y}{\partial y}\right) - gik\xi_y \frac{dv}{dy} \qquad (5\text{-}73)$$

and we arrive at the differential equation which describes the variation of ξ_y in the y direction:

$$k^2\left(v\omega^2 + g \frac{dv}{dy}\right)\xi_y = \omega^2 \frac{\partial}{\partial y}\left(v \frac{\partial \xi_y}{\partial y}\right) \qquad (5\text{-}74)$$

Assume that the fluid is of uniform density v_0 separated by a thin current-carrying layer from the vacuum at $y = 0$. (Note that this assumption does not exclude the presence of a magnetic field in the fluid interior.) For $y > 0$, $dv/dy = 0$, and

$$k^2 \xi_y = \frac{\partial^2}{\partial y^2} \xi_y \qquad (5\text{-}75)$$

with the solution (which does not diverge for $y \to \infty$)

$$\xi_y = \xi_y{}^0 e^{-ky} e^{ikx} \qquad (5\text{-}76)$$

Since $dv/dy = v_0\delta(y)$, one may integrate (5-74) across the boundary from

$y = -\varepsilon$ to $y = \varepsilon$ $(\varepsilon \to 0)$, to obtain

$$k^2 g v_0 \xi_y = -\omega^2 k v_0 \xi_y \tag{5-77}$$

and the dispersion relation

$$\omega^2 = -kg \tag{5-78}$$

leading to unstable solutions for every wave number, as expected. In fact the larger k is (the smaller the wavelength), the faster the instability grows. This is the *Kruskal-Schwarzschild instability*. Its significance lies not so much in the unstable behavior of a fluid supported against a gravitational field but in the instability exhibited by systems supported by a magnetic field against inertial forces. Clearly a fluid accelerated by a magnetic field, with the acceleration vector pointing toward the fluid (see Fig. 5-7a) can be viewed

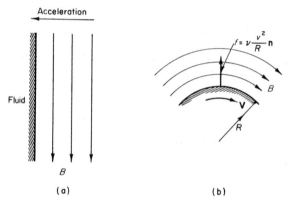

(a)   (b)

FIG. 5-7. The fluid–magnetic field interface is unstable if the fluid elements are accelerated toward the fluid interior.

from the moving system as being subjected to an equivalent gravitational field and must be unstable. A fluid streaming along slightly inward-curving field lines (Fig. 5-7b) is another example, where the centrifugal force plays the role of the gravitational field.

The case where the magnetic field (light fluid) is up and the fluid down is of course again stable, and exhibits oscillations around equilibrium. The proof is simple and left as an exercise (Exercise 5-8).

When the ripples are not parallel to the magnetic field lines the growth of the instability is reduced (Exercise 5-9). This can be expected, since the stretched field lines exert a stabilizing force, especially for short wavelengths. This is of no help, however, since it only means that the parallel flutes will soon outgrow all others. If, on the other hand, one introduces a shear in the magnetic field in such a way that the field changes direction across the boundary, all ripples result in field-line stretching, and the fastest growth rate can be effectively reduced (see Exercise 5-10).

Finally we turn to the stability analysis of a simplified version of the pinch configuration, where the inside is current- and field-free and the confining field is generated by a thin surface current flowing in the z direction. The equation of motion of the fluid reduces to

$$-v\xi\omega^2 = \gamma p \nabla \nabla \cdot \xi \tag{5-79}$$

One introduces cylindrical coordinates to fit the boundary conditions and the trial solution

$$\xi(r,\varphi,z) = \xi(r)e^{i(m\varphi+kz)} \tag{5-80}$$

[This choice is justified by $\mathbf{F}(\partial/\partial z) = (\partial/\partial z)\mathbf{F}$ and $\mathbf{F}(\partial/\partial \varphi) = (\partial/\partial\varphi)\mathbf{F}$ (Exercise 5-11) or *a posteriori* by the fact that this trial function leads to solutions.] Inserting (5-80) in (5-79) leads to the component equations

$$-v\xi_z\omega^2 = \gamma pik\nabla \cdot \xi \tag{5-81}$$

$$-v\xi_\varphi\omega^2 = \gamma p \frac{im}{r}\nabla \cdot \xi = -\frac{m\omega^2 v}{rk}\xi_z \tag{5-82}$$

and

$$-v\xi_r\omega^2 = \gamma p \frac{d}{dr}\nabla \cdot \xi = -\frac{\omega^2 v}{ik}\frac{d}{dr}\xi_z \tag{5-83}$$

Since ξ_φ and ξ_r are expressed as functions of ξ_z, it suffices to solve for the latter component only. The differential equation for ξ_z can be obtained if one expands $\nabla \cdot \xi$ in (5-81) and inserts ξ_φ and ξ_r from (5-82) and (5-83).

$$-v\omega^2\xi_z = \gamma p \left[\frac{1}{r}\frac{d}{dr}r\frac{d}{dr}\xi_z - \frac{m^2}{r^2}\xi_z - k^2\xi_z\right] \tag{5-84}$$

or

$$\frac{d^2\xi_z}{dr^2} + \frac{1}{r}\frac{d\xi_z}{dr} + \left(\alpha^2 - \frac{m^2}{r^2}\right)\xi_z = 0 \tag{5-85}$$

where

$$\alpha^2 = \frac{\omega^2 v}{\gamma p} - k^2 \tag{5-86}$$

Equation (5-85) is *Bessel's differential equation*, with the (nonsingular) solution of $J_m(\alpha r)$, which yields the full form of ξ_z,

$$\xi_z = A_{m,k}J_m(\alpha r)e^{i(m\varphi+kz)} \tag{5-87}$$

where the constants (one for each pair of m and k) will be determined with the help of the boundary conditions.

Outside the plasma the equilibrium magnetic field is

$$B_\varphi = B_0(r_0/r) \qquad (5\text{-}88)$$

where B_0 is the magnitude of the field at the surface of cylinder $r = r_0$. In equilibrium the pressure-balance equation

$$p = B_0^2/2\mu_0 \qquad (5\text{-}89)$$

holds. The perturbation of the plasma gives rise to the additional vacuum magnetic field $\delta \mathbf{B}$. Since in the vacuum there is no current, and one neglects the displacement current (this is equivalent to neglecting the small radiation pressure), $\nabla \times \delta \mathbf{B} = 0$ and the magnetic field can be derived from a scalar potential as

$$\delta \mathbf{B} = \nabla \psi \qquad (5\text{-}90)$$

where, since $\nabla \cdot \delta \mathbf{B} = 0$, ψ obeys the Laplace equation

$$\nabla^2 \psi = 0 \qquad (5\text{-}91)$$

The trial solution

$$\psi = \psi(r) e^{i(kz + m\varphi)} \qquad (5\text{-}92)$$

leads immediately to the differential equation

$$\frac{1}{r}\frac{d}{dr}\left(r \frac{d\psi}{dr}\right) + \left(-k^2 - \frac{m^2}{r^2}\right)\psi = 0 \qquad (5\text{-}93)$$

The solutions are Bessel and Neumann functions with the imaginary argument ikr. It is somewhat more convenient instead to use the hyperbolic Bessel functions I and K with real arguments. The solution is

$$\psi = C_{m,k} K_m(kr) e^{i(m\varphi + kz)} \qquad (5\text{-}94)$$

because $I(kr)$ diverges at $r \to \infty$. The field components are

$$\delta B_r = Ck K_m'(kr) e^{i(m\varphi + kz)} \qquad (5\text{-}95a)$$

$$\delta B_\varphi = C\frac{im}{r} K_m(kr) e^{i(m\varphi + kz)} \qquad (5\text{-}95b)$$

$$\delta B_z = Cik K_m(kr) e^{i(m\varphi + kz)} \qquad (5\text{-}95c)$$

where we dropped the subscripts from the constant.

$\delta \mathbf{B}$ has to satisfy the boundary conditions at the new fluid-vacuum interface. First we require that the magnetic field be tangential to the boundary. For this end we need the surface normal. The equation of the boundary is in linearized form (see Fig. 5-8):

$$r = r_0 + \xi_r = r_0 + \frac{\alpha}{ik} A J_m'(\alpha r_0) e^{i(kz + m\varphi)} \qquad (5\text{-}96)$$

where ξ_r was obtained from (5-83) and (5-87). If the equation of a surface is written in the form $\varphi(\mathbf{r}) = 0$, it is known that the vector $\nabla\varphi$ is perpendicular

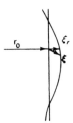

FIG. 5-8. Surface perturbation of a cylindrical fluid.

to this surface. Writing (5-96) in the form

$$\varphi = r_0 + r_1 e^{i(m\varphi + kz)} - r = 0 \tag{5-97}$$

the vector

$$\nabla\varphi = \left(-1, \frac{imr_1}{r_0} e^{i(m\varphi + kz)}, ikr_1 e^{i(m\varphi + kz)}\right) \tag{5-98}$$

is a surface normal. The magnetic field on the vacuum side of the boundary can be obtained from (5-28) and the remark following (5-28),

$$\mathbf{B}(\mathbf{r}, t) = \mathbf{B}(\mathbf{r}_0) + \delta\mathbf{B} + (\boldsymbol{\xi} \cdot \nabla)\mathbf{B}(\mathbf{r}_0) = B_0\boldsymbol{\varphi} + \delta\mathbf{B} - \xi_r \frac{B_0}{r_0} \boldsymbol{\varphi} \tag{5-99}$$

where use has been made of (5-88). The boundary condition becomes, neglecting second-order quantities,

$$\nabla\varphi \cdot \mathbf{B} = -\delta B_r + \frac{im}{r_0} B_0 r_1 e^{i(kz + m\varphi)} = 0 \tag{5-100}$$

Substituting δB_r from (5-95a) and r_1 from (5-96) one obtains

$$\frac{C}{A} = \frac{B_0}{r_0} \frac{\alpha m}{k^2} \frac{J_m'(\alpha r_0)}{K_m'(kr_0)} \tag{5-101}$$

The other boundary condition comes from the first-order pressure-balance equation (5-32), which gives for our case

$$\frac{v\omega^2 \xi_z}{ik} = \frac{B_0}{\mu_0} \left[\delta B_\varphi - \xi_r \frac{B_0}{r_0}\right] \tag{5-102}$$

Substituting the values of ξ_z, δB_φ, and ξ_r at the boundary gives

$$\frac{v\omega^2}{ik} J_m(\alpha r_0) = \frac{B_0}{\mu_0 r_0} \left[\frac{C}{A} im K_m(kr_0) + iB_0 \frac{\alpha}{k} J_m'(\alpha r_0)\right] \tag{5-103}$$

Inserting C/A from (5-101) and using (5-89) yields the dispersion relation

$$-\frac{\omega^2 r_0 v}{\alpha 2 p}\frac{J_m(\alpha r_0)}{J_m{}'(\alpha r_0)} = 1 + \frac{m^2}{k r_0}\frac{K_m(k r_0)}{K_m{}'(k r_0)} \qquad (5\text{-}104)$$

where α is related to ω and k through (5-86). It is easy to see that (5-104) leads to instabilities. Take, for instance, perturbations independent of φ, where $m = 0$. If one varies $-\omega^2$ from 0 to $+\infty$, the left side of the equation varies continuously from 0 to $+\infty$, taking on the value 1 somewhere. However, a negative value of ω^2, satisfying the dispersion relation, indicates an instability. Hence for $m = 0$, any k value is unstable. The investigation of modes with $m \neq 0$ shows the appearance of further unstable solutions, although now some values of k lead to stable oscillations.

The physical mechanisms of these instabilities are easy to find. An $m = 0$ perturbation causes periodic constrictions on the column as on Fig. 5-9a.

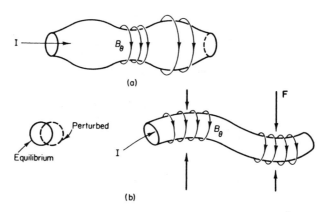

(a)

Perturbed

Equilibrium

(b)

FIG. 5-9. Some characteristic instabilities of the pinch: (a) sausage, (b) kink.

Since the magnetic field is larger at the constriction, the perturbation tends to grow and the column will look like a sausage. The $m = 0$ instability is often referred to as the *sausage-type instability*.

Another dangerous instability is the $m = 1$ or *kink instability*. The result of an $m = 1$, arbitrary k, perturbation is a screw-like surface, while we plotted on Fig. 5-9b the combination of an $m = 1, k$ with the $m = 1, -k$ mode. The magnetic pressures are such as to cause the kink to grow. The understanding of the physical mechanism of these instabilities suggests a cure. We shall return to this question soon.

A similar computation can be carried out for the θ pinch with a sharp boundary. This calculation is left as an exercise to the reader (Exercise 5-12).

The resulting dispersion relation is

$$- \frac{\omega^2 v}{2p} (k\alpha)^{-1} \frac{J_m(\alpha r_0)}{J_m{}'(\alpha r_0)} \frac{K_m{}'(kr_0)}{K_m(kr_0)} = 1 \qquad (5\text{-}105)$$

and it leads to no unstable solutions.

5–4. Some Consequences of the Energy Principle

While the normal-mode analysis is well suited for the detailed investigation of specific configurations, the energy principle is a powerful tool for seeking general stability criteria. Also, in many cases one is not interested in all the ω's of a system, but wants to find out, with as little effort as possible, whether a system is stable or not. In some of these cases instability is easy to prove: One introduces a suitable perturbation, usually led by physical reasoning, which decreases the potential energy of the system, and instability is established. Stability is, of course, more difficult to prove: One must show that no perturbation can lead to a negative δW. This is achieved by minimizing δW with respect to all possible perturbations. If δW_{min} is positive, the system is stable.

Some useful comparison theorems follow immediately from the form of δW.

1. Since γ appears only in a positive-definite term in the energy integral it follows that the larger γ is, the more stable the system. An incompressible fluid, $\gamma \to \infty$, gives at least as much stability as a compressible one in the same configuration. A fluid with pressure independent of volume, $\gamma \to 0$, is the most unstable.

2. Consider a fluid-vacuum system. If one replaces the vacuum region with a zero-pressure fluid, the latter system is at least as stable as the former one. The proof is simple and is left to the reader (Exercise 5-13).

3. Take a fluid, surrounded by a vacuum, which is in turn enclosed by perfectly conducting walls. Moving the walls so as to include more vacuum volume cannot increase, but may decrease, the stability of the system.

This can be seen if one considers that the force on the walls resulting from the magnetic pressure $\delta B^2/2\mu_0$ is always directed outward from the vacuum region. Consequently by moving the walls so as to include a larger vacuum region (think of elastic walls), the perturbation magnetic field does work on the wall currents, reducing the energy content of the fluid-vacuum system.

These theorems can either be used to show the direction a system should be modified to get more stability, or they enable one to get necessary, or sufficient, criteria for stability through the investigation of a modified system. We shall make use of them later in this section.

We proceed now to prove the instability of some systems, by introducing the concept of *interchange instability*. Consider two thin adjacent flux tubes in the fluid. A possible perturbation is one that interchanges these tubes without disturbing the rest of the system. If this lowers the potential energy of the system, it must be unstable.

The change in magnetic energy due to interchange of the tubes follows from (4-31),

$$\delta W_m = \delta \int \frac{B^2}{2\mu_0} A\, dl = \delta \frac{\phi^2}{2\mu_0} \int \frac{dl}{A} \tag{5-106}$$

where one integrates along the flux tube and A is the cross-sectional area. Adopting the notation of Fig. 5-10,

$$\delta W_m = \frac{1}{2\mu_0}\left[\phi_2{}^2\left(\int \frac{dl_1}{A_1} - \int \frac{dl_2}{A_2}\right) + \phi_1{}^2\left(\int \frac{dl_2}{A_2} - \int \frac{dl_1}{A_1}\right)\right]$$

$$= \frac{1}{2\mu_0}(\phi_2{}^2 - \phi_1{}^2)\left(\int \frac{dl_1}{A_1} - \int \frac{dl_2}{A_2}\right) = -\frac{\delta\phi^2}{2\mu_0}\delta\int \frac{dl}{A} \tag{5-107}$$

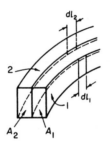

Fig. 5-10. Interchange of flux tubes.

In the case of an incompressible fluid $(\gamma \to \infty)$ the other term in the potential energy is zero. According to our first comparison theorem, however, if one can prove instability for an incompressible fluid, the compressible fluid is at least as unstable. Consequently, to demonstrate instability it suffices to prove that δW_m is negative for an incompressible fluid of the same configuration. Incompressibility implies that the two tubes have equal volume. In this case $A_1\, dl_1 = A_2\, dl_2$ and (5-107) can be written

$$\delta W_m = \frac{1}{2\mu_0}(\phi_2{}^2 - \phi_1{}^2)\int \left[1 - \left(\frac{dl_2}{dl_1}\right)^2\right]\frac{dl_1}{A_1} \tag{5-108}$$

Take, for example, a conductive fluid confined in a mirror geometry (such as Fig. 4-7a). Instead of the sharp boundary between fluid and magnetic field we assume the existence of a thin transition region. If we take tube 2 to be on the outer (magnetic field) side and tube 1 to be on the inner (fluid) side, it follows immediately that $dl_2 > dl_1$, and if the magnetic field increases rapidly enough (thin transition layer), $\phi_2 > \phi_1$. Consequently $\delta W_m < 0$ and the configuration is unstable. The cusp geometries of Fig. 4-7b and c prove stable against this type of perturbation since here $dl_2 < dl_1$. Equation (5-108) shows that a field-free fluid confined by an external magnetic field is always unstable if $dl_2 > dl_1$ or if the field lines on the surface are concave to the fluid. This also follows from the more formal theory developed in Sec. 5-2 (see Exercise 5-14). This can also be viewed by looking at the ripples (or flutes) parallel to the field, produced on the fluid-vacuum interface by a perturbation (Fig. 5-11).

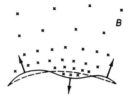

FIG. 5-11. Flute instability.

The protruding flutes push against a weaker field, while the regions newly occupied by magnetic field contain shortened field lines, hence increased magnetic pressure. The result is an exponential growth of the perturbation. The opposite is expected in a cusp geometry. In fact the cusp geometry is absolutely stable, as shown in Exercise 5-14.

It should be noted that for a fluid completely penetrated by the external field, one may find $\phi_2 < \phi_1$ in a mirror geometry (it is certainly true for the vacuum field), and there is no interchange instability for an incompressible fluid. Nevertheless, instability can be proved if one drops the $\gamma \rightarrow \infty$ assumption.

To compute the change in the pressure term in (4-31) we take for simplicity two infinitesimally short tube sections of lengths dl_1 and dl_2 and perform integration along the tubes afterward. The change in the corresponding energy is

$$\delta W_p = \frac{1}{\gamma - 1} \delta(pV) = \frac{1}{\gamma - 1} \delta \frac{pV^\gamma}{V^{\gamma-1}}$$

$$= \frac{1}{\gamma - 1} \left\{ (pV^\gamma)_2 \left(\frac{1}{V_1^{\gamma-1}} - \frac{1}{V_2^{\gamma-1}} \right) + (pV^\gamma)_1 \left(\frac{1}{V_2^{\gamma-1}} - \frac{1}{V_1^{\gamma-1}} \right) \right\}$$

$$= \frac{1}{\gamma - 1} [(pV^\gamma)_2 - (pV^\gamma)_1] \left(\frac{1}{V_1^{\gamma-1}} - \frac{1}{V_2^{\gamma-1}} \right) \tag{5-109}$$

where use has been made of the adiabatic equation of state. Since

$$\left(\frac{1}{V_2^{\gamma-1}} - \frac{1}{V_1^{\gamma-1}} \right) = \delta V^{1-\gamma} = \frac{1-\gamma}{V^\gamma} \delta V \tag{5-110}$$

it follows that

$$\delta W_p = \frac{\delta(pV^\gamma)\,\delta V}{V^\gamma} \tag{5-111}$$

Interchange two flux tubes which contain equal flux. If $\phi_2 = \phi_1$, $\delta W_m = 0$, and the only energy change results from δW_p. The system is then certainly unstable if

$$\delta(pV^\gamma)\,\delta V = \delta V(\delta p\, V^\gamma + \gamma p V^{\gamma-1}\,\delta V) < 0 \tag{5-112}$$

Near the fluid boundary $p \to 0$ and the first term dominates the second. Since $\delta p < 0$, instability develops if

$$\delta V > 0 \tag{5-113}$$

or, in the integrated form, if

$$\delta \int A\,dl = \phi\,\delta \int \frac{dl}{B} > 0 \tag{5-114}$$

Considering a mirror configuration, where the fluid is penetrated by the magnetic field (the field is approximately the same as it was in the absence of the fluid), it is clear at once that inequality (5-114) is satisfied, hence the system is unstable. Again no instability is found for the cusp geometry.

An even more general theorem can now be proved: *Geometries in which the magnetic field lines curve toward the fluid along the entire fluid boundary are interchange-unstable, provided adjacent field lines in the direction of the pressure-gradient exist.* (We shall return later to the problem of the existence of adjacent field lines.) For such a system the field lines lengthen as one approaches the boundary, hence the integral in (5-108) is negative. Instability due to the lessening of magnetic energy follows if $\phi_2 > \phi_1$ or if

$$\frac{B_2}{B_1} > \frac{A_1}{A_2} = \frac{dl_2}{dl_1} \tag{5-115}$$

After integration the condition for magnetic interchange instability follows:

$$\delta \int \frac{dl}{B} < 0 \tag{5-116}$$

On the other hand, if inequality (5-116) is not satisfied, (5-114) is [except for $\delta\int(dl/B) = 0$], which proves our theorem. The special case of $\delta\int B^{-1}\,dl = 0$ can be treated separately by interchanging flux tubes containing neither the same flux nor equal volumes, only to find that this case presents no exception (Exercise 5-15).

The susceptibility of mirror geometry to interchange instabilities led to modifications of this arrangement. For instance, a combination of a mirror geometry with a line cusp, the cusp mirror, or the stabilized mirror (Fig. 5-12)

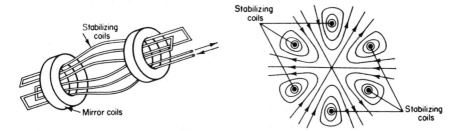

FIG. 5-12. Stabilized cusp-mirror geometry.

prevents interchange instabilities by introducing an "outwardly increasing" magnetic field at the boundary with a cusp-like curvature.

The main drawback of a cusp-like geometry when used for the confinement of a plasma is the high particle-loss rate at the line cusp. This can be circumvented by application of the so-called "multipole geometry" (Fig. 5-13) in

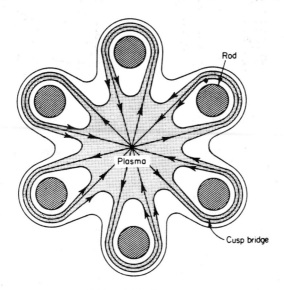

FIG. 5-13. Multipole geometry.

which the line cusps are bridged around the conductors, eliminating leakage. Although the "bridges" themselves would be unstable at the outer boundary, just like a mirror machine, the combined concave and convex regions when designed properly can lead to $\delta W > 0$ in both (5-108) and (5-114). In fact the (rather lengthy) formal analysis shows that complete stability can be achieved with such a configuration.

These stability criteria can of course also be applied to other configurations. For the pinch without an internal field, e.g., δW_m is negative by (5-108), hence instability is proved without the elaborate calculations performed in the previous section. Field-free fluids are obtainable only in the unrealistic case in which only surface currents are present. For any distribution of axial volume currents, however, the pinch configuration is still unstable, as field lines curve "the wrong way" at the boundary. By the same token, the hard-core pinch or "unpinch," where the fluid assumes the shape of a cylindrical shell with the return current flowing in a metallic rod along the cylinder axis (Fig. 5-14a), is stable against interchange instabilities. A more elaborate

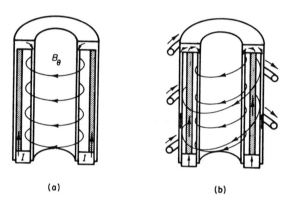

(a)   (b)

FIG. 5-14. Two versions of the hard-core pinch.

calculation shows that such configurations exhibit general hydromagnetic stability (Exercise 5-17).

Another way of avoiding interchange instabilities is based on the requirement that to be interchanged the flux tubes need to be adjacent. By introducing shear into the magnetic field two field lines adjacent in a small region are far removed elsewhere, which prevents their interchange by an infinitesimal perturbation. Such is the case, e.g., in the stellarator configurations shown in Fig. 4-6, where the pitch of the field-line spirals varies with the axial distance.

Another version of the hard-core pinch utilizes the same principle (Fig. 5-14b). Here the plasma, confined to a cylindrical shell, and the outer cylindrical conductor carry the current in one direction while the return current is carried by the inner conductor. So far this configuration is not in equilibrium (the plasma shell expands), and it is also interchange-unstable at the outer plasma boundary. Both ills may be cured by a superimposed $B_z(r)$ field. The field lines become spirals whose pitch increases with r, which excludes interchange modes. In fact, one can prove on the basis of the energy principle that such a configuration exhibits general hydromagnetic stability (Exercise 5-18).

5-5. Application of the Energy Principle

We turn now to the formal application of the energy principle. This consists of finding the displacement field which minimizes δW. Inserting this $\xi(\mathbf{r})$ into δW we obtain $\delta W_{\min}$, whose sign determines the stability of the system. Since δW is a homogeneous quadratic form in ξ, one can make $|\delta W|$ arbitrarily large by choosing $|\xi|$ large enough, or zero, by choosing $\xi(\mathbf{r}) \equiv 0$. To make minimization meaningful one has to introduce some normalization condition for ξ, e.g., $\int \xi^2 \, d\tau = 1$, to keep δW bounded. Often one can find a more convenient normalization condition which leads to analytical simplification of the problem at hand.

An alternative method of prescribing the distribution of ξ at some boundary is sometimes more convenient to use. In this case one must minimize δW separately for any admissible prescription of ξ at the boundary. This is usually achieved by Fourier-analyzing an arbitrary ξ at the boundary and minimizing for each mode. Both methods will be illustrated by an example.

Consider a force-free magnetic field enclosed by a perfectly conducting rigid wall. (A wall is necessary to take up the forces, since an over-all force-free system does not exist, according to the virial theorem derived in Sec. 3-5). We assume that the system is filled with hydromagnetic fluid, so δW_s and δW_v do not appear. We shall derive a sufficient stability criterion for such a general system. To simplify the form of δW_F we make the system less stable by choosing $\gamma = 0$. If this system is stable, so is ours, according to the first comparison theorem. Since the fluid is force-free, $\nabla p = 0$ and

$$\nabla \times \mathbf{B} = \alpha \mathbf{B} \tag{4-52}$$

As a further simplification we restrict ourselves to $\alpha = $ const fields. Consequently,

$$\delta W = \frac{1}{2\mu_0} \int [Q^2 + \alpha \mathbf{B} \cdot (\xi \times \mathbf{Q})] \, d\tau$$

$$= \frac{1}{2\mu_0} \int [(\nabla \times \mathbf{R})^2 - \mathbf{R}\alpha \cdot \nabla \times \mathbf{R}] \, d\tau \tag{5-117}$$

where the notation

$$\mathbf{R} = \xi \times \mathbf{B} \tag{5-118}$$

has been introduced. Since δW depends on ξ only through $\mathbf{R}$, it is sufficient if we minimize with respect to $\mathbf{R}$. One has to note, however, that while for any possible ξ there is an $\mathbf{R}$, any $\mathbf{R}$ with a component along the field has no corresponding ξ. Hence only $\mathbf{R}$ fields perpendicular to $\mathbf{B}$ can be used.

As a convenient normalization condition for ξ (or $\mathbf{R}$) one may choose

$$-\frac{1}{2\mu_0} \int \alpha \mathbf{R} \cdot \nabla \times \mathbf{R} \, d\tau = \text{const} \tag{5-119}$$

This condition represents a constraint on $\mathbf{R}$. Following the rules of the variational calculus, we multiply this constant by the undetermined Lagrange multiplier λ and add it to (5-117). Instead of δW we minimize,

$$I = \int \left[(\nabla \times \mathbf{R})^2 - (\lambda + 1)\alpha \mathbf{R} \cdot \nabla \times \mathbf{R} \right] d\tau = \int L \, d\tau \qquad (5\text{-}120)$$

I has an extremum where L satisfies the Euler equation

$$\frac{\partial L}{\partial R_i} = \frac{d}{dx_k} \frac{\partial L}{\partial R_{i,k}} \qquad (5\text{-}121)$$

where

$$R_{i,k} = \frac{\partial R_i}{\partial x_k} \qquad (5\text{-}122)$$

For instance,

$$\frac{\partial L}{\partial R_x} = \frac{\partial}{\partial x} \frac{\partial L}{\partial R_{x,x}} + \frac{\partial}{\partial y} \frac{\partial L}{\partial R_{x,y}} + \frac{\partial}{\partial z} \frac{\partial L}{\partial R_{x,z}} \qquad (5\text{-}123)$$

Each term can be calculated readily:

$$\frac{\partial L}{\partial R_x} = -(\lambda + 1)\alpha(\nabla \times \mathbf{R})_x \qquad (5\text{-}124)$$

$$\frac{\partial L}{\partial R_{x,x}} = 0 \qquad (5\text{-}125)$$

$$\frac{\partial L}{\partial R_{x,y}} = -2(\nabla \times \mathbf{R})_z + (\lambda + 1)\alpha R_z \qquad (5\text{-}126)$$

$$\frac{\partial L}{\partial R_{x,z}} = 2(\nabla \times \mathbf{R})_y - (\lambda + 1)\alpha R_y \qquad (5\text{-}127)$$

and

$$\frac{\partial}{\partial y} \frac{\partial L}{\partial R_{x,y}} + \frac{\partial}{\partial z} \frac{\partial L}{\partial R_{x,z}} = -2[\nabla \times (\nabla \times \mathbf{R})]_x + (\lambda + 1)\alpha(\nabla \times \mathbf{R})_x \qquad (5\text{-}128)$$

Similar equations hold for the other components. The Euler equations therefore become, in vector form,

$$\nabla \times \nabla \times \mathbf{R} = (\lambda + 1)\alpha \, \nabla \times \mathbf{R} \qquad (5\text{-}129)$$

An $\mathbf{R}$ satisfying this differential equation and normalized according to (5-119) minimizes (or maximizes) δW. To get the corresponding values of δW one calculates (5-117) for a value of $\mathbf{R}$ which satisfies (5-129). To this end we use the identity (A-10),

$$(\nabla \times \mathbf{R}) \cdot (\nabla \times \mathbf{R}) = \mathbf{R} \cdot \nabla \times (\nabla \times \mathbf{R}) - \nabla \cdot [(\nabla \times \mathbf{R}) \times \mathbf{R}] \qquad (5\text{-}130)$$

to write, with the help of (5-129),

$$\alpha \mathbf{R} \cdot \nabla \times \mathbf{R} = \frac{1}{\lambda + 1} \{ (\nabla \times \mathbf{R})^2 + \nabla \cdot [(\nabla \times \mathbf{R}) \times \mathbf{R}] \} \qquad (5\text{-}131)$$

We now insert this expression into (5-117). The $\nabla \cdot$ in (5-131) results in a surface integral over the surface of the wall. Since at this wall both $\boldsymbol{\xi}$ and $\mathbf{B}$ must be tangential, $\mathbf{R}$ points in the direction of the surface normal. Consequently the surface integral vanishes. The remaining integral reduces to

$$\delta W_{\text{extr}} = \frac{1}{2\mu_0} \frac{\lambda}{\lambda + 1} \int Q^2 \, d\tau \qquad (5\text{-}132)$$

where the values of λ can be determined as eigenvalues of (5-129), which can also be written

$$(1/\alpha) \nabla \times \mathbf{Q} = (\lambda + 1)\mathbf{Q} \qquad (5\text{-}133)$$

The values of δW_{extr} contain the absolute minimum of δW. δW_{extr} can only be negative if there is an eigenvalue of (5-129) where $-1 < \lambda < 0$. In the absence of such a λ the configuration is stable. When solving (5-133) it should be remembered that $\mathbf{R}$ is restricted to $\mathbf{R} \cdot \mathbf{B} = 0$.

The problem of stability has now been reduced to finding the eigenvalues of an operator. This operator equation, however, is far simpler than the equation of motion, owing to the judicious choice of the normalization condition. It is interesting to note that the choice

$$\int v\xi^2 \, d\tau = \text{const} \qquad (5\text{-}134)$$

for normalization results in the Euler equation identical with the equation of motion, where the Lagrange multiplier λ replaces ω^2. If no normalization condition is imposed, the Euler equation is simply

$$\mathbf{F}(\boldsymbol{\xi}) = 0 \qquad (5\text{-}135)$$

The proof in a somewhat extended form (α is not constant and compressibility is taken into account) is left as an exercise to the reader (Exercise 5-20), with the remark that it is generally true for arbitrary fields. The proof for the general case is rather tedious.

If a vacuum region is also present, the minimization of δW_F is not sufficient. As an illustration of the method for such a case, we are going to treat the so-called stabilized pinch configuration.

If one accepts the physical explanation for the pinch instability as given in Sec. 5-3, one can easily devise countermeasures to fight these instabilities. The attempt of the external B_θ field to "strangle" the cylinder at the constrictions of the $m = 0$ modes can be counteracted by freezing an axial

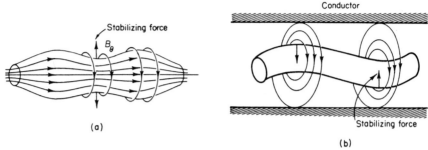

FIG. 5-15. Stabilized pinch: (a) outward pressure of trapped axial magnetic field hinders sausage instability; (b) stabilization of kinks using outer conductor.

magnetic field into the column. The "kink," on the other hand, may be stabilized by the use of an external, coaxial, conducting cylinder, which compresses the B_θ field of the approaching column and thereby keeps the kink away from the wall (Fig. 5-15).

We set out therefore to investigate the stability of a cylinder of radius r_0, with an external pinch field $B_\theta = B_0(r_0/r)$, a uniform axial vacuum field $B_z^v = b_v$, an internal stabilizing field $B_z^{in} = b_i$ and a perfectly conducting hollow cylinder of internal radius $R_0 = \Lambda r_0$. We investigate first the fluid part of the energy integral

$$\delta W_F = \frac{1}{2} \int \left[\frac{Q^2}{\mu_0} + \gamma p (\nabla \cdot \xi)^2 \right] d\tau_F \qquad (5\text{-}136)$$

If this expression can be made negative with a perturbation that does not effect the boundary $[(\mathbf{n} \cdot \xi) = 0]$ the configuration is certainly unstable, since such displacement fields leave $\delta W_s = \delta W_v = 0$. However, as the integrand of (5-136) is positive-definite, instability can only result from the perturbation of the boundary. The procedure is then to minimize δW_F and δW_v for a prescribed perturbation of the boundary $\mathbf{n} \cdot \xi$. If this is done for all $\mathbf{n} \cdot \xi$'s possible and in every case δW was found positive, the system is stable. (Of course δW_s is determined by $\mathbf{n} \cdot \xi$.)

Leaving the prescription of $\mathbf{n} \cdot \xi$ for a later time, we calculate the Euler equation to minimize (5-136). No normalization is necessary now. Writing

$$\mathbf{Q} = \nabla \times (\xi \times \mathbf{B}) = (\mathbf{B} \cdot \nabla)\xi - \mathbf{B}(\nabla \cdot \xi) - (\xi \cdot \nabla)\mathbf{B} \qquad (5\text{-}137)$$

the Lagrangian density becomes

$$L = \frac{1}{2\mu_0} \left[B_i \frac{\partial \xi_k}{\partial x_i} - B_k \frac{\partial \xi_i}{\partial x_i} - \xi_i \frac{\partial B_k}{\partial x_i} \right] Q_k + \tfrac{1}{2}\gamma p \left(\frac{\partial \xi_i}{\partial x_i} \right) \nabla \cdot \xi$$

$$= \frac{1}{2\mu_0} \left[B_k \xi_{i,k} Q_i - B_l \xi_{i,k} Q_l \delta_{ik} - \xi_i \frac{\partial B_k}{\partial x_i} Q_k \right] + \tfrac{1}{2}\gamma p \, \delta_{ik} \xi_{i,k} \nabla \cdot \xi \qquad (5\text{-}138)$$

where some of the summation indices have been changed. Furthermore,

$$\frac{\partial L}{\partial \xi_i} = -\frac{1}{\mu_0} \frac{\partial B_k}{\partial x_i} Q_k \tag{5-139}$$

and

$$\frac{\partial}{\partial x_k} \frac{\partial L}{\partial \xi_{i,k}} = \frac{1}{\mu_0} \left[\frac{\partial}{\partial x_k} (B_k Q_i) - \frac{\partial}{\partial x_i} (B_k Q_k) \right] + \gamma p \frac{\partial}{\partial x_i} \nabla \cdot \xi \tag{5-140}$$

Here use has been made of $\partial Q^2 = 2Q \, \partial Q$ and $\partial (\nabla \cdot \xi)^2 = 2\nabla \cdot \xi \, \partial \nabla \cdot \xi$, where ∂ represents differentiation with respect to any variable. The Euler equation becomes, finally,

$$\frac{1}{\mu_0} \left[\frac{\partial B_k}{\partial x_i} Q_k + \frac{\partial B_k}{\partial x_k} Q_i + B_k \frac{\partial Q_i}{\partial x_k} - \frac{\partial B_k}{\partial x_i} Q_k - B_k \frac{\partial Q_k}{\partial x_i} \right] + \gamma p \frac{\partial}{\partial x_i} \nabla \cdot \xi = 0 \tag{5-141}$$

or, since $\partial B_k / \partial x_k = \nabla \cdot \mathbf{B} = 0$ and because of the cancellation of the first and fourth term, (5-141) becomes

$$\frac{B_k}{\mu_0} \left(\frac{\partial Q_i}{\partial x_k} - \frac{\partial Q_k}{\partial x_i} \right) + \gamma p \frac{\partial}{\partial x_i} \nabla \cdot \xi = \frac{1}{\mu_0} (\nabla \times \mathbf{Q}) \times \mathbf{B} + \gamma p \, \nabla \nabla \cdot \xi = 0 \tag{5-142}$$

which is just $\mathbf{F}(\xi) = 0$, as expected on the basis of Exercise 5-20. One again neglects the displacement current and derives the perturbed vacuum magnetic field from a scalar potential

$$\delta \mathbf{B} = \nabla \phi \tag{5-143}$$

and writes

$$\delta W_v = \frac{1}{2\mu_0} \int (\nabla \phi)^2 \, d\tau_v \tag{5-144}$$

where the Euler equation yields immediately, with $\phi_{,k} = \partial \phi / \partial x_k$,

$$\frac{\partial}{\partial x_k} \frac{\partial}{\partial \phi_{,k}} \frac{1}{2} \phi_{,k} \phi_{,k} = \frac{\partial}{\partial x_k} \nabla \phi = \nabla^2 \phi = 0 \tag{5-145}$$

This result is also known from electro- and magnetostatics. It states that the vacuum electric (or magnetic) field which contains the least energy is the one which can be deduced from a scalar potential satisfying the Laplace equation. (The minimizing vacuum field is actually the only one possible. Since $\nabla \cdot \delta \mathbf{B} = 0$, $\nabla^2 \phi = 0$.)

The next step is the insertion of the minimizing solutions into the energy expression. To this end we use the identities

$$(\nabla \cdot \xi)(\nabla \cdot \xi) = -\xi \, \nabla \nabla \cdot \xi + \nabla \cdot (\xi \, \nabla \cdot \xi) \tag{5-146}$$

and

$$Q^2 = \mathbf{Q} \cdot \nabla \times (\xi \times \mathbf{B}) = \nabla \cdot [(\xi \times \mathbf{B}) \times \mathbf{Q}] + [\mathbf{B} \times (\nabla \times \mathbf{Q})] \cdot \xi \tag{5-147}$$

in (5-136). If ξ satisfies the Euler equation (5-142), the fluid energy reduces to a surface integral,

$$\delta W_F^{\min} = \frac{1}{2} \int \left[\frac{1}{\mu_0} (\xi \times \mathbf{B}) \times \mathbf{Q} + \gamma p \xi \nabla \cdot \xi \right] d\mathbf{S} = \frac{1}{2} \int \left(\gamma p \nabla \cdot \xi - \frac{\mathbf{B} \cdot \mathbf{Q}}{\mu_0} \right) \xi \cdot d\mathbf{S} \tag{5-148}$$

since $\mathbf{B} \cdot d\mathbf{S} = 0$. A similar transformation can be performed for the vacuum field. One uses

$$\nabla \phi \cdot \nabla \phi = \nabla \cdot (\phi \nabla \phi) - \phi \nabla^2 \phi \tag{5-149}$$

and (5-145) in (5-144) to obtain

$$\delta W_v^{\min} = \frac{1}{2\mu_0} \int \phi \nabla \phi \cdot d\mathbf{S} \tag{5-150}$$

Here the surface normal $d\mathbf{S}$ points away from the vacuum (into the fluid), while it points away from the fluid in (5-148). All three parts of $\delta W_{\min}$ are now reduced to integrals extended over the fluid-vacuum interface. Now one has to investigate the behavior of these expressions for all possible perturbations of the boundary. Since inspection of the surface terms does not lead to an immediate answer, one proceeds to solve the Euler equations. One introduces again the trial solution $\xi = \xi(\mathbf{r}) \, e^{i(m\varphi + kz)}$ into (5-142). Scalar multiplication of (5-142) with $\mathbf{B}$ yields

$$\gamma p B i k \nabla \cdot \xi = 0 \qquad \text{or} \qquad \nabla \cdot \xi = 0 \tag{5-151}$$

This does not hold for $k = 0$, which must be investigated separately. Using a vector analytical identity, and the constancy of $\mathbf{B}$ in the fluid, (5-142) reduces to

$$\mathbf{B} \times (\nabla \times \mathbf{Q}) = \nabla(\mathbf{B} \cdot \mathbf{Q}) - Bik\mathbf{Q} = 0 \tag{5-152}$$

Since $\mathbf{Q} = \nabla \times (\xi \times \mathbf{B})$ the $\nabla \cdot$ of (5-152) becomes

$$\nabla^2(\mathbf{B} \cdot \mathbf{Q}) = 0 \tag{5-153}$$

Hence the displacement field ξ, to minimize δW_F with given boundary displacement, is such that $\mathbf{B} \cdot \mathbf{Q}$ satisfies the Laplace equation. The solutions of this equation in cylindrical coordinates have been obtained previously:

$$\mathbf{B} \cdot \mathbf{Q} = A I_m(kr) e^{i(m\varphi + kz)} \tag{5-154}$$

because the K function diverges for $r \to 0$. The vacuum potential satisfies the same equation but only in the region $r_0 < r < R_0$; hence

$$\phi = C I_m(kr) e^{i(m\varphi + kz)} + D K_m(kr) e^{i(m\varphi + kz)} \tag{5-155}$$

At the external conductor the normal component of $\delta\mathbf{B}$ vanishes:

$$\left(\frac{\partial\phi}{\partial r}\right)_R = CkI_m{}'(kR_0)e^{i(m\varphi+kz)} + DkK_m{}'(kR_0)e^{i(m\varphi+kz)} = 0 \qquad (5\text{-}156)$$

At the fluid-vacuum interface the surface normal is again given by (5-98) and the magnetic field by (5-99) with $b_v\hat{z}$ added. The condition that the field be tangential becomes

$$\nabla\varphi \cdot \mathbf{B} = -\delta B_r + \frac{imr_1}{r_0}B_0 + ikr_1b_v = 0 \qquad (5\text{-}157)$$

Substituting $\delta B_r = \partial\phi/\partial r$ from (5-155),

$$CkI_m{}'(kr_0) + DkK_m{}'(kr_0) = ikr_1b_v + \frac{imr_1}{r_0}B_0 \qquad (5\text{-}158)$$

From (5-156) and (5-158) the two constants can be evaluated to yield

$$C = \frac{i[b_v + (mB_0/kr_0)]K_m{}'(kR_0)r_1}{K_m{}'(kR_0)I_m{}'(kr_0) - K_m{}'(kr_0)I_m{}'(kR_0)} \qquad (5\text{-}159)$$

and

$$D = -\frac{i[b_v + (mB_0/kr_0)]I_m{}'(kR_0)r_1}{K_m{}'(kR_0)I_m{}'(kr_0) - K_m{}'(kr_0)I_m{}'(kR_0)} \qquad (5\text{-}160)$$

The third constant A is also determined by the surface displacement amplitude r_1.

$$\mathbf{Q} = \nabla \times (\boldsymbol{\xi} \times \mathbf{B}) = ikB\boldsymbol{\xi} \qquad (5\text{-}161)$$

since $\nabla \cdot \boldsymbol{\xi} = 0$ and $\mathbf{B} = \text{const}$. Equations (5-152) and (5-161) can be combined into

$$\boldsymbol{\xi} = -(kB)^{-2}\nabla(\mathbf{B} \cdot \mathbf{Q}) \qquad (5\text{-}162)$$

and the displacement of the boundary gives

$$r_1e^{i(m\varphi+kz)} = \xi_r(r_0) = -(kb_i)^{-2}AkI_m{}'(kr_0)e^{i(m\varphi+kz)} \qquad (5\text{-}163)$$

so that

$$A = -\frac{kb_i{}^2}{I_m{}'(kr_0)}r_1 \qquad (5\text{-}164)$$

It remains only to write down the minimized energy expressions. From (5-148), (5-151), (5-154), (5-163), and (5-164),

$$\delta W_F^{\min} = \frac{r_0\pi L}{2\mu_0}\frac{kb_i{}^2}{I_m{}'(kr_0)}I_m(kr_0)r_1{}^2 \qquad (5\text{-}165)$$

where L is the length of the column. (One should remember that the exponentials represent sine or cosine functions, so the square integrated over a surface is half the surface area.) The vacuum contribution is obtained by a similar substitution in (5-150) and integration:

$$\delta W_v^{min} = \frac{r_0 \pi L k}{2\mu_0} \left(b_v + \frac{m B_0}{k r_0} \right)^2 \frac{I_m'(kR_0)K_m(kr_0) - K_m'(kR_0)I_m(kr_0)}{K_m'(kR_0)I_m'(kr_0) - I_m'(kR_0)K_m'(kr_0)} r_1^2 \quad (5\text{-}166)$$

Finally the surface energy is to be determined. The equilibrium quantities obey the pressure-balance equation

$$p + \frac{b_i^2}{2\mu_0} = \frac{b_v^2}{2\mu_0} + \frac{B_0^2}{2\mu_0} \quad (5\text{-}167)$$

and

$$\left[\nabla \left(p + \frac{B^2}{2\mu_0} \right) \right] = -\frac{B_0^2}{\mu_0 r_0} \, \mathbf{n} \quad (5\text{-}168)$$

where **n** is the surface normal pointing into the vacuum. The surface energy is, therefore, from (5-44),

$$\delta W_s = -\frac{B_0^2 \pi L}{2\mu_0} r_1^2 \quad (5\text{-}169)$$

a negative-definite expression. Investigation of the other two terms show that they are always positive.

The sum of (5-165), (5-166), and (5-169) gives δW^{min}, the minimum potential energy change compatible with the fluid-surface perturbation characterized by the r_1, m, k triplet. Since any perturbation of the boundary can be Fourier-analyzed into such $r_1 e^{i(m\varphi + kz)}$ components, it is clear that the necessary and sufficient condition for stability is the nonnegative value of δW^{min} for any m and k (the amplitude enters only as r_1^2, as seen).

The results of a numerical analysis of the above expressions are plotted in Fig. 5-16. Owing to (5-167), for any physical system the inequality $b_i^2 \leq b_v^2 + B_0^2$ must hold. The curves marked with given values of m and Λ mark the b_i/B_0 and b_v/B_0 values, where instability sets in for the m, Λ pair. $m = 2, 3, \ldots$ modes turn out to be more stable than the $m = 0$ and $m = 1$ modes. Hence a given configuration is stable if the b_i/B_0 and b_v/B_0 points lie above both the $m = 0$ and $m = 1$ curves corresponding to the Λ value of the configuration. We see that large b_i/b_v and small Λ are good for stability. (The latter also follows from the third comparison theorem.) For $b_v = 0$, one can stabilize the column with $\Lambda \leq 5$, but increasing b_v reduces this value quickly.

It remains to treat the $k = 0$ case. This is the subject of Exercise 5-21. It is easy to see that no instability arises for this perturbation. Unfortunately such an idealized configuration with infinitely thin surface currents cannot be realized experimentally. While similar experimental arrangements resulted in increased stability, the finite conductivity of the fluid causes the internal and external fields to intermix, destroying stability. After a time the axial current becomes uniform and $b_i = b_v$. A calculation of stability shows that such a configuration is unstable.

We perform this calculation in the absence of conducting walls when $R_0 \to \infty$. Further simplification is introduced by taking the long-wavelength limit $kr_0 \ll 1$ since more detailed calculation shows that this is the regime where instability develops. In this limit $K_m'/K_m = -m/kr_0$, $I_m'/I_m = m/kr_0$ ($m > 0$), and (5-165), (5-166), (5-169) reduce to

$$\delta W^{\min} = \frac{r_1^2}{2\mu_0}\, \pi L\, \frac{k^2 r_0^2}{m} \left[B_z^2 + \left(B_z + \frac{mB_0}{kr_0} \right)^2 - \frac{mB_0^2}{k^2 r_0^2} \right] \quad (5\text{-}170)$$

where $B_z = b_v = b_i$. The details are left for Exercise 5-23. This expression can be negative for $m = 1$ (but not for $m > 1$) if $B_z + B_0/kr_0 < 0$. If B_z, $B_0 > 0$, this happens when $k < 0$. For stability one requires therefore that $B_0/B_z < |k|r_0$. For a finite column $|k|$ cannot be smaller than $2\pi/L$. Hence the configuration is stable if $B_0/B_z < 2\pi r_0/L$. This is the *Kruskal-Shafranov condition*.

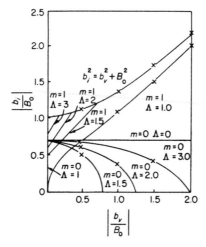

Fig. 5-16. Region of stability of stabilized pinch. [From R. J. Tayler, *Proc. Phys. Soc.* (*London*) **B70**, 1049 (1957).]

Returning to the method used in determining $\delta W^{\min}$, attention should be called to the fact that there was no need for the application of the boundary condition describing the balance of surface forces (5-32). This is due to the fact that this condition was used in deriving δW_s and is already incorporated in (5-44). This feature of the present form of the energy principle is an added advantage of this method of stability analysis.

5-6. Tearing Modes

If the infinite conductivity assumption is relaxed, another set of instabilities, not covered in the previous treatment, is obtained.

Consider a magnetic field with shear, embedded in a conductive fluid with large but finite conductivity. In a slab geometry the magnetic field has the form $\mathbf{B} = B_x(y)\mathbf{e}_x + B_z(y)\mathbf{e}_z$. There is free energy in the system; by eliminating, e.g., B_x, the shear would be removed and the magnetic energy of the system lowered. The last term in (4-11a) describes the diffusion of the magnetic field, and this diffusion would lead after a long time to the elimination of shear. There are, however, modes that speed up this process considerably by greatly enhancing locally the value of $\nabla^2 B$.

Consider an electromagnetic perturbation with a periodic x,z dependence $\exp[i(k_x x + k_z z)]$. The wave vector $\mathbf{k}$ can always be chosen in such a way that at some value of y, say at $y = 0$, $\mathbf{k} \cdot \mathbf{B} = 0$. The field configuration in the xy plane is shown in Fig. 5-17 with dotted lines. Consider now a periodic perturbation added with the resulting field also shown in the figure. As a result of the

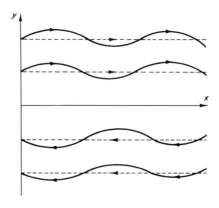

Fig. 5-17. Magnetic field line components in the xy plane for a sheared field. Dashed lines illustrate unperturbed lines; solid lines illustrate perturbed $\mathbf{B}$ lines.

perturbation, a field component B_y has been added. If the slab is bounded by conducting walls at, say $y = \pm a$, $B_y(\pm a) = 0$, and the possible B_y profiles are plotted in Fig. 5-18. If one now permits magnetic field diffusion, this diffusion will be strongest in a diffusive layer around $y = 0$, where $\nabla^2 B_y$ is largest. It may be seen from (4-11a) that if $\partial^2 B_y/\partial y^2 < 0$, $\partial B_y/\partial t < 0$ (curve 1), and the perturbation will be diffusion damped. If, however, $\partial^2 B_y/\partial y^2 > 0$, $\partial B_y/\partial t > 0$, (curve 2), the perturbation grows. The growing perturbation leads to more field line diffusion, and so on, resulting in an exponentially growing instability.

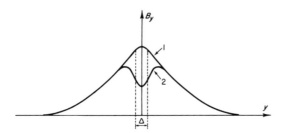

FIG. 5-18. B_y profiles for magnetic tearing. Curve 1, stable; curve 2, unstable.

This *tearing instability* develops in general if the field lines can rearrange themselves in such a way as to increase $\nabla^2 \mathbf{B}$ in some region of the fluid to produce enhanced field diffusion, and if this field diffusion leads to further local enhancement of $\nabla^2 \mathbf{B}$.

If the growth rate is again designated by γ, it is clear that the diffusive layer must be thinner than the skin depth $\delta = (\eta/\mu_0\gamma)^{1/2}$, where $\eta = \sigma^{-1}$ is the resistivity. Since η is small, it is also reasonable to consider cases when the skin depth is small compared to the slab size a. Thus one arrives at an ordering $\Delta \ll \delta \ll a$, where Δ is the thickness of the diffusive layer.

For an analytic treatment of the problem we use the linearized version of (4-11a)

$$\dot{\mathbf{B}}_1 = \nabla \times (\mathbf{v}_1 \times \mathbf{B}_0) + \frac{\eta}{\mu_0} \nabla^2 \mathbf{B}_1 \qquad (5\text{-}171)$$

and taking the curl of the linearized (4-10)

$$\nu\mu_0 \nabla \times \mathbf{v}_1 = \nabla \times [(\mathbf{B}_0 \cdot \nabla)\mathbf{B}_1 + (\mathbf{B}_1 \cdot \nabla)\mathbf{B}_0] \qquad (5\text{-}172)$$

where $\mathbf{v}_0 = 0$ and the fluid is incompressible with $\nabla \cdot \mathbf{v} = 0$, $\nu = $ const. Since all quantities vary as $\exp[\gamma t + i(k_x x + k_z z)]$, the y component of (5-171) becomes

$$\gamma b = iFv + \frac{\eta}{\mu_0}(b'' - k^2 b) \qquad (5\text{-}173)$$

where $F = \mathbf{k} \cdot \mathbf{B}$, $b = B_{1y}$, $v = v_{1y}$, $\mathbf{B} = \mathbf{B}_0$, and prime designates d/dy. Differentiating the x component of (5-172) by z and the z component by x and subtracting results in

$$i\mu_0 \nu\gamma(v'' - k^2 v) = -Fb'' + F''b + k^2 Fb \qquad (5\text{-}174)$$

Eliminating v between these two equations yields the fourth-order equation

$$\delta^2 b^{IV} - 2\delta^2 \frac{F'}{F} b''' - \left(1 + 2k^2\delta^2 + \delta^2 \frac{F''}{F} - 2\delta^2 \frac{F'^2}{F^2} + \frac{F^2}{\mu_0 v\gamma^2}\right)b''$$

$$+ 2\frac{F'}{F}(1 + k^2\delta^2)b' - \left[2\left(\frac{F'}{F}\right)^2(1 + k^2\delta^2)\right.$$

$$\left. - \left(\frac{F''}{F} + k^2\right)\left(1 + k^2\delta^2 + \frac{F^2}{\mu_0 v\gamma^2}\right)\right]b = 0 \qquad (5\text{-}175)$$

This equation can be simplified if one considers wavelengths, comparable to the slab width, ka of order unity, and one assumes that the shear length is comparable to a. Since $a \gg \delta$, it follows that $(k\delta)^2 \ll 1$ and $\delta^2 F''/F \ll 1$, leading to

$$\delta^2 b^{IV} - 2\delta^2 \frac{F'}{F} b''' - \left(1 - 2\delta^2 \frac{F'^2}{F^2} + \frac{F^2}{\mu_0 v\gamma^2}\right)b'' + 2\frac{F'}{F}b'$$

$$+ \left[\left(k^2 + \frac{F''}{F}\right)\left(1 + \frac{F^2}{\mu_0 v\gamma^2}\right) - 2\left(\frac{F'}{F}\right)^2\right]b = 0 \qquad (5\text{-}176)$$

This equation is approximately solved in the following fashion. The region $|y| < a$ is divided into a large outer region $y_0 < |y| < a$ and a thin inner region $|y| < y_0$ that contains the diffusion layer. One determines y_0 from the condition that the last term in the coefficient of b be negligible in the outer region. Notice that this term diverges at $y = 0$, where $F = 0$, and decreases further out. One estimates $F'' \approx F/L^2$, where L is the shear length and $F \approx kB_0 y/L$ for $y \ll L$, where B_0 is some characteristic value of B. Demanding that $F''F/\mu_0 v\gamma^2 \gg (F'/F)^2$ at y_0 leads to $y_0 \gg L\sqrt{\gamma/kv_A}$, where $v_A^2 = B_0^2/\mu_0 v$. It follows that $F^2/\mu_0 v\gamma^2)_{y_0} \approx (k^2 v_A^2 y_0^4/\gamma^2 L^4)(L^2/y_0^2) \gg 1$ since both terms are much larger than one. Finally, one expects that in the outer region b varies on a scale a, so $b'/b \approx a^{-1}$. So in the exterior region (5-176) becomes a second-order equation

$$b'' - \left(k^2 + \frac{F''}{F}\right)b = 0 \qquad (5\text{-}177)$$

Note that this is the same equation one would have obtained by simply setting the inertial term in the equation of motion (5-174) equal to zero ($v = 0$). Hence in the outer region the fluid slowly moves from equilibrium to equilibrium. This sets an upper limit to the growth rate since $L \gg y_0 \gg L\sqrt{\gamma/kv_A}$, so $\gamma \ll kv_A$. The permissible growth rate is also bounded from below since $\delta \ll a$, $\gamma \gg \eta/\mu_0 a^2$.

One now proceeds to solve the problem in the following manner. In the outer region one solves (5-177) with the boundary conditions at $y = \pm a$. Such solutions when extended to $y = 0$ (where they are not valid) do not have matching first derivatives. One defines the normalized jump in the first derivative

$$\Delta' = \frac{b'(\varepsilon) - b'(-\varepsilon)}{b(0)} \tag{5-178}$$

where ε is small. If this quantity is positive, the curve, after the inner solution has also been constructed, looks like curve 2 in Fig. 5-18 and an unstable solution is possible. This can only happen when $|F''/F| \gg k^2$ in at least part of the slab. Now one has to construct the inner solution and match it to the outer one. For $y \ll y_0$ the last term dominates in the coefficient of b in (5-176) and one gets, approximating F'/F by y^{-1},

$$\delta^2 b^{IV} - 2\delta^2 \frac{1}{y} b''' - \left(1 - 2\frac{\delta^2}{y^2} + \frac{k^2 v_A^2 y^2}{\gamma^2 L^2}\right) b'' + \frac{2}{y} b' - \frac{2b}{y^2} = 0 \tag{5-179}$$

Introducing the dimensionless length $x = y/\delta$, this expression becomes

$$b^{IV} - 2\frac{b'''}{x} - g(x)b'' + \frac{2}{x} b' - \frac{2b}{x^2} = 0 \tag{5-180}$$

where

$$g(x) = 1 - \frac{2}{x^2} + \frac{k^2 v_A^2 \delta^2}{\gamma^2 L^2} x^2 = 1 - \frac{2}{x^2} + \sigma x^2 \tag{5-181}$$

When (5-180) is differentiated one obtains a fifth-order equation, which, when combined with (5-180) again, reduces to

$$z''' - (1 + \sigma x^2)z' - 4\sigma x z = 0 \tag{5-182}$$

where $z = b''$. This equation is still complicated but asymptotic solutions for large x, in the form $z \sim x^{-4}$ and $z \sim \exp(\pm\frac{1}{2}\sqrt{\sigma}x^2)$ can be found.

The strategy employed in fitting the inner and outer solutions is as follows. The former varies on a very short length scale $\Delta \ll a$, the characteristic length of the outer solution. Hence the first derivative of the outer solution can be taken to be a constant on the Δ length scale. To fit the two, the first derivative of the inner solution must also approach a constant; hence $z = b'' \to 0$ in the vicinity of y_0. Consequently, z must vanish asymptotically. Since, as we have seen, two out of the three asymptotic solutions of (5-182) vanish, this condition can always be satisfied. Numerical solutions of (5-182) give the characteristic scale length.

The growth rate can be estimated by integrating $\partial^2 A_z/\partial y^2 \approx \nabla^2 A_z = -\mu_0 J_z$ (where A is the perturbed vector potential) across the diffusion layer.

$$\left[\frac{\partial A_z}{\partial y}\right]_{-y_0}^{+y_0} = -\mu_0 \langle J_z \rangle \Delta \approx \mu_0 \eta^{-1} \gamma A_z \Delta = \frac{\Delta}{\delta^2} A_z \qquad (5\text{-}183)$$

since $\mathbf{E} = -\mathbf{A} = -\gamma \mathbf{A}$. Here $\langle J \rangle$ is the average current density in the current-carrying diffusion layer of thickness Δ. Since A_z is proportional to b (see Exercise 5-25)

$$\Delta' = \frac{[\partial b/\partial y]}{b} = \frac{[\partial A_z/\partial y]}{A_z} = \frac{\Delta}{\delta^2} \qquad (5\text{-}184)$$

Consequently, when Δ' is known from the outer solution, and $\Delta(\gamma)$ from the inner one, the growth rate can be determined.

As an example consider the case when the dominant asymptotic solution is $z \sim \exp(-\frac{1}{2}\sqrt{\sigma}x^2)$. The characteristic layer width is then determined from $\sigma x^4 = 1$ or

$$\frac{\Delta}{\delta} = \sigma^{-1/4} = \left(\frac{\gamma L}{kv_A\delta}\right)^{1/2} \qquad (5\text{-}185)$$

This yields for the growth rate

$$\gamma = (\Delta')^{4/5}\left(\frac{\eta}{\mu_0}\right)^{3/5}\left(\frac{kv_A}{L}\right)^{2/5} \qquad (5\text{-}186)$$

In general, Δ' is found by numerical integration of (5-177), a much easier task than integrating the full fourth-order equation. Some special cases can be treated analytically (see Exercise 5-26).

The practically important examples involve cylindrical geometries. In a tokomak, magnetic shear is inevitable and tearing modes have been observed in experiments.

5-7. Summary

A magnetohydrostatic system, if it is to be useful, in addition to being in equilibrium, must also satisfy the far more stringent condition of stability. This is the case either if any small displacement of fluid elements with the "frozen" magnetic field results in oscillations about the equilibrium or if the potential energy content (4-31) of the system has a minimum at the equilibrium. This latter (energy) condition can be formulated in the following way: The change of potential energy due to a fluid displacement $\xi(\mathbf{r})$,

$$\delta W = \delta W_F + \delta W_s + \delta W_v$$

must be nonnegative for any choice of the (small) displacement field $\xi(\mathbf{r})$. The forms of the fluid, surface, and vacuum terms are given in (5-43), (5-44), and (5-45), respectively. The practical use of the energy criterion can take various forms. One may find by inspection that some displacements lead to a negative δW, hence the system is unstable, or one minimizes δW in terms of $\xi(\mathbf{r})$ and determines the sign of $\delta W^{\min}$. To keep δW bounded in the process, a constraint should be imposed on ξ. This may take the form of a condition like $\int \xi^2 \, d\tau = \text{const}$ or one may prescribe ξ at the fluid-vacuum boundary and minimize δW with respect to all possible prescribed surface displacements, which usually takes the form of Fourier surface modes with different wave numbers.

By applying either of these (the equation of motion or energy) methods to a fluid-vacuum system, the solutions involve constants of integration. These must be determined by the use of the fluid-vacuum, vacuum-vessel boundary conditions. The energy method has the advantage that the boundary condition on the fluid-vacuum interface, the one prescribing the continuity of the total pressure, is already included in δW_s, and the number of constants of integration is reduced by one.

Applying these methods we found that a plasma in a gravitational field suspended by a magnetic field is unstable (Kruskal-Schwarzschild)—a mirror or pinch configuration likewise. A θ pinch is neutrally stable, while a cusp configuration is stable. In short:

The pinch configuration can be stabilized by an added (suitably chosen) axial field inside the fluid and a cylindrical conductive wall placed close enough to the fluid column. Other stable configurations can also be constructed, such as the unpinch, the multipole arrangement, the cusp-mirror, etc.

The equation of motion of a hydromagnetic system is also useful for studying small oscillations around an equilibrium. This way we obtained the equations of the Alfven and magnetoacoustic waves in an infinite fluid immersed in a uniform magnetic field.

In the case of finite but small resistivity, sheared magnetic field lines can produce tearing mode instabilities. These come about by enhanced local diffusion of magnetic field lines, due to the rearrangement of the field in the entire fluid, leading to further field perturbation and hence more diffusion.

It should be kept in mind that the hydromagnetic stability of a configuration (the stability of an idealized perfectly conducting fluid) does not imply the stability of a geometrically analogous plasma system. We will discuss later some of the ways (other than hydromagnetic) in which a plasma can misbehave.

EXERCISES

5-1. Examine the equations of a mechanical system with two degrees of freedom for small displacements about an equilibrium. How should one set up the coordinate system so that the motion along the x and y axes becomes uncoupled?

5-2. How many different types of equilibria are there in a system with n degrees of freedom? How many of them are stable?

5-3. Prove that $\mathbf{Q}$ in (5-20) is the change of local magnetic field due to the perturbation.

5-4. Prove that the work done against the surface currents by displacing the fluid-vacuum boundary is given in (5-44).

5-5. What is the explicit component form of (5-56) and the corresponding dispersion relation?

5-6. Use the result of Exercise 5-5 to investigate plane waves propagating in an arbitrary direction in a perfectly conducting uniform fluid penetrated by a homogeneous magnetic field. Prove there are three wave modes corresponding to a given wave vector $\mathbf{k}$, and analyze their properties.

5-7. Prove that the first-order gravitational force is given by $\mathbf{F}_g = \nabla \cdot (v\boldsymbol{\xi}) \nabla \phi$, where ϕ is the gravitational potential.

5-8. Prove that the inverse Kruskal-Schwarzschild configuration, with the magnetic field above the fluid, exhibits stable oscillations with the frequency $\omega = (kg)^{1/2}$.

5-9. Investigate the Kruskal-Schwarzschild instability in the more general case where the ripples are not parallel to the magnetic field.

(a) Prove that ξ_y obeys, instead of (5-74), the differential equation

$$k^2\left(v\omega^2 - \frac{(\mathbf{k}\cdot\mathbf{B})^2}{\mu_0} + g\frac{dv}{dy}\right)\xi_y = \frac{\partial}{\partial y}\left\{\left[v\omega^2 - \frac{(\mathbf{k}\cdot\mathbf{B})^2}{\mu_0}\right]\frac{\partial\xi_y}{\partial y}\right\}$$

(b) Prove that in the case of a field-free fluid fluid separated by a thin surface current from a uniform field, the dispersion equation becomes

$$\omega^2 = -kg + \frac{(\mathbf{k}\cdot\mathbf{B}_0)^2}{\mu_0 v_0}$$

(*Hint*: To obtain the dispersion relation you will need $\partial\xi_y/\partial y$ on the vacuum side of the current layer, where it is not defined. Assume that for $y < 0$ the half-space is filled with a weightless and pressureless fluid.)

5-10. Modify the previous problem by introducing a uniform magnetic field $\mathbf{B}_f$ inside the fluid ($y > 0$) not parallel with the vacuum field $\mathbf{B}_v$.

(a) Prove that the dispersion relation now becomes

$$\omega^2 = -kg + \frac{1}{\mu_0 v_0} [(\mathbf{k} \cdot \mathbf{B}_f)^2 + (\mathbf{k} \cdot \mathbf{B}_v)^2]$$

(b) What is the growth rate of the "fastest" instability if $\mathbf{B}_f$ and $\mathbf{B}_v$ are at right angles? How does this compare with the growth rate without magnetic shear?

5-11. Prove that for $\mathbf{F} = \gamma p \, \nabla\nabla \cdot$ and a uniform pressure in (5-79), F commutes with both $\partial/\partial z$ and $\partial/\partial \varphi$.

5-12. Derive the dispersion relation (5-105) for a θ pinch with a sharp boundary without an internal magnetic field.

5-13. Prove comparison theorem 2 in Sec. 5-4.

5-14. Consider a field-free fluid with sharp boundaries. Prove that
(a) $\delta W_F \geqq 0$, $\delta W_V \geqq 0$.
(b) δW_s is positive (negative) if the field lines are convex (concave) to the fluid.
(c) A suitable choice of surface perturbation (short wavelength and fluid volume kept constant) results in $|\delta W_s| \gg |\delta W_v|$ while leaving $\delta W_F = 0$.
Hence stability is determined by the sign of δW_s and convex surfaces (cusp) are stable, concave ones (mirror, pinch, etc.) unstable.

5-15. Demonstrate that a magnetofluid in a magnetic field with the field lines curving toward the fluid is interchange-unstable even if $\delta \int dl/B_0 = 0$. (*Hint:* Interchange two flux tubes with the reciprocal average cross sections, defined by $\langle A^{-1} \rangle = (1/L) \int A^{-1} \, dl$ equal.)

5-16. Consider a pinch geometry with a uniform current density. On the basis of the sign of $\delta \int dl/B$, to which class of interchange instability is this configuration subject? On the basis of Exercise 5-15 prove instability directly.

5-17. Apply the energy principle to the unpinch configurations, and show its stability. No surface currents need be assumed, so $\delta W_s = 0$ and the vacuum region can be replaced by a zero-pressure plasma according to comparison theorem 2. The remaining δW_F can be calculated with $\xi(r,\theta,z) = \xi(r)e^{i(kz+m\theta)}$ and minimized with respect to $\xi_\theta(r)$ and $\xi_z(r)$. Substituting the minimizing values of ξ_θ and ξ_z into δW_F, one finds that the latter is always positive for the unpinch.

5-18. Use the same method as in Exercise 5-17 to investigate the hydromagnetic stability of the hard-core pinch configuration of Fig. 5-14b. Prove that an arrangement where the plasma current density $J_z(r)$ is everywhere opposite to the total current enclosed $I_z(r)$ (as in the case in the hard-core configuration discussed) is hydromagnetically stable. [W. B. Thompson, "An Introduction to Plasma Physics." Macmillan (Pergamon), New York, 1962.]

5-19. Prove that in a general cylindrical configuration $B(0, B_\varphi, B_z)$ with the normal-mode displacements $\xi(r) \exp[i(kz + m\varphi)]$, the modes with $(mB_\varphi/r) + kB_z = 0$ result in uniform displacement of a field line (the same $\xi_r, \xi_\varphi, \xi_z$ along the line, that is). These correspond, therefore, to interchange modes.

5-20. Consider a force-free field $[\nabla \times \mathbf{B} = \alpha(\mathbf{r})\mathbf{B}, \mathbf{B} \cdot \nabla\alpha = 0]$ with conducting boundaries. Prove that
   (a)   The Euler equation of the energy integral is $\mathbf{F}(\xi) = 0$.
   (b)   The Euler equation with the normalization condition $\int v\xi^2 \, d\tau = $ const results in the equation of motion where the Lagrange multiplier λ plays the role of ω^2.

5-21. Prove that the energy principle applied to a stabilized pinch yields for a $k = 0$ perturbation:
   (a) $\delta W_F^{extr} = 0$.
   (b)   No instability.

5-22. Use the method of normal-mode analysis for the treatment of the stabilized pinch. The stability condition, of course, must be the same as obtained from the energy principle.

5-23. Derive (5-170) and obtain the Kruskal-Shafranov condition for stability. Using this result, under what conditions is the $kr_0 \ll 1$ expansion justified?

5-24. Derive (5-174).

5-25. Prove that for the tearing mode calculation one may write $\Delta' = [\partial A_z/\partial y]_{-y_0}^{+y_0} A_z^{-1}$.

5-26. Calculate Δ' for a magnetic field profile $F = y$ when $|y| < 1, F = 1$ and when $y > 1, F = -1$ for $y < -1$, and let $a \to \infty$. What is the range of wave numbers for which the configuration is tearing mode unstable?

VI

Plasma in the Steady State

6-1. Electric Fields in Plasmas

In this chapter we return to the discussion of plasmas, keeping their multi-charged-particle aspect in the foreground, and use the self-consistent field approach to investigate the macroscopic behavior of steady and quasi-steady plasmas. By examining specific problems of practical interest, we shall make use of the general results developed in Chapters II and III. The plasmas under discussion are free of collisions, but the particles do interact via long-range collective electromagnetic fields.

At first we consider the problem of maintaining electric fields inside a plasma in thermal equilibrium. The field is produced by an object, e.g., a sphere immersed in the plasma and kept at the potential V_0 with respect to the plasma, or rather with respect to regions of the plasma far away from the sphere. In the absence of the sphere, the plasma contains a uniform density of positive and negative charges n_+ and n_-, respectively, and it is neutral, $q_+ n_+ + q_- n_- = 0$. For the sake of simplicity we take a hydrogen or deuterium plasma with $q_+ = -q_- = e$.

The qualitative effect of the presence of the charged sphere on the plasma is easy to estimate. A positively charged sphere will attract the electrons and repel ions, and vice versa. This results in a "screening" effect of the plasma weakening the field farther away from the sphere. The resulting electric field is produced by both the sphere and the plasma and it can be described by the *Poisson equation*,

$$\nabla^2 V = -\frac{e}{\varepsilon_0}(n_+ - n_-) \qquad \text{for} \quad r > R \tag{6-1}$$

where R is the radius of the sphere. Since the plasma is in thermal equilibrium, the particle densities at a point with the potential V are given by

$$n^+ = n_0 e^{-eV/kT} \tag{6-2a}$$

and

$$n^- = n_0 e^{\, eV/kT} \tag{6-2b}$$

Here it has been tacitly assumed that the number of particles of both kinds to be found in a spherical shell of volume $4\pi r^2 \, \Delta r$ is large enough for statistical treatment, and the potential does not vary appreciably over Δr, $[(\partial V/\partial r) \, \Delta r \ll kT/e]$.

Equations (6-1), (6-2a), and (6-2b) constitute the complete set of self-consistent equations for this problem. Indeed, by combining them, one arrives at

$$\nabla^2 V = - \frac{en_0}{\varepsilon_0} (e^{-eV/kT} - e^{eV/kT}) \tag{6-3}$$

which can be solved to yield $V(r)$. Let us assume that

$$eV_0/kT \ll 1 \tag{6-4}$$

the Taylor expansion of the exponentials, neglecting higher-order terms, yields

$$\frac{1}{r^2} \frac{\partial}{\partial r} r^2 \frac{\partial V}{\partial r} - \frac{2e^2 n_0}{\varepsilon_0 kT} V = 0 \tag{6-5}$$

Defining

$$\frac{\varepsilon_0 kT}{2e^2 n_0} = L^2 = \frac{\lambda_D^{\,2}}{2} \tag{6-6}$$

it can be seen by direct substitution that the solution of (6-5) is

$$V = A \frac{e^{-r/L}}{r} \tag{6-7}$$

The boundary condition ($V = V_0$ at $r = R$) yields for the value of the constant,

$$A = RV_0 e^{R/L} \tag{6-8}$$

which leads to

$$V = V_0 \frac{R}{r} \exp\left(-\frac{r-R}{L}\right) \tag{6-9}$$

The solution (6-9) verifies our expectation. In the presence of a plasma the potential falls off more rapidly than $1/r$. Owing to the exponential factor, the potential drops e^{-1}-fold at the *shielding distance* L. $\lambda_D = 2^{1/2}L$ is called the *Debye radius* or *Debye length*, and it is one of the characteristic lengths of the plasma. This length is usually rather short in plasmas produced in the laboratory. For example, in the case of a plasma with the density $n_0 = 10^{12} \text{cm}^{-3} = 10^{18} \text{m}^{-3}$, and a temperature $T = 10^4$ °K, $\lambda_D \approx 7 \times 10^{-6}$ m = 7 μ, but even for $T = 10^8$ °K the Debye length is only ≈ 0.7 mm. Thus a plasma provides a very effective shielding mechanism against electric fields.

The same process also prevents the development of any appreciable charge imbalance in parts of a plasma. The removal of charges of one sign (e.g., electrons) in a region of a plasma would create local electric fields, which in turn would act immediately to restore equilibrium.

An appreciable charge imbalance in the plasma as a whole is also very difficult to maintain. Consider a spherical plasma with radius $r = 10^{-1}$ m, of the same density as in the previous example. If only 1 per cent of the electrons (or ions) were removed, the potential in the sphere would rise to roughly half a million volts.

For the aforementioned reasons, we shall usually assume charge neutrality in plasmas over regions larger than the Debye length, therefore to a high degree of approximation,

$$\frac{|n_+ - n_-|}{n} \ll 1 \tag{6-10}$$

It should be noted that in deriving the shielding formula (6-9) use has been made of the assumption of thermal equilibrium. In plasmas with nonthermal velocity distributions, electric fields can be maintained over distances considerably larger than λ_D. However, the range of validity of (6-10) is much larger, since, as mentioned, an appreciable charge imbalance would give rise to enormous electric fields for the particle densities customary in laboratory plasmas.

A plasma (now a nonequilibrium one) placed in a magnetic field also reacts in a characteristic fashion to electric fields. Consider a plasma of uniform density n filling the space in the region $x_1 < x < x_2$, placed in the homogeneous magnetic field $\mathbf{B} = B\mathbf{e}_3$. The particles move in this field in circles, and we choose $|x_2 - x_1| \gg R$, the ion gyration radius. This enables one to treat the plasma as if the particles were located at their guiding center, e.g., use the "guiding-center plasma" approximation. The circulating particles, of course, produce currents which give rise to magnetic fields, but we assume that they are small compared to B. This imposes a limitation on both n and the kinetic temperature of the plasma, and will be examined later.

Consider now two condenser plates placed at $x < x_1$ and $x > x_2$, respectively (Fig. 6-1). At $t = 0$ a voltage is switched on slowly enough so that the guiding-center approximation holds for the motion of individual particles. The electric field $\mathbf{E}$ causes a motion of the particles in the y direction with the drift velocity $\mathbf{w}^E = (\mathbf{E} \times \mathbf{B})/B^2$. Since, however, $\mathbf{E}$ varies in time, so does $\mathbf{w}^E$, and this, according to (2-166), results in an additional polarization drift:

$$\mathbf{w}^p = \frac{m}{qB^2} \dot{\mathbf{E}} \tag{6-11}$$

This drift is in the x-direction, with a different sign for positive and negative particles, giving rise, thereby, to polarization of the plasma. Since positive charges move in the direction of, and negative charges against, a growing field, the two oppositely charged layers emerge at x_1 and x_2 and set up an

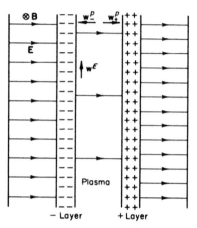

FIG. 6-1. Polarization of guiding-center plasma between condenser plates.

electric field opposing the impressed field. This behavior is similar to the reaction of a dielectric in response to the switching on of an electric field. Let us consider the "polarization current" density

$$\mathbf{J}_p = \sum_\sigma q_\sigma n \mathbf{w}_\sigma{}^p = (m_+ + m_-)n(\dot{\mathbf{E}}/B^2) \tag{6-12}$$

Inserting this in Maxwell's equation one obtains

$$(\nabla \times \mathbf{H})_\perp = \mathbf{J}_p + \varepsilon_0 \frac{\partial \mathbf{E}_\perp}{\partial t} = \left(\frac{m_+ + m_-}{B^2} n + \varepsilon_0\right)\dot{\mathbf{E}}_\perp \tag{6-13}$$

taking into account in the notation that this is true only for vector components perpendicular to $\mathbf{B}$. Defining the *dielectric permeability* of the magnetoplasma by

$$\kappa_\perp = \frac{m_+ + m_-}{\varepsilon_0 B^2} n + 1 \tag{6-14}$$

Maxwell's equation can be written

$$(\nabla \times \mathbf{H})_\perp = \varepsilon_0 \kappa_\perp \, \partial \mathbf{E}_\perp/\partial t \tag{6-15}$$

The same dielectric constant is obtained if the ratio of the normal components of the electric field inside and outside the plasma is computed, provided (and only if) the initial conditions yield no polarization for zero electric field. In this case at time t both polarization layers contain the charge per unit surface area:

$$Q = \int J_p \, dt = (m_+ + m_-)n(E/B^2) \tag{6-16}$$

giving rise to the polarization field $E_p = Q/\varepsilon_0$, which leads again to (6-14).

The formal analogy between dielectric materials and magnetoplasmas can be carried even further. The former when polarized contain a polarization energy (the work done by the field while polarizing the dielectric) whose density is

$$W_p = (\kappa - 1)\varepsilon_0(E^2/2) \tag{6-17}$$

where E is the field inside the material. The latter picks up kinetic energy from the field during polarization, with the density

$$W_k = \tfrac{1}{2}n(m_+ + m_-)(w^E)^2 = \tfrac{1}{2}n(m_+ + m_-)(E^2/B^2) \tag{6-18}$$

Using the definition of $\kappa_\perp$, (6-14) leads to

$$W_k = (\kappa_\perp - 1)\varepsilon_0(E^2/2) \tag{6-19}$$

in analogy with (6-17).

The first term of (6-14) can be rewritten

$$\frac{(m_+ + m_-)n}{\varepsilon_0\mu_0} \bigg/ \frac{B^2}{\mu_0} = \frac{(m_+ + m_-)nc^2}{2(B^2/2\mu_0)} \tag{6-20}$$

which is just one-half the rest-mass energy density divided by the magnetic energy density. This is usually a large number for laboratory plasmas [take, e.g., $n = 10^{18}$ m^{-3}, $B = 10^{-1}$ webers m^{-2}, and obtain from (6-20) $10^4 \div 10^6$ for light elements]. Thus it is often a good approximation to take

$$\kappa_\perp = \frac{m_+ n}{B^2 \varepsilon_0} \tag{6-21}$$

recognizing also that $m_+ \gg m_-$. The large value of $\kappa_\perp$ results in an almost perfect cancellation of the perpendicular component of the external electric field inside the plasma by polarization.

It will be seen in the next chapter that a plasma possesses the characteristic frequencies

$$(\omega_p^+)^2 = \frac{q^2 n}{\varepsilon_0 m_+} \quad \text{and} \quad (\omega_p^-)^2 = \frac{q^2 n}{\varepsilon_0 m_-} \tag{6-22}$$

the ion and electron plasma frequencies, respectively. Using $\omega_c^+ = (e/m_+)B$ and (6-22) in (6-21) one obtains

$$\kappa_\perp = (\omega_p^+/\omega_c^+)^2 \tag{6-23}$$

In the surface layers produced by polarization, only one kind of charge is

present, and the inequality (6-10) breaks down. The layer thickness can be
calculated from (6-11), considering the distance a particle on the outer surface
of the layer travels during the process of polarization,

$$D = \int w^p \, dt = \frac{m_+}{e \, B^2} \, E_{\text{ext}} \qquad (6\text{-}24)$$

where E_{ext} is the external field. Since the electric field in the layer exceeds the
plasma field, w^p is also larger in the layer, while the particle density is correspondingly smaller. It is easy to see that (Exercise 6-2)

$$\frac{n_{\text{layer}}}{n_{\text{plasma}}} = \frac{E_{\text{pl}}}{E_{\text{ext}}} = \frac{1}{\kappa_\perp} \qquad (6\text{-}25)$$

Using the value of $\kappa_\perp$ from (6-21) and the Debye-length formula (6-6) (with
n_{pl}), one obtains from (6-24)

$$D = \kappa_\perp^2 \frac{qE}{kT} \lambda_D^2 = \kappa_\perp^2 \frac{\lambda_D^2}{\delta} \qquad (6\text{-}26)$$

where a charged particle has to be moved a distance δ along the electric field
to pick up roughly the mean thermal energy. If $\delta < \lambda_D \kappa_\perp^2$, the layer thickness
exceeds the Debye length. Since the plasma is not in thermal equilibrium (if
it were it would not be confined without material walls), this is not surprising.

A word of caution is in order concerning the limitation of applicability of
the effective dielectric constant concept. Obviously, this is valid only in the
plane perpendicular to the magnetic field. It has already been mentioned that
in order to obtain the right jump of the normal electric field component
across the boundary (or equivalently to satisfy $\nabla \cdot \varepsilon_0 \kappa_\perp \mathbf{E} = \rho$), proper initial
conditions have to be chosen. The positioning of the condenser plates is also
critical. If they are in touch with the plasma, while (6-14) still holds and leads
formally to the same dielectric constant, no charged surface layers can form
as they are carried away by the electrodes, and the field is consequently not
weakened in the plasma interior.

6-2. Plasma Motion in Magnetic Fields

As a consequence of applying the crossed magnetic and electric fields, the
plasma as a whole is set in motion with the drift velocity $\mathbf{w}^E$. We have seen
in the example in the last section that the electric field is greatly reduced inside
the plasma, owing to the dielectric properties of plasmas. This can be avoided
if the plasma is in touch with the electrodes, and such an arrangement can
serve to move a plasma across a magnetic field. Often two concentric cylinders
are used as electrodes, with the plasma rotating between the cylinders. Such

an arrangement can even be used as a capacitor, since the kinetic energy of the moving plasma can be extracted when shorting the electrodes through a load. This can be seen as follows.

When charging the capacitor, an electric field is set up between the plates accelerating the particles to their $\mathbf{w}^E$ velocity. If the voltage source is disconnected, the electric drift velocity decreases, giving rise to an inertial drift $\mathbf{w}^p$ and a depolarization current $\mathbf{J}^p$ flowing into the electrodes, to charge them up, so as to maintain the electric field. If $\mathbf{J}_p$ flows through a load, the voltage on the capacitor gradually decreases to zero (or further if the load is inductive) and the plasma drift changes correspondingly. The stored energy density is again given by (6-17) (see Exercise 6-3).

A system consisting of two parallel " rail " electrodes, with a dc magnetic field producing a $\mathbf{w}^E$ drift along the electrodes, can serve as a plasma accelerator (Fig. 6-2). If the plasma is injected into the system when the electric field is low, it is accelerated with the electric field increasing in time, provided, of course, that it is in touch with the electrodes, so that no charge separation can develop.

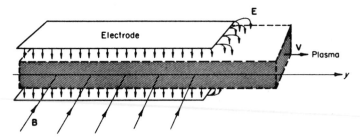

FIG. 6-2. Plasma emerging from an accelerator.

It is of interest to investigate what happens at the " muzzle " of this " plasma gun " at $y > 0$. Will the plasma continue to move across the magnetic field or will it stop at the end of the electrodes? A single charged particle would simply decelerate as the electric field decreases to zero, but the collective action of particles in the plasma leads again to a qualitatively different behavior. The decelerating particles undergo an inertial or polarization drift resulting in the formation of charged surface layers in the electrodeless $y > 0$ region. The electric field set up by these layers facilitates the continued motion of the rest of the plasma across the magnetic field. The calculation goes as follows.

A particle in the plasma experiences two drifts, both in the perpendicular direction:

$$w_y = \frac{E_x}{B_z} = \frac{E}{B} \tag{6-27}$$

and

$$w_x = \frac{m}{qB}\frac{dw_y}{dt} = \frac{m}{qB^2}\frac{dE}{dt} \tag{6-28}$$

The electric field consists of two parts: the stray field of the electrodes and the polarization field of the plasma:

$$E = E_s + E_p \tag{6-29}$$

The latter can be computed with the help of Poisson's equation. Equating the current density with the rate of change of surface charge density,

$$\varepsilon_0 \frac{dE_p}{dt} = -J_p = -qn(w_x{}^+ - w_x{}^-) \approx -\frac{m_+ n}{B^2}\frac{dE}{dt} \tag{6-30}$$

These are the self-consistent equations, (6-27) and (6-28) giving the motion of particles in given fields while (6-30) yields the fields that arise as a result of particle motion. Inserting E from (6-29) and using our definition of $\kappa_\perp$ leads to

$$\kappa_\perp \frac{dE_p}{dt} = -(\kappa_\perp - 1)\frac{dE_s}{dt} \tag{6-31}$$

This means that the decrease of the stray field is upset by a proportional increase of the polarization field. For $\kappa_\perp \gg 1$,

$$\frac{dE_p}{dt} = -\frac{dE_s}{dt} \quad \text{or} \quad \frac{dE}{dt} = 0 \tag{6-32}$$

the electric field inside the "dielectric" is not weakened at all after it leaves the electrodes and w_y is also unchanged. If, however, $\kappa_\perp \ll 1$, the plasma behaves as a collection of single particles and it is unable to cross the field.

It should not be thought, however, that the plasma can move indefinitely across a magnetic field. Particles drifting out into the surface layer experience a smaller electric field and hence a smaller drift than the bulk of the plasma and are consequently left behind. Since the total electric field in the plasma cannot change, the particles lost from the surface layer are continuously replaced from the plasma interior. Thus the plasma velocity remains unchanged while the mass is gradually decreasing.

Let us turn now to plasma motion along the magnetic field lines. The case of a homogeneous field is uninteresting; particles move along the field as in the absence of the field. The situation is different if the field lines have a curvature.

Consider a plasma being driven into a curved magnetic field, with velocity $v_\parallel$ (Fig. 6-3). A single particle moving with a velocity $v_\parallel$ parallel to the field, and gyrating with the velocity $v_\perp$ perpendicular to it, would follow the field lines in the zero-order approximation while experiencing a centrifugal and a

$\mathbf{V}_\perp B$ drift in the first-order approximation. Since many particles are present, and electrons and ions drift in opposite directions, an electric field builds up inside the plasma giving rise to the electric drift. Since this is a zero-order quantity, its time rate of change gives rise to an inertial drift, resulting in polarization currents.

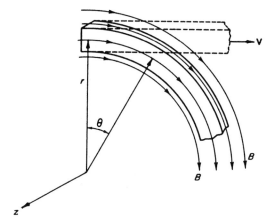

FIG. 6-3. Plasma motion in a curved magnetic field.

The velocity of the guiding center of each particle is constant along the field lines, as there are neither electric fields nor magnetic field gradients in this direction, while the guiding center drifts perpendicular to the magnetic field with the velocity

$$\mathbf{w}_\perp = \frac{\mathbf{E} \times \mathbf{B}}{B^2} + \frac{\mu_m}{qB^3}\left[\mathbf{B} \times \nabla \frac{B^2}{2}\right] + \frac{m}{qB^2}[\mathbf{B} \times \dot{\mathbf{w}}^0] \qquad (2\text{-}75)$$

The last term is the inertial drift caused by the inertial forces arising in the coordinate system moving with the zero-order drift velocity,

$$\mathbf{w}^0 = v_{\parallel}\frac{\mathbf{B}}{B} + \frac{\mathbf{E} \times \mathbf{B}}{B^2} \qquad (6\text{-}33)$$

The inertial drift splits into two parts. The centrifugal drift is, from (2-66),

$$\mathbf{w}^c = \frac{m}{q}\frac{v_{\parallel}^2}{B^4}\mathbf{B} \times (\mathbf{B} \cdot \nabla)\mathbf{B} \qquad (6\text{-}34)$$

while the polarization drift, coming from the inertial forces due to changes in the electric drift, is

$$\mathbf{w}^p = \frac{m}{qB^2}\mathbf{B} \times \frac{d}{dt}\frac{\mathbf{E} \times \mathbf{B}}{B^2} \qquad (6\text{-}35)$$

We have assumed that the plasma currents do not influence the imposed magnetic field. Therefore, $\nabla \times \mathbf{B} \approx 0$ and

$$\nabla \frac{B^2}{2} = (\mathbf{B} \cdot \nabla)\mathbf{B} \tag{6-36}$$

Equation (2-75) in this case can be put in the following form:

$$\mathbf{w}_\perp = \frac{\mathbf{E} \times \mathbf{B}}{B^2} + \frac{m}{qB^4}\left(v_\parallel^2 + \frac{v_\perp^2}{2}\right)\mathbf{B} \times (\mathbf{B} \cdot \nabla)\mathbf{B} + \frac{m}{qB^2}\mathbf{B} \times \frac{d}{dt}\frac{\mathbf{E} \times \mathbf{B}}{B^2} \tag{6-37}$$

If one sets up a cylindrical coordinate system (Fig. 6-3), the plasma drift velocities are as follows: The plasma moves along the field lines with

$$w_\theta = v_\parallel \tag{6-38}$$

The first term in (6-37) does not cause charge separation; the second term gives rise to particle drifts in the z direction resulting in an electric field in the same direction. As a result of this, the plasma as a whole experiences an electric drift in the r direction. Finally, the last term is perpendicular to both B and the time rate of change of the electric drift which points in the r direction. Hence the last term is again a z drift. Consequently,

$$\begin{aligned}
w_z &= \frac{m}{qB^4}\left(v_\parallel^2 + \frac{v_\perp^2}{2}\right)[\mathbf{B} \times (\mathbf{B} \cdot \nabla)\mathbf{B}]_z + \frac{m}{qB^2}\left[\mathbf{B} \times \frac{d}{dt}\frac{\mathbf{E} \times \mathbf{B}}{B^2}\right]_z \\
&= -\frac{m}{qrB}\left(v_\parallel^2 + \frac{v_\perp^2}{2}\right) + \frac{m}{qB}\frac{d}{dt}\left(\frac{E}{B}\right)
\end{aligned} \tag{6-39}$$

while

$$w_r = E/B \tag{6-40}$$

For the sake of simplicity we take an infinitely long plasma slab with a rectangular cross section and uniform density n. To make the equations self-consistent we again use the fact that the electric field is produced by the drift

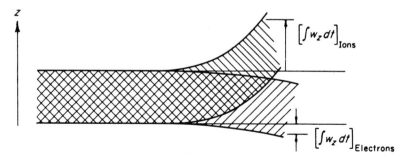

Fig. 6-4. Polarization of plasma entering a region with curved field lines.

in the z direction, resulting in the appearance of surface charges on the side-walls of the slab (Fig. 6-4):

$$\frac{dE_z}{dt} = -\frac{e\,n(w_z{}^+ - w_z{}^-)}{\varepsilon_0} \approx -\frac{e\,nw_z{}^+}{\varepsilon_0} \tag{6-41}$$

Using (6-41) in (6-39) one obtains

$$w_z = -\frac{m}{erB}\left(v_\parallel{}^2 + \frac{v_\perp{}^2}{2}\right) - \frac{mnw_z}{\varepsilon_0 B^2} - \frac{mE}{eB^3}\frac{dB}{dt} \tag{6-42}$$

where m stands for m_+. This can be written as

$$\kappa_\perp w_z = -\frac{m}{erB}\left(v_\parallel{}^2 + \frac{v_\perp{}^2}{2}\right) - \frac{mE}{eB^3}\frac{dB}{dt} \tag{6-43}$$

For $m_+ n/\varepsilon_0 B^2 \gg 1$,

$$\frac{ew_z n}{\varepsilon_0 B} + \frac{1}{r}\left(v_\parallel{}^2 + \frac{v_\perp{}^2}{2}\right) + \frac{E}{B^2}\frac{dB}{dt} = 0 \tag{6-44}$$

or, by using (6-41),

$$\frac{1}{B}\frac{dE}{dt} - \frac{E}{B^2}\frac{dB}{dt} = \left(v_\parallel{}^2 + \frac{v_\perp{}^2}{2}\right)\frac{1}{r} \tag{6-45}$$

This can be written finally:

$$\frac{d}{dt}\left(\frac{E}{B}\right) = \frac{dw_r}{dt} = \frac{v_\parallel{}^2 + (v_\perp{}^2/2)}{r} \tag{6-46}$$

Under experimental conditions the energy associated with the mass flow is usually much larger than the thermal energy of the plasma, or $v_\parallel{}^2 \gg v_\perp{}^2$. In this case, (6-46) yields for the differential equation of the radial velocity,

$$\frac{dw_r}{dt} = \frac{v_\parallel{}^2}{r} \tag{6-47}$$

This is the equation of a particle in uniform motion along a straight line, tangential to the magnetic field lines at the point of entry.

In the other limiting case of very tenuous plasma $\kappa_\perp - 1 \ll 1$. This case leads to essentially independent particles following the curved field lines. Note that in the case where thermal motion is appreciable in the dense plasma,

$$\frac{dw_r}{dt} > \frac{v_\parallel{}^2}{r} \tag{6-48}$$

Consequently, the plasma column bends away from the field lines.

Another case of plasma motion along the magnetic field is shown in

Fig. 6-5. Here the magnetic field intensity varies along the plasma path, but there is no appreciable curvature of the field lines.

Consider a charge-neutral homogeneous plasma moving along this field with velocity v_0 in the uniform field region, while each particle gyrates

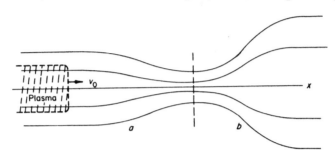

FIG. 6-5. Plasma moving along field lines enters a region of constriction.

around the field lines due to its (thermal) velocity $v_\perp$. We assume for the sake of simplicity that each electron has the same $v_\perp^-$ and each ion the same $v_\perp^+$, corresponding to the same kinetic temperature, so that

$$[\tfrac{1}{2}m_+(v_\perp^+)^2]_0 = [\tfrac{1}{2}m_-(v_\perp^-)^2]_0 \tag{6-49}$$

where the subscript 0 refers to the homogeneous field region.

In the region of constriction (region a) the motion along the field is slowed down as in a mirror machine. Since v_0 is the common longitudinal velocity for electrons as well as ions, ions, because of their larger mass, have a larger longitudinal energy and penetrate the constriction farther than electrons with the same velocity. As a result of this charge separation, an electric field with the potential $V(x)$ is set up; this helps the electrons to overcome the barrier, but slows down the ions. Owing to (6-49) and the definition of the magnetic moment (2-14),

$$\mu_m^+ = \mu_m^- = \mu_m \tag{6-50}$$

Since both μ_m and the total energy of each particle are conserved, the energy conservation law takes the form

$$\tfrac{1}{2}m_+(v_\|^+)^2 + \mu_m B + q_+ V = E^+ \tag{6-51a}$$

for ions, and

$$\tfrac{1}{2}m_-(v_\|^-)^2 + \mu_m B + q_- V = E^- \tag{6-51b}$$

for electrons in a hydrogen plasma. It can be seen that $\mu_m B$ acts as a potential energy. Since this potential barrier is the same for both types of particles and $E^+ > E^-$ obviously in the absence of charge separation, ions reach regions of higher B. Because of the electrostatic coupling between the two components, there are only two possibilities: The plasma passes through the constriction as a whole, or it is totally reflected. The potential field is defined by Poisson's equation:

$$\nabla^2 V = -\frac{(n_+ - n_-)e}{\varepsilon_0} \tag{6-52}$$

Consider the case of transmission. Since—except for an unimportant azimuthal drift—the particles follow the field lines, the cross-sectional plasma area is inversely proportional to the magnetic field. This leads to the equations of continuity:

$$n_+ v_\parallel{}^+ = \frac{B}{B_0} I_0 \tag{6-53a}$$

and

$$n_- v_\parallel{}^- = \frac{B}{B_0} I_0 \tag{6-53b}$$

where the particle flux density I_0 in the homogeneous field region is common to both species. Equations (6-51) to (6-53) constitute the complete set of self-consistent equations. By eliminating all variables but V, one obtains

$$\nabla^2 V = -\frac{e}{\varepsilon_0} \frac{B}{B_0} I_0 \left\{ \left[\frac{2}{m_+} (E^+ - \mu_m B - eV) \right]^{-1/2} \right.$$
$$\left. - \left[\frac{2}{m} (E^- - \mu_m B + eV) \right]^{-1/2} \right\} \tag{6-54}$$

With I_0, the perpendicular kinetic temperature, and the variation of $B(x)$ given, (6-54) can be integrated.

A necessary, but not sufficient condition for transmission can be obtained by adding (6-51a) to (6-51b):

$$\tfrac{1}{2} m_+ (v_\parallel{}^+)^2 + \tfrac{1}{2} m_- (v_\parallel{}^-)^2 + 2\mu_m B = E^+ + E^- \tag{6-55}$$

If the maximum field in the throat is B_{max}, it is required that

$$E^+ + E^- > 2\mu_m B_{max} \tag{6-56}$$

In this case, at least, the ions reached the critical cross section, although $\tfrac{1}{2} m_- (v_\parallel{}^-)^2$ might still be negative, meaning that the electrons are reflected before reaching the throat. In this case the ions will also be dragged back after a short penetration. A sufficient and necessary criterion for transmission depends on the details of field variation and can be obtained only by solving (6-54).

The same equation applies also for region *b*. Here the particles move "downhill" in the $\mu_m B$ potential with electrons preceding the ions, as long as $B < B_0$.

6-3. Plasma Confinement by Magnetic Fields

It has been assumed so far that the reaction of plasma currents on the magnetic field can be neglected. In the following we shall do just the opposite: investigate the contribution of the plasma currents to the magnetic field and neglect electrostatic effects. This latter assumption is justified by considering configurations with charge neutrality.

We start by considering an initially uniform magnetic field $\mathbf{B}_0$ pointing in the z direction. Particles injected into this field will execute the well-known helical motion, giving rise to a "diamagnetic" field opposing $\mathbf{B}_0$. As a result, the uniformity of the field will be destroyed and also the particle paths will cease to be helical, taking on drift motion caused by the field inhomogeneity. If the injected particle distribution is independent of z as well as y, and the density declines symmetrically toward $+x$ and $-x$, we expect a configuration like the one depicted in Fig. 6-6 to arise. The decrease of the

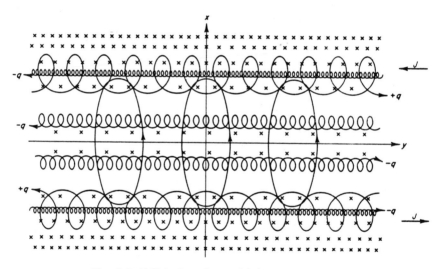

FIG. 6-6. Collisionless plasma slab in a magnetic field.

magnetic field inside the plasma is associated with a magnetic field gradient drift in the y direction. Of course, the particle currents have to produce the right self-consistent fields to maintain the same particle motion. Note that the plasma currents can be divided into two categories: diamagnetic currents arising from particle gyration, and paramagnetic currents strengthening the field due to the field gradient drift (see Fig. 6-6).

To derive the self-consistent particle-field equation, we divide particles into groups, denoting particles of the same species which have the same velocity components v_x, v_y at the same coordinate x with index j. The particle speed v_j

(in the absence of electric fields) is a constant of motion. The magnetic field intensity $B(x)$ (a function to be determined) bends the particle paths into curves with a local radius of curvature

$$R_j(x) = \frac{m_j}{q_j} \frac{v_j}{B(x)} \tag{6-57}$$

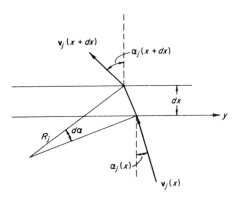

FIG. 6-7. Illustration to aid calculation of equations of motion.

Adopting the notation of Fig. 6-7 one obtains

$$R_j \, d\alpha_j = \frac{dx}{\cos \alpha_j} \tag{6-58}$$

where, of course, $\alpha_j = \alpha_j(x)$. Substituting R_j from (6-57) yields

$$\frac{m_j v_j}{q_j B} \cos \alpha_j \, d\alpha_j = dx \tag{6-59}$$

This differential equation fully describes the motion of particles of group j in a given field $B(x)$. To find out what this field is we need the current density

$$J_y(x) = -\sum_j q_j n_j(x) v_j \sin \alpha_j(x) \tag{6-60}$$

since all groups contribute to the current; n denotes particle density. Through $\nabla \times \mathbf{H} = \mathbf{J}$ this current is responsible for the variation of the field

$$-(\nabla \times \mathbf{H})_y = \frac{1}{\mu_0} \frac{dB}{dx} = \sum_j q_j n_j v_j \sin \alpha_j \tag{6-61}$$

The requirement that the particle flux is conserved for each group in the steady state can be written

$$n_j v_j \cos \alpha_j = \lambda_j \tag{6-62}$$

Equations (6-59), (6-61), and (6-62) are self-consistent and complete. Since every particle moves once in the positive and once in the negative x direction, belonging to a different group in each case, the number of groups is always even, say $2N$. Thus the number of unknown $n_j(x)$ functions, as well as the number of the $\alpha_j(x)$'s, is also $2N$. Therefore, the number of functions to be determined, including $B(x)$, equals $4N + 1$. Since both (6-59) and (6-62) represent $2N$ equations, these, with the additional equation (6-61), just suffice to determine the unknown functions. The v_j's and λ_j's are prescribed numbers, the latter actually square pulse functions.

Combining (6-59), (6-61), and (6-62) one obtains

$$\frac{1}{\mu_0} dB = \sum_j \frac{m_j v_j \lambda_j}{B} \sin \alpha_j \, d\alpha_j \tag{6-63}$$

This is just the differential form of the pressure-balance equation. The x-x component of the pressure tensor, as defined in the laboratory system, is

$$P_{xx} = \sum_j m_j n_j v_j^2 \cos^2 \alpha_j = \sum_j m_j \lambda_j v_j \cos \alpha_j \tag{6-64}$$

and its differential is

$$dP_{xx} = -\sum_j m_j \lambda_j v_j \sin \alpha_j \, d\alpha_j \tag{6-65}$$

This leads to the more familiar form of (6-63),

$$\frac{B \, dB}{\mu_0} + dP_{xx} = 0 \tag{6-66}$$

or

$$\frac{\partial}{\partial x} \left(\frac{B^2}{2\mu_0} + P_{xx} \right) = 0 \tag{6-67}$$

the only surviving component of the tensor equation (3-45a).

The contribution of a particle species to the plasma current is closely related to its partial pressure. To prove this we differentiate (6-62):

$$dn_j v_j \cos \alpha_j - n_j v_j \sin \alpha_j \, d\alpha_j = 0 \tag{6-68}$$

and calculate the current density resulting from all the groups of ions with charge e and mass m:

$$J_y^+ = -e \sum_j {}^+ n_j v_j \sin \alpha_j = -e \sum_j {}^+ v_j \cos \alpha_j \frac{dn_j}{d\alpha_j} \tag{6-69}$$

To introduce the variable x instead of α one uses (6-59) and obtains

$$J_y^+ = -\frac{m^+}{B} \sum_j {}^+ v_{jx}^2 \frac{dn_j}{dx} \tag{6-70}$$

It follows from (6-62) that

$$d(v_{jx}^2 n_j) = \lambda_j^2\, d\frac{1}{n_j} = -\frac{\lambda_j^2}{n_j^2}\, dn_j = -v_{jx}^2\, dn_j \tag{6-71}$$

and the partial current density becomes

$$J_y^{+} = \frac{m^{+}}{B}\frac{d}{dx}\sum{}^{+}n_j v_{jx}^2 = \frac{1}{B}\frac{d}{dx}P_{xx}^{+} \tag{6-72a}$$

The electronic current density is, similarly,

$$J_y^{-} = \frac{1}{B}\frac{d}{dx}P_{xx}^{-} \tag{6-72b}$$

The results so far are quite general. They can be simplified further if the assumption is made that the field variation as well as the variation of guiding-center density is small within the cyclotron radii of particles. In this case,

$$\sum_j v_{jx}^2 = \sum_j u_{jx}^2 = \tfrac{1}{2}\sum u_j^2 \tag{6-73}$$

where $\mathbf{u}$ is the gyrating velocity vector. The pressure is now a scalar and (6-67) becomes, integrating,

$$\frac{B^2}{2\mu_0} + p = \frac{B_0^2}{2\mu_0} \tag{6-74}$$

where B_0 is the field intensity in regions unoccupied by plasma. The phenomenon that the magnetic field is weakened by particle currents inside the plasma is sometimes referred to as *plasma diamagnetism*, and its magnitude is characterized by the ratio of plasma to magnetic pressure:

$$\beta = \frac{p}{B^2/2\mu_0} \tag{6-75a}$$

The scalar pressure in two dimensions is related to the kinetic temperature by

$$p = nkT \tag{6-76}$$

If the electronic and ionic kinetic temperatures are equal, the partial pressures are identical and, from (6-72),

$$J_y^{+} = \frac{kT}{B}\frac{dn^{+}}{dx} \tag{6-77a}$$

and

$$J_y^{-} = \frac{kT}{B}\frac{dn^{-}}{dx} \tag{6-77b}$$

hence the contributions to the current are also identical. In a nonhydrogen (or deuterium) plasma, where the ionic charge is Z times the electronic charge $Zn^+ = n^-$ and

$$J^- = ZJ^+ \qquad (6\text{-}77\text{c})$$

The same procedure can be followed with other geometries. An important example is the so-called θ pinch, where a plasma of cylindrical symmetry is confined in an asymptotically uniform B_z field produced by a cylindrical coil. The name comes from the fact that the plasma currents flow in the θ direction. The self-consistent equations are now (see Exercise 6-6)

$$d\alpha_j = \left(\frac{q_j B}{m_j v_j} - \frac{\sin \alpha_j}{r} \right) \frac{dr}{\cos \alpha_j} \qquad (6\text{-}78)$$

where r is the radial coordinate,

$$dB/dr = \mu_0 \sum_j q_j n_j v_j \sin \alpha_j \qquad (6\text{-}79)$$

and

$$r n_j v_j \cos \alpha_j = \lambda_j \qquad (6\text{-}80)$$

The equations again yield a pressure-balance equation, which assumes the following form for cylindrical geometry (see Exercise 3-2):

$$\frac{B}{\mu_0} \frac{dB}{dr} + \frac{dP_{rr}}{dr} + \frac{P_{rr} - P_{\theta\theta}}{r} = 0 \qquad (6\text{-}81)$$

where

$$P_{rr} = \sum_j n_j m_j v_j{}^2 \cos^2 \alpha_j \qquad (6\text{-}82)$$

and

$$P_{\theta\theta} = \sum_j n_j m_j v_j{}^2 \sin^2 \alpha_j \qquad (6\text{-}83)$$

(see Exercise 6-7). It is easy to see that the outward-directed $-P_{\theta\theta}/r$ is just the centrifugal force density, which has to be balanced by additional magnetic and plasma pressure gradients. The P_{rr}/r term comes from the distortion of the planar geometry, which results in a purely geometrical radial decrease of P_{rr} (a negative dP_{rr}/dr) which has to be compensated for by this additional term (see Exercise 6-9). For a scalar pressure, these additional terms cancel and we obtain again the simple equation (6-74) and the diamagnetic effect can again be characterized by β.

In previous sections it was assumed that the contribution of plasma currents to the magnetic field is negligible while $\kappa_\perp \gg 1$. We see now that this assumption is justified for a large class of plasmas. The first requirement can be formulated as $\beta \ll 1$. Since

$$\beta = \frac{\frac{1}{2} \sum^{\pm} nm \langle v_\perp{}^2 \rangle}{B^2/2\mu_0} \qquad (6\text{-}75\text{b})$$

the ratio of kinetic energy density to magnetic energy density is roughly $\langle v_\perp^2 \rangle^+/c^2$ times smaller than $\kappa_\perp$. Except for plasmas in which even the ionic component is relativistic, $\beta \ll \kappa_\perp$; consequently, there is a large class of plasmas with a high dielectric constant but not showing any appreciable diamagnetism.

Next we consider the pinch configuration (Fig. 4-4). Here the plasma current flows in the z direction, giving rise to a θ magnetic field which serves to confine the plasma. No imposed external magnetic field is necessary. We employ here the alternative method, based on the Boltzmann equation, instead of the equivalent equations of motion. The distribution function is a function of the constants of motion only (3-60):

$$f = f(\mathcal{H}, P_\theta, P_z) \tag{6-84}$$

where P_θ and P_z are constants due to symmetry and the Hamiltonian

$$\mathcal{H} = \frac{1}{2m}\left[P_r^2 + \left(\frac{P_\theta}{r}\right)^2 + (P_z - qA_z)^2 \right] + qV \tag{6-85}$$

The vector potential has a z component only and the potential V arises from charge separation. Every nonconstant quantity depends on r only. Note that (6-84) already contains the equations of motion and we have only to add

$$\nabla^2 A_z = \frac{1}{r}\frac{d}{dr}\left(r\frac{dA_z}{dr} \right) = -\mu_0 e \int (f^+ - f^-)v_z \, d^3P \tag{6-86}$$

and

$$\frac{1}{r}\frac{d}{dr}\left(r\frac{dV}{dr} \right) = -\frac{e}{\varepsilon_0} \int (f^+ - f^-) \, d^3P \tag{6-87}$$

to make the system self-consistent. For simplicity we again take a hydrogenic plasma, and chose to work in an $(\mathbf{r}, \mathbf{P})$ phase space instead of the equivalent $(\mathbf{r}, \mathbf{v})$ one. Insertion of (6-84) into (6-86) and (6-87) leads to two equations from which $A_z(r)$ and $V(r)$ can be determined. Our result, obtained so quickly, is completely equivalent to results obtained earlier in this section by lengthier, although physically more transparent, considerations. To arrive at a concrete field distribution, the functions $f^\pm$ in (6-84) have to be specified. This is equivalent to specifying all the v_j's and λ_j's in the previous cases. If we choose a Maxwellian distribution, no current and no confinement results. Let us take instead the simple

$$f = \frac{a}{2\pi mkT}\exp\left[-\frac{\mathcal{H}}{kT} \right] \delta(P_z) \tag{6-88}$$

distribution function for both electrons and ions. Since

$$P_z = mv_z + qA_z \tag{6-89}$$

the $\delta(P_z)$ term yields a particle current in the z direction. Of course, any other distribution function of the type (6-84) is equally justified, but this has been chosen for its simplicity and closeness to thermal equilibrium. Inserting (6-88) in (6-86) and (6-87) the integrals on the right side can be carried out. The volume element of phase space is $dr\, d\theta\, dz\, dP_r\, dP_\theta\, dP_z$. Since, however, the volume element in configuration space is $r\, dr\, d\theta\, dz$, the remaining momentum space element should be written $d^3P = (1/r)\, dP_r\, dP_\theta\, dP_z$. The P_z part of the integral in (6-86) is trivial and the remainder yields

$$q \iint \frac{a}{2\pi mkT}\left(-\frac{qA_z}{m}\right)\exp\left[-\frac{1}{2mkT}\left(P_r^2 + \left(\frac{P_\theta}{r}\right)^2 + q^2 A_z^2\right) - \frac{qV}{kT}\right] dP_r\, d\frac{P_\theta}{r}$$

$$= -\frac{q^2 A_z}{m}\, a\, \exp\left(-\frac{q^2 A_z^2}{2mkT} - \frac{qV}{kT}\right) \tag{6-90}$$

The dependence on q^2 shows that electrons and ions produce currents in the same direction, as expected. The integral in (6-87) yields for the particle densities

$$n^+(r) = a^+ \exp\left(-\frac{e^2 A_z^2}{2m^+ kT} - \frac{eV}{kT}\right) \tag{6-91a}$$

and

$$n^-(r) = a^- \exp\left(-\frac{e^2 A_z^2}{2m^- kT} + \frac{eV}{kT}\right) \tag{6-91b}$$

One can simplify the equations by assuming approximate charge neutrality $n^+ \approx n^-$ and by taking advantage of the fact that $m^+ \gg m^-$. Since $a^+ = a^-$, this leads to

$$\frac{e^2 A_z^2}{2m^- kT} \approx 2\frac{eV}{kT} \tag{6-92}$$

Inserting (6-90) into (6-86) the ion current can be neglected, since it is only m^-/m^+ times the electron current. Equation (6-86) becomes, therefore,

$$A_z'' + \frac{A_z'}{r} - \frac{\mu_0 e^2 A_z a}{m^-}\exp\left(-\frac{e^2 A_z^2}{4m^- kT}\right) = 0 \tag{6-93}$$

This is a nonlinear differential equation, which can be solved numerically to yield $A_z(r)$.

Figures 6-8 and 6-9 show the results of solving the coupled equations (6-86) and (6-87) for both the vector and scalar potential. One can see from Fig. 6-9 that the assumption of charge neutrality used to arrive at the single equation (6-93) is really justified.

Another example is the so-called *Bennett pinch*. Here the geometry is the same as before, with a stream of electrons and a stream of ions moving through each other along the z axis. In the center-of-mass system (roughly moving with the ion stream, since $m^+ \gg m^-$) complete space-charge neutralization is assumed to take place. The magnetic field in this system is again described by (6-86), but the ion contribution to the current can be neglected,

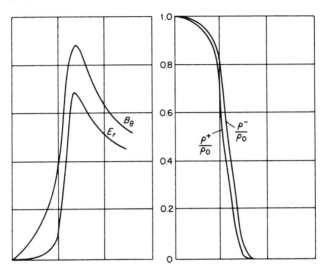

FIG. 6-8. Electric and magnetic field of a pinch geometry. [After E. S. Weibel, *Phys. Fluids* **2**, 52 (1959).]

FIG. 6-9. Computed charge-density distribution of a pinch geometry. [After E. S. Weibel, *Phys. Fluids* **2**, 52 (1959).]

for the aforementioned reason. Let us now transform our coordinate system to move with the center of mass of the electrons, which moves with velocity $\mathbf{V}$ compared to the ions. An observer in this system observes a radial electric field, related to the magnetic field through

$$\mathbf{E} = \mathbf{V} \times \mathbf{B} \tag{6-94a}$$

or

$$E_r = -V_z B_\theta \tag{6-94b}$$

and a scalar potential related to the vector potential through

$$\phi = -V A_z \tag{6-95}$$

since $E_r = -\partial\phi/\partial r$ and $B_\theta = -\partial A_z/\partial r$. (For deeper reasons see Exercise 6-12.) This electric field is such as to confine the electrons in their rest frame. One now assumes a Maxwellian distribution for the electrons in this coordinate system. This distribution depends on $\mathscr{H}$ only, hence it satisfies (6-84). For the density distribution it follows immediately that

$$n = n_0 \exp(-e\phi/kT) = n_0 \exp(eVA_z/kT) \tag{6-96}$$

and the current density in the center-of-mass system produced by the electron stream is

$$\mathbf{J} = e n_0 V \exp\left(\frac{eVA_z}{kT}\right) = -\frac{1}{\mu_0}\frac{1}{r}\frac{d}{dr}\left(r\frac{dA_z}{dr}\right) \tag{6-97}$$

This equation can be solved analytically with the boundary conditions $A(0) = 0$ and $A'(0) = B(0) = 0$, and it yields

$$A_z = -\frac{2kT}{eV}\log\left(1 + \frac{1}{8}\frac{e^2 n_0 V^2 \mu_0}{kT} r^2\right) \tag{6-98}$$

One can see via substitution that (6-98) satisfies (6-97) (see Exercise 6-13). The magnetic field distribution can be obtained by differentiation:

$$B_\theta = -\frac{\partial A_z}{\partial r} = \frac{\tfrac{1}{2}eVn_0\mu_0 r}{1 + \tfrac{1}{8}(e^2 n_0 V^2 \mu_0/kT)r^2} \tag{6-99}$$

6-4. The Plasma–Magnetic Field Boundary

The pressure-balance equation presents a limitation to the maximum plasma pressure one can confine by a given magnetic field. The confinement is, of course, more economical the more magnetic field pressure can be replaced in the plasma interior by plasma pressure. In this respect the goal is to achieve a high-β confinement. The best one can do in this respect is to create regions where the magnetic field is completely absent, and the plasma pressure p equals the magnetic pressure of the external confining field. We set out to investigate now whether such an arrangement is possible, that is, whether a self-consistent solution of this kind exists.

In the course of the calculation a distribution function has to be chosen for the plasma interior. Since collisions are always present, every distribution function tends to become Maxwellian. A completely Maxwellian plasma cannot be confined by magnetic fields—as proved earlier. The question then arises: Can it be Maxwellian in the field-free region only?

We now set up our problem in the following form: A geometry like the first one treated in the previous section is investigated; the confining magnetic field points in the z direction, while every quantity is independent of y and z. The field becomes asymptotically uniform for large positive x values where the plasma pressure is zero, while the magnetic field is zero at large negative x values and the plasma Maxwellian. We expect to find something like Fig. 6-10, where particles emerging from the plasma move in a curved path in the magnetic field and re-enter the plasma. The field which these particles produce must be self-consistent. We wish to assume, furthermore, that all particles found in this boundary layer came from the Maxwellian plasma interior. This means we'exclude "trapped particles" drifting in the boundary layer, like those in Fig. 6-6.

We seek to avoid complications arising from charge separation. The simplest way to do this is to work with a "positronium plasma" model, where $m^+ = m^-$. Since the plasma interior is in thermal equilibrium, $T^+ = T^-$.

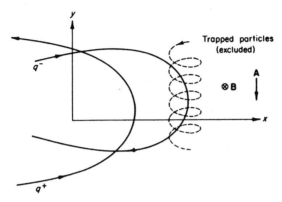

FIG. 6-10. Particle trajectories in a plasma–magnetic field boundary.

The constants of motion are: $\mathscr{H}$, P_y, P_z. Accordingly, the distribution function has the form

$$f(\mathbf{r},\mathbf{v}) = f(\mathscr{H}, P_y, P_z) \qquad (6\text{-}100)$$

where $\mathscr{H}$ and P_y contain $A_y = A$ as a variable. Self-consistency requires that

$$\frac{d^2A}{dx^2} = -2\mu_0 q \int f v_y \, d^3P \qquad (6\text{-}101)$$

where the factor 2 arises from the additive contribution of the two particle species. When discussing the pinch configuration in the previous section, we were free to choose any distribution function that depended on the constants of motion only. Here f, in addition to having the form (6-100), has to satisfy the boundary conditions and the requirement of exclusion of trapped particles. These, with (6-100) and (6-101), determine the problem uniquely.

We proceed to solve the problem by analyzing phase space. Let us again choose $(\mathbf{r},\mathbf{P})$ space. [For the equivalent solution in $(\mathbf{r},\mathbf{v})$ space see Exercise 6-17.] Owing to the chosen symmetries, the six-dimensional phase space reduces to a more transparent three-dimensional system. Since nothing varies along y and z, these coordinates can be omitted. Furthermore, since $P_z = mv_z = $ const is known, and this motion is not coupled to either the magnetic field, or the particle motion in the xy plane, the P_z axis can also be left out of our coordinate system, which reduces finally to an $x - P_x - P_y$ system (see Fig. 6-11).

It was shown in Sec. 3-4 that the constants of motion define subspaces in phase space, lowering the dimensionality of the problem by one. The $P_y =$ const and $\mathscr{H} =$ const subspaces are, therefore, two-dimensional surfaces in our three-dimensional phase space. In fact, $P_y =$ const defines planes parallel

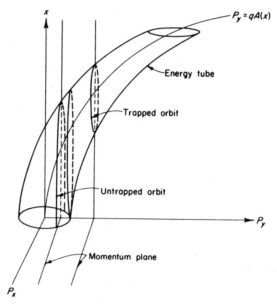

FIG. 6-11. Constant-of-motion surfaces and curves in phase space.

to the $x - P_x$ plane. The $\mathscr{H} =$ const surfaces are also simple, defined by the equation

$$P_x^2 + (P_y - qA)^2 = 2m\mathscr{H} - P_z^2 = 2m\mathscr{H}_\perp = \text{const} \qquad (6\text{-}102)$$

The shape of this surface is not known, since $A_y(x)$ is unknown as yet, but its characteristics are easily evaluated. Its intersection with any $x =$ const plane is a circle with radius $r = (2m\mathscr{H}_\perp)^{1/2}$, its center displaced by $qA(x)$ in the P_y direction (see Fig. 6-12). Since A is assumed to increase monotonically with increasing x from zero at $x \to -\infty$, the $\mathscr{H}_\perp =$ const tubes look like the ones in Fig. 6-11. The intersections of the $P_y =$ const and $\mathscr{H} =$ const surfaces are the phase paths of the particles, each one of them being populated by a uniform phase point (or, briefly, particle) density (see Sec. 3-4). Since the intersections of subspaces yielded the one-dimensional phase paths, we see that all the constants of motion have been found. (Note that this was achieved by finding three cyclic variables, y, z, and t, leading to the elimination of these variables and yielding three constants of motion: the canonically conjugate P_y, P_z, and $\mathscr{H}$.)

The phase paths on Fig. 6-11 fall into two categories: those extending to $x = -\infty$ (or at whatever point $A = 0$), and those being limited on both sides in x. The latter paths correspond to trapped orbits, which we do not populate with particles, while the particle density on the former ones is

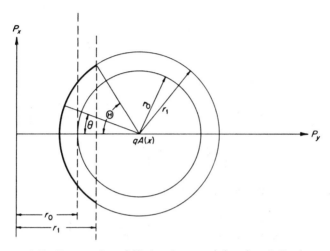

FIG. 6-12. Construction of filled and empty regions in a P_xP_y plane.

determined by its value at $x \to -\infty$. Phase space is thus divided into filled and empty regions. Introducing polar coordinates in the $x = $ const planes according to Fig. 6-12, it is found that the arc $-\Theta < \theta < +\Theta$ is filled while the rest of the circle is empty. Here

$$\cos \theta = \frac{qA - P_y}{r} = -\frac{mv_y}{r} \tag{6-103}$$

and

$$\cos \Theta = \frac{qA - r}{r} = \frac{qA}{r} - 1 \tag{6-104}$$

At $x \to -\infty$ the distribution function is Maxwellian, hence it depends only on $\mathcal{H} = \mathcal{H}_\perp + P_z^2/2m$. Therefore, for a fixed value of P_z, the particle density in phase space is a constant on an $r = $ const circle at $x \to -\infty$. On the other hand, the particle density does not vary on a phase path either. It follows that on the filled arc of $r = $ const the particle density is the same as on the corresponding circle at $-\infty$ and in our case a constant. Hence the distribution function in any $x = $ const plane is a truncated Maxwellian distribution. It is easy to see that the boundary between filled and empty regions is a parabola (see Exercise 6-16). The integral on the right side of (6-101) is now easily carried out,

$$\int f v_y \, d^3 P = -\frac{1}{m} \int\limits_{-\Theta}^{+\Theta} \int\limits_{r_0}^{\infty} \int\limits_{-\infty}^{+\infty} f(\mathscr{H}) r^2 \cos\theta \, dP_z \, dr \, d\theta$$

$$= -\frac{1}{m} \int\limits_{-\Theta}^{+\Theta} \int\limits_{r_0}^{\infty} f_\perp(r) r^2 \cos\theta \, dr \, d\theta \qquad (6\text{-}105)$$

where we introduced

$$f_\perp(r) = \int f(\mathscr{H}_\perp + P_z^2/2m) \, dP_z \qquad (6\text{-}106)$$

Integration over θ yields, using (6-104),

$$\int\limits_{-\Theta}^{+\Theta} \cos\theta \, d\theta = [\sin\Theta - \sin(-\Theta)] = 2\left[1 - \left(\frac{qA}{r} - 1\right)^2\right]^{1/2}$$

$$= 2\left(2\frac{qA}{r} - \frac{q^2 A^2}{r^2}\right)^{1/2} \qquad (6\text{-}107)$$

This leads to the differential equation

$$\frac{d^2 A}{dx^2} = +\frac{4\mu_0 q}{m} \int\limits_{r_0}^{\infty} f_\perp(r) r^2 \left(2\frac{qA}{r} - \frac{q^2 A^2}{r^2}\right)^{1/2} dr \qquad (6\text{-}108)$$

The integral is to be carried out over real values of the integral, yielding for the lower limit

$$r_0 = qA/2 \qquad (6\text{-}109)$$

Evidently, as seen from Fig. 6-12, in an $x = \text{const}$ plane circles with radii $r < r_0$ are completely empty. This means physically that particles with very small energy cannot penetrate far into the magnetic field, whatever the direction of their velocity vectors in the field-free region. Equation (6-109) can also be derived using the notion of effective potential (see Exercise 6-15).

The solution $A(x)$ of (6-108) can be obtained by numerical integration. It is not obvious beforehand that it exists. Therefore, we investigate the asymptotic solution for $A \to 0$. In this case $(qA/r)^2 \ll qA/r$ and one obtains

$$\frac{d^2 A}{dx^2} = 4\mu_0 \frac{q}{m} \int\limits_{r_0}^{\infty} f_\perp(r) r^2 \left(\frac{2qA}{r}\right)^{1/2} dr = \alpha A^{1/2} \qquad (6\text{-}110)$$

where the value of the constant α can be determined by integration if necessary. Equation (6-110) can be readily solved, yielding

$$A = (\alpha/12)^2 (x - x_0)^4 \qquad (6\text{-}111)$$

It may seem that this solution is in contradiction with our assumption $A \to 0$ if $x \to -\infty$. There is no reason, however, why one could not build up the solution by adding two solutions in the following manner. Let

$$A = 0, \quad f(\mathscr{H}) \text{ (Maxwellian)} \qquad \text{for} \quad x < x_0$$

and

$$A = \text{solution of (6-108)} \qquad \text{for} \quad x > x_0$$

(6-112)

At $x = x_0$ these solutions have to be pieced together. As can be seen from (6-111), A, as well as its first three derivatives, vanishes at $x = x_0$ from both sides, which is more than necessary (fitting the function and the first derivative would suffice for a second-order differential equation). The physically significant quantities, the magnetic field and current density, the first and second derivatives of A, respectively, are also both zero at the boundary. The value of x_0 can be chosen arbitrarily.

Note that the combination of the collisionless Boltzmann equation with the field equations is nonlinear ($\partial f/\partial v$ contains the field variables and it is also multiplied by $E + v \times B$). This can also be seen from (6-108). It is known that the solutions of nonlinear equations are not uniquely determined by the boundary conditions; this is how the undetermined x_0 enters our solution.

On the magnetic field side (A large), r_0 is large, and since $f_\perp(r)$ drops off rapidly at large values of r, the integral in (6-108) becomes small. This leads to the asymptotic solution of A:

$$A = C_1 x + C_2$$

(6-113)

which leads to the desired $A' = B_0$ constant magnetic field at large values of x.

The second-order differential equation (6-108) can easily be integrated once. Multiplying both sides by dA/dx and integrating with respect to x, one obtains

$$\frac{1}{2\mu_0}\left(\frac{dA}{dx}\right)^2 + \frac{1}{m}\int \left\{ (r - qA)(2rqA - q^2A^2)^{1/2} \right.$$
$$\left. + r^2 \left[\frac{\pi}{2} + \sin^{-1}\left(1 - \frac{qA}{r}\right)\right]\right\} f_\perp(r)r\, dr = \frac{B_0^2}{2\mu_0} \quad (6\text{-}114)$$

The proof is left to the reader (Exercise 6-18). It is easy to see that this is the expression of the pressure-balance equation, with the first term $(1/2\mu_0)(dA/dx)^2 = B^2/2\mu_0$ and the second term P_{xx} (Exercise 6-19).

Using the technique of "piecing solutions together," various field configurations can be produced (see Fig. 6-13). If the solution of (6-108) is used by itself, one arrives at a plasma that is Maxwellian only in the field-free plane $x = x_0$ confined by a positive magnetic field $+B_0$ to the right and a negative $-B_0$ on the left. The change in sign of B at x_0 can be readily seen looking at (6-111). In this case, one of the confining fields (e.g., $+B_0$) must be produced externally, while the other one is produced by the plasma itself.

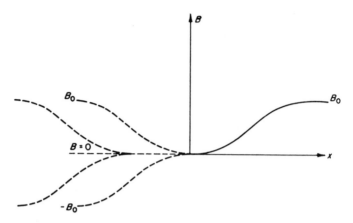

FIG. 6-13. Some possible high-β confined-slab configurations in asymptotically uniform magnetic fields.

Other solutions, such as that of a plasma slab confined on both sides by uniform magnetic fields, can also be obtained with ease.

6-5. The E Layer

We have termed the weakening of the magnetic field inside a plasma by particle currents "plasma diamagnetism," and characterized its magnitude by the factor β. The similarity between plasmas and dielectric materials, however, is rather superficial, and holds only in the limiting case of scalar pressure and nearly circular particle motion. The exact statement one can make about the "diamagnetic effect" is the reference to the full form of the pressure-balance equation (3-45a). In the case discussed in the last section, for instance, we have seen that even "complete diamagnetism" can be achieved, with the complete disappearance of the magnetic field in the plasma interior. In another case, briefly mentioned in the same section, the magnetic field actually changed sign across the plasma boundary, a case which could be arbitrarily characterized by a negative diamagnetic constant, if one wanted to stick to this concept.

We are going to examine now another example where the concept of diamagnetism and "β" breaks down. Consider a uniform magnetic field $\mathbf{B} = B_0\mathbf{k}$ with particles injected uniformly along the z and θ directions, forming a thin solenoidal layer (Fig. 6-14). Each particle has the same velocity v and the radius of the solenoid is the cyclotron radius, $R_c = mv/qB$. Injecting more particles into this layer, the self-magnetic field B^s, produced by the solenoid, becomes appreciable, resulting in the weakening of the initially established internal field. The question arises: How many particles can be accommodated in such a layer, and how does the internal field change during this process of filling up the layer? In particular: Is the complete elimination of the internal field, and perhaps also its reversal, possible?

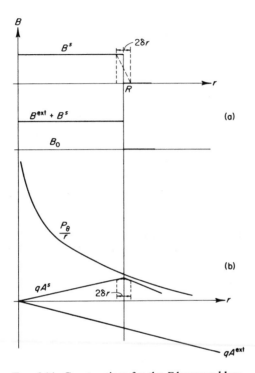

FIG. 6-14. Constructions for the E-layer problem.

In solving this problem the Boltzmann equation could be used, just as in the previous section. However, since the particle distribution function is singular, it is easier to resort to the equation of motion directly. Owing to the cylindrical symmetry of the system, the Hamiltonian method with the effective potential can be used. Space-charge effects will again be neglected (think, for example, of a positronium plasma sheath).

The magnetic field consists of two parts—external and self-field:

$$B = B^{\text{ext}} + B^s \tag{6-115}$$

where $B^{\text{ext}} = B_0$ is uniform and

$$B^s = \begin{cases} qNv\mu_0 & \text{for} \quad r < R \\ 0 & \text{for} \quad r > R \end{cases} \tag{6-116}$$

where N is the surface charge density of the layer and R is its radius. The vector potential $A_\theta = A$ can be calculated from the flux ϕ,

$$A^{ext} = \frac{\phi^{ext}}{2\pi r} = \frac{B_0}{2} r \tag{6-117}$$

and

$$A^s = \begin{cases} \dfrac{\mu_0 q N v}{2} r & \text{for } r < R \\[2ex] \dfrac{\mu_0 q N v R^2}{2r} & \text{for } r > R \end{cases} \tag{6-118}$$

The radius R is as yet unknown. It must be derived from the self-consistency requirement, namely, the particles moving in the combination of imposed and self-field have to move just on the circle with the radius $r = R$. In other words, the effective potential (2-143)

$$\psi = \frac{1}{2m}\left(\frac{P_\theta}{r} - qA^{ext} - qA^s\right)^2 \tag{6-119}$$

has a minimum at $r = R$. The fields and the various terms appearing in ψ are plotted in Fig. 6-14. Note that the self-field opposes the external field and also that P_θ and qA^{ext} have necessarily opposite signs for particles encircling the axis (see Sec. 2-6).

In the absence of a self-field, the minimum of the effective potential is where the slopes of P_θ/r and qA^{ext} are equal, where

$$-\frac{P_\theta}{r^2} = \frac{q}{2} B_0 \tag{6-120}$$

and since

$$P_\theta = (mv + qA^{ext})r = \left(mv + q\frac{B_0}{2} r\right)r \tag{6-121}$$

the potential minimum is at

$$r = \left|\frac{mv}{qB}\right| = R_c \tag{6-122}$$

as expected. The presence of a self-field, however, complicates matters slightly. By minimizing ψ at $r = R$, we need the derivative of A^s at the same point where it is double-valued. The values are

$$\frac{\partial A^s}{\partial r} = \begin{cases} \dfrac{\mu_0 q N v}{2} & \text{at } \lim_{\delta r \to 0}(R - \delta r) \\[2ex] -\dfrac{\mu_0 q N v}{2} & \text{at } \lim_{\delta r \to 0}(R + \delta r) \end{cases} \tag{6-123}$$

and the arithmetical average is zero. If the layer has a small but finite thickness $2\,\delta r$ around R with a uniform charge distribution, the field increases linearly, the flux cubically, and the vector potential varies parabolically inside the layer. Figure 6-14b shows the behavior of qA^s, which has a genuine extremum at $r = R$ in this case. It can be said, therefore, that

$$(\partial A^s/\partial r)_{r=R} = 0 \tag{6-124}$$

The requirement that ψ be a minimum at R leads to

$$\frac{P_\theta}{R^2} + q\left(\frac{\partial A^{\text{ext}}}{\partial r}\right)_R + q\left(\frac{\partial A^s}{\partial r}\right)_R = \frac{P_\theta}{R^2} + \frac{qB_0}{2} = 0 \tag{6-125}$$

the same as (6-120). It follows that particles with the same P_θ move in the same orbit in the field B_0, irrespective of the self-field. This does not mean, however, that particles with identical P_θ move with the same velocity as well. Expressing P_θ in terms of the velocity, one obtains

$$P_\theta = [mv + qA^{\text{ext}}(R) + qA^s(R)]R \tag{6-126}$$

Inserting this in (6-125) and using the expressions for the vector potentials,

$$\frac{mv}{R} + \frac{qB_0}{2} + \frac{q^2\mu_0 Nv}{2} + \frac{qB_0}{2} = 0 \tag{6-127}$$

from which the radius of the solenoid becomes

$$R = - \frac{mv}{qB_0 + (Nq^2v\mu_0/2)} \tag{6-128}$$

Since qB_0 was chosen negative, $v = v_\theta$, and R is positive. One can see immediately that to keep R finite the inequality

$$|B_0| > \left|\frac{Nqv\mu_0}{2}\right| = \frac{|B^s|}{2} \tag{6-129}$$

must be satisfied, which means that the self-field cannot exceed twice the value of the external field. If the self-field equals the external field in magnitude, the solenoid interior is exactly field-free. If B^s is somewhat larger (but less than twice as large) the inequality (6-129) is still satisfied and the field is reversed inside the solenoid. If B^s exceeded the limit imposed by (6-129), the internal fields were stronger than the external. For $N = 0$, (6-128) goes over into the familiar formula of the cyclotron radius, while the increase of N results in an increase of R, keeping all the other variables constant.

Introducing the internal field,

$$B_{\text{in}} = B_0 + \mu_0 qNv \tag{6-130}$$

and inserting this into (6-128), one obtains

$$R = -\frac{mv}{q\frac{1}{2}(B_0 + B_{in})} \tag{6-131}$$

Consequently, the solenoid radius is the cyclotron radius for particles moving in the average field $\frac{1}{2}(B_0 + B_{in})$. It is interesting to note that this relationship can also be derived at once from the pressure-balance equation (6-81). Since

$$P_{rr} = 0 \tag{6-132}$$

and

$$P_{\theta\theta} = Nmv^2\,\delta(r - R) \tag{6-133}$$

(6-81) can be integrated over the layer, yielding

$$\frac{B_0{}^2}{2\mu_0} - \frac{B_{in}^2}{2\mu_0} - \frac{Nmv^2}{R} = 0 \tag{6-134}$$

Writing (6-134) in the form

$$\frac{(B_0 + B_{in})(B_0 - B_{in})}{2} = \mu_0 Nvq\,\frac{mv}{qR} \tag{6-135}$$

and using (6-130) leads, indeed, to (6-131).

The Astron thermonuclear machine is based on the confinement of ions by an E layer. Energetic (relativistic) electrons are injected into a magnetic field terminated by magnetic mirrors to form an E layer (Fig. 6-15), creating a

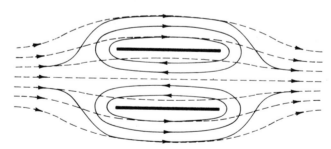

FIG. 6-15. The E layer.

magnetic bottle of closed field lines with the ions confined in the bottle.

Another interesting application arises if one considers a collisionless θ pinch. The magnetic field in a θ-pinch device is usually generated by discharging a capacitor through a single-turn coil. As the system rings, the magnetic field reverses at successive half-cycles. The motion of a single particle in a sign-changing spacially uniform magnetic field was investigated in Sec. 2-7 and the trajectory shown in Fig. 2-21. When many particles are present, one expects a strong diamagnetic effect to arise from the spiralling particles after field reversal. A self-consistent treatment of this problem is left as an exercise (Exercise 6-26). A more thorough analysis reveals that with a sufficient number of particles, a surface layer is formed after field reversal just strong enough to eliminate (approximately) but never to reverse the internal field. This is quite contrary to what one expects of a hydromagnetic fluid of high conductivity. There surface currents arise just to preserve the internal flux independent of the external field. This discrepancy is not surprising, however, since any analogy between a collisionless plasma and a hydromagnetic fluid breaks down if the particle motion is nonadiabatic. Remember that the CGL or double-adiabatic approximation was based on the guiding-center approximation.

6-6. Plasma Confinement by High-Frequency Fields

As was seen in Chapter II, particles in the rf field of a standing electromagnetic wave experience an average force acting in the direction of the negative field gradient. This also means that a plasma in a nonuniform rf field tends to be confined in regions of minimum field strength. Here again, however, one must inquire as to the reaction of the oscillating particles to the electromagnetic field, and solve the self-consistent equations.

We shall look for steady-state solutions, which means in this case that the density of oscillation centers, $n^+(r)$ and $n^-(r)$, respectively, and the field amplitudes, $\mathbf{E}_0(r)$ and $\mathbf{B}_0(r)$, are stationary. It is also reasonable to assume that due to electrostatic forces in the plasma, $n^+ \approx n^- = n$. The oscillating charges give rise to the plasma current density

$$\mathbf{J} = n \sum_{\pm} q\dot{\mathbf{r}}_1 = -n \sum_{\pm} \frac{q^2}{m\omega^2} \dot{\mathbf{E}} \approx -\frac{q^2 n}{m^-\omega^2} \dot{\mathbf{E}} \qquad (6\text{-}136)$$

where (2-190) and $m^- \ll m^+$ have been used. The oscillation-center drifts do not contribute to the current in the steady state, since $\langle \dot{\mathbf{r}}_0 \rangle = 0$. From Maxwell's equation,

$$\nabla \times \mathbf{H} = \varepsilon_0 \dot{\mathbf{E}} - \frac{q^2 n}{m^-\omega^2} \dot{\mathbf{E}} = \varepsilon_0 \left[1 - \left(\frac{\omega_p^-}{\omega} \right)^2 \right] \dot{\mathbf{E}} \qquad (6\text{-}137)$$

where the definition of the electron plasma frequency (6-22) was used. Just as the circulating particles in the guiding-center approximation weakened the imposed magnetic field, the oscillating charges—being 180 degrees out of phase with the electric field—weaken the displacement current and thereby the high-frequency magnetic field. The quantity

$$\kappa_\omega = 1 - (\omega_p^-/\omega)^2 \qquad (6\text{-}138)$$

can also be considered as the high-frequency dielectric constant of the plasma. It is frequency-dependent and, of course, different from the dielectric constant obtained for the guiding-center plasma.

Using the other field equation and dropping the index one obtains

$$\nabla \times (\nabla \times E) = \nabla(\nabla \cdot E) - \nabla^2 E = -\mu_0 \varepsilon_0 (1 - \omega_p{}^2/\omega^2)\ddot{E} \qquad (6\text{-}139)$$

Let us consider only fields which do not give rise to space-charge accumulation in the plasma, hence $\nabla \cdot E = 0$. In this case (6-139) reduces to the wave equation

$$\nabla^2 E = \frac{\kappa}{c^2}\ddot{E} \qquad (6\text{-}140)$$

For standing waves,

$$E(r,t) = E_0(r)e^{i\omega t} \qquad (6\text{-}141)$$

one obtains

$$\nabla^2 E_0 + (\kappa/c^2)\omega^2 E_0 = 0 \qquad (6\text{-}142)$$

For $\omega > \omega_p$, κ is positive and one obtains wave-like solutions in the plasma. In the opposite case, $\omega < \omega_p$, $\kappa < 0$ however, only exponentially decreasing solutions exist, with the characteristic penetration depth

$$\delta = c/\omega(-\kappa)^{1/2} \qquad (6\text{-}143)$$

Equation (6-142) describes the electromagnetic field if the particle density n is known. One also needs to know the particle density as a function of the fields to arrive at a solvable set of equations. One can either resort to the equations of motion directly (see Exercise 6-27) or use macroscopic equations. Equating the force density with the pressure gradient

$$mn\ddot{r}_0 = \nabla \cdot \overset{\leftrightarrow}{p} \qquad (6\text{-}144)$$

yields, with the help of (2-195), the pressure-balance equation

$$\frac{nq^2}{m\omega^2} \nabla \frac{E_0{}^2}{4} + \nabla \cdot \overset{\leftrightarrow}{p} = 0 \qquad (6\text{-}145)$$

where $\overset{\leftrightarrow}{p}$ is the time-averaged pressure tensor. One again needs an equation of state relating $\overset{\leftrightarrow}{p}$ to n. Unfortunately, the difficulties of finding such a relationship are even more formidable here than in the case of the guiding-center fluid. Those particles are prevented from moving across the magnetic field and carrying energy with them, but particles in an rf field are free to move, giving rise to heat transport. The effect of the strong magnetic field, causing a nearly isotropic pressure perpendicular to the field, is also absent here. On the other hand, since the particles are free to move in the plasma (actually they oscillate in the effective potential well Φ), they are likely to exchange energy by occasional collisions and arrive at a common temperature T and an isotropic velocity distribution perpendicular to the direction of oscillation, that is, to E. If the field configuration is such that the E lines are straight and parallel (plane waves, cylindrical waves with axial E, etc.), one writes

$$p_\perp = 2nkT \tag{6-146}$$

where the factor 2 is due to the presence of the two particle species. This yields, when put into (6-145),

$$\varepsilon_0 \left(\frac{\omega_p}{\omega}\right)^2 \nabla_\perp \frac{E_0{}^2}{4} + \nabla_\perp(2nkT) = 0 \tag{6-147}$$

where $\perp$ refers now to the direction perpendicular to the *electric* field. Equations (6-142) and (6-147) contain only the two variables n and E_0, and they constitute the complete set of self-consistent equations. The relationship between electric field intensity and particle density follows from (6-147) directly:

$$\frac{q^2}{8kTm\omega^2} \nabla_\perp E_0{}^2 + \frac{1}{n} \nabla_\perp n = 0 \tag{6-148}$$

and can be written

$$\frac{q^2}{8kTm\omega^2} \nabla_\perp E_0{}^2 + \nabla_\perp \log n = 0 \tag{6-149}$$

which can be integrated directly to yield

$$n = n_0 \exp\left(-\frac{q^2 E_0{}^2}{8kTm\omega^2}\right) \tag{6-150}$$

Inserting (6-150) into (6-142) leads to a nonlinear differential equation in E_0 which can only be integrated numerically. For a cylindrical geometry, this equation is identical mathematically to (6-93) except for the absence of two components, and the solutions are the same as the ones plotted in Fig. 6-9.

It should be mentioned that the main difficulties in using high-frequency plasma confinement are of an economic and technical nature. The skin losses arising on the conductor surfaces for high frequencies would outweigh the gains in fusion energy. The technical factor involves the relative weakness of the rf effective potential. The confinement of a plasma with the kinetic temperature $T \approx 10$ keV, with a high-frequency field of $\omega \approx 10^{10}$, requires a field strength of the order of $E \approx 10^7$ volts m^{-1}, as calculated from (2-199). This estimate is based on the following confinement mechanism: The rf field confines the electrons, while each electron can be thought of as being electrostatically coupled with one ion, since these are almost unaffected by the field.

6-7. Quasi-Steady Processes

Frequently one wants to know what happens if some of the parameters defining the state of the plasma are slowly varied in time. This variation is assumed to be so slow that at any time the plasma is in one of the steady states. In the course of time, however, it can pass from one state into another one characterized by substantially different parameters. These processes will be called *adiabatic* or *quasi-steady processes*.

At any time during such a process the plasma is described by the steady-state equations. The problem is to find out *which* new state the plasma goes into from a given one under the given variation of some parameters. This can be done if one knows a sufficient number of constants and adiabatic invariants of the particle motion.

As an example we investigate a cylindrical guiding-center plasma confined by a uniform external magnetic field under the influence of the slow variation of the confining field. In such a field the particle motion can be considered as two-dimensional (the motion along the field lines plays no role in the process), and the adiabatic invariant is the magnetic moment of each particle.

Consider an arbitrary two-dimensional geometry (Fig. 6-16). In the steady state, particles drift along $B = $ const lines, while their gyration around the guiding center generates the diamagnetic currents. The slow variation of the confining field results in the additional $\mathbf{E} \times \mathbf{B}/B^2$ drift (inertial drifts are absent in an adiabatic process). The change of flux inside a closed line defined by guiding centers initially lying on a $B = $ const line is

$$\frac{d\phi}{dt} = \frac{\partial}{\partial t} \int \mathbf{B} \cdot d\mathbf{S} + \oint \mathbf{B} \cdot \left(\frac{\mathbf{E} \times \mathbf{B}}{B^2} \times d\mathbf{s}\right)$$

$$= -\oint \mathbf{E} \cdot d\mathbf{s} + \oint \mathbf{E} \cdot d\mathbf{s} = 0 \tag{6-151}$$

Applying this result to the plasma boundary, one finds that the magnetic flux is "frozen" into the plasma.

Let us assume that the external field is increasing. To satisfy (6-151), the plasma area must decrease, so that particles drift inward. Since the flux contained inside the previously defined loop stays constant, the magnetic field in the guiding-center frames also increases. Introducing the transverse kinetic temperature $T = T_\perp = \langle \frac{1}{2}mv_\perp^2 \rangle$, because of the constancy of μ_m,

$$\frac{T(\xi)}{T(\xi_0)} = \frac{B(\xi)}{B(\xi_0)} \tag{6-152}$$

where ξ is the momentary guiding-center position. It follows from the conservation of flux in each strip in Fig. 6-16 that the change in volume (strip

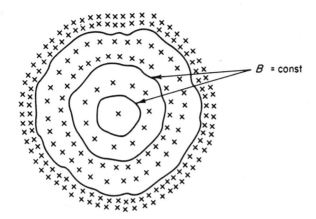

FIG. 6-16. Cross section of plasma column confined by a B_z field.

surface × length along **B**) is

$$\frac{V(\xi)}{V(\xi_0)} = \frac{B(\xi_0)}{B(\xi)} \tag{6-153}$$

or for the guiding-center density

$$\frac{n(\xi)}{n(\xi_0)} = \frac{B(\xi)}{B(\xi_0)} \tag{6-154}$$

The combination of (6-152) and (6-154) leads to the equation of state

$$T(\xi) = T(\xi_0) \frac{n(\xi)}{(n\xi_0)} \tag{6-155}$$

This is the adiabatic equation of state of a two-dimensional fluid ($\gamma = 2$). Since $p = p_\perp \sim nT$,

$$p(\xi) = p(\xi_0) \frac{n(\xi)T(\xi)}{n(\xi_0)T(\xi_0)} \tag{6-156}$$

and one obtains

$$p(\xi) = p(\xi_0) \frac{B^2(\xi)}{B^2(\xi_0)} \tag{6-157}$$

The magnetic field distribution can only be obtained by solving the self-consistent equations. At $t = 0$ the field $B(\xi_0)$ is related to the external field $B(\infty,0)$ through the pressure-balance equation

$$p(\xi_0) + \frac{B^2(\xi_0)}{2\mu_0} = \frac{B^2(\infty,0)}{2\mu_0} \tag{6-158}$$

The same equation holds at a time t at the new guiding-center position ξ:

$$p(\xi) + \frac{B^2(\xi)}{2\mu_0} = \frac{B^2(\infty,t)}{2\mu_0} \qquad (6\text{-}159)$$

Inserting (6-157) in (6-159) and using (6-158), it follows that

$$\frac{B(\xi)}{B(\xi_0)} = \frac{B(\infty,t)}{B(\infty,0)} \qquad (6\text{-}160)$$

This means that a v-fold increase of the external field results in a corresponding field increase in the coordinate system of each particle (but not in a fixed frame). By virtue of the constancy of μ_m, each particle experiences a v-fold energy increase, a similar increase of local particle density, and a v^2-fold increase in pressure.

Our results—adiabatic equations of state, flux conservation, etc.—are all consequences of the hydromagnetic properties of a two-dimensional guiding-center plasma (see Sec. 3-6).

In the general case in which the guiding-center approximation is not applicable, things are much more complicated. In the problem of planar symmetry discussed in Sec. 6-3, for example, there are two constants of motion, P_z and P_y, and an adiabatic invariant $\int P_x \, dx$, but the solution of the problem leads to complicated integral equations. The trouble lies in the fact that the microscopic problem—single particles moving self-consistently—cannot be simply reduced to macroscopic equations, dealing with densities kinetic temperatures, etc., as in the hydromagnetic case.

The difficulties are quite similar in the treatment of the pinch geometry in the nonhydromagnetic case. The distribution function (6-84) does not satisfy the time-dependent Boltzmann equation, even for slow time variation. While P_θ and P_z are constants, $\mathscr{H}$ no longer is, and if the distribution function looked like (6-88) at $t = 0$, it might change its entire character if a parameter, e.g., the pinch current, is changing. In the guiding-center limit, however, this problem can also be treated on a hydromagnetic basis.

The E-layer configuration discussed in Sec. 6-5 can be treated without difficulty for slow variations of the external field. Since P_θ is the common and strictly constant canonical momentum of all particles, and since with the slow field variation oscillations in the r direction do not arise in the potential well, the problem remains one-dimensional. The variation in the external field results in a corresponding change in R described by (6-125). Since the total number of particles remains constant, $N = N_0(R_0/R)$ and the varying particle velocity can be readily calculated from (6-128) (see Exercise 6-28).

6–8. Summary

A plasma in thermal equilibrium tends to shield electric fields with the characteristic Debye shielding distance

$$\lambda_D = \left(\frac{\varepsilon_0 kT}{q^2 n_0}\right)^{1/2}$$

If the plasma is situated in a magnetic field and has free plasma-vacuum boundaries (no thermal equilibrium), it behaves like a dielectric with the dielectric constant

$$\kappa_\perp = \frac{m_+ + m_-}{\varepsilon_0 B^2} n + 1$$

with respect to an electric field perpendicular to the magnetic field. This quantity is usually large enough to effectively shield the plasma interior from electric fields. This property of the guiding-center plasma is responsible for the fact that a collisionless plasma can propagate across a magnetic field over distances far exceeding the cyclotron radii of particles. The plasma when injected into a magnetic field with curved field lines, along the lines, will tend to move on a straight line instead of following field lines, as single particles would do.

A guiding-center plasma also tends to be diamagnetic, as a result of currents created by the gyrating particles. It is characterized by the quantity

$$\beta = \frac{p}{B^2/2\mu_0}$$

where p and B are the local pressure and magnetic field, respectively. β is usually much smaller than $\kappa_\perp$; hence a plasma of moderate density can have a high dielectric constant, but much higher densities are required for effective diamagnetism. For straight and parallel field lines the pressure-balance equation becomes

$$\frac{B^2}{2\mu_0} + p = \frac{B_0^2}{2\mu_0}$$

where B_0 is the external field and

$$\beta = (B_0/B)^2 - 1$$

In the pinch configuration no external field is applied; the field is generated by the particle currents. Several solutions to such a general configuration can be constructed in terms of self-consistent solutions of the collisionless Boltzmann equation, the constants of motion, and the field equations. An interesting solution is the Bennett pinch, where each species is in thermal equilibrium in its own frame and the two systems are in relative motion, generating the pinch current.

When the guiding-center approximation is abandoned, one can find solutions to the plasma–magnetic field boundary problem, the E layer, etc. One may have, for example, a plasma entirely free of magnetic field and in thermal equilibrium, separated by a self-consistent boundary layer from a uniform magnetic field. Solutions in which the field reverses over the boundary can also be found.

One may also confine a plasma (provided it is not too hot or dense) by a high-frequency field. The treatment of particle motion is based on the oscillation-center approximation solved self-consistently with the equations for high-frequency fields.

For slow variation of the fields (compared to cycles in particle motion) a quasi-steady-state analysis can be used to calculate the motion of guiding centers, boundaries, heating, etc. Such a calculation was carried out for a guiding-center plasma situated in a magnetic field with straight and parallel field lines.

EXERCISES

6-1. Derive the polarization current density (6-12) from the first moment of the Boltzmann equation (3-24) considering that the force density is $n\langle \mathbf{F} \rangle = \mathbf{J} \times \mathbf{B}$ and the mean velocity $\mathbf{u} = \mathbf{w}^E$.

6-2. Show that the density of the polarization charge layer is decreased $\kappa_\perp$-fold compared to the plasma density.

6-3. Calculate the distribution of $\mathbf{w}^E$ and the plasma energy stored in a cylindrical plasma capacitor.

6-4. Owing to the rotation of particles around the axis of a cylindrical plasma capacitor, a centrifugal drift arises.
(a) Calculate its value for ions and electrons.
(b) Compute the current produced by this drift and estimate its effect on distorting the imposed magnetic field.

6-5. In a two-stage plasma accelerator, a plasma drifts in the y direction between two charged rails as in Fig. 6-2, except that B decreases with increasing y, yielding a growing E/B drift velocity. Describe the process
(a) if the plasma is in touch with the electrodes,
(b) if the plasma is not in touch with the electrodes.

6-6. Derive equations (6-78) to (6-80) for the θ pinch.

6-7. Prove the pressure-balance equation (6-81), using (6-78) to (6-80).

6-8. Prove that equations (6-78) to (6-81) go over into the corresponding equations for the planar geometry if $r \to \infty$.

6-9. Show that in a cylindrical geometry for particles traveling radially in the absence of fields, there is a "geometric" decrease of radial pressure $dP_{rr}/dr = -P_{rr}/r$.

6-10. Prove from (6-93) that the pressure-balance equation for the pinch configuration is

$$\frac{d}{dr}\left[(n_+ + n_-)kT + \frac{B_\theta{}^2}{2\mu_0}\right] + \frac{B_\theta{}^2}{\mu_0 r} = 0$$

6-11. Prove that for a pinch configuration not neglecting charge separation, the pressure-balance equation is like the one obtained in Exercise 6-10, except that $B_\theta{}^2$ has to be replaced everywhere by $B_\theta{}^2 - E_r{}^2/c^2$.

6-12. The reason why the observer moving with the electron beam in the Bennett pinch sees a confining electric potential lies in relativistic effects. Prove that the density of a stream is smallest in a co-moving system. This is why the observer moving with the electrons sees the positive ions while the observer moving with the ions sees the electrons predominate.

6-13. Prove that (6-98) satisfies (6-97).

6-14. Consider an electron and ion stream with planar symmetry in the yz plane moving through each other with a relative velocity V in the y direction. In the center-of-mass system, charge neutrality prevails, while in the electron frame the distribution is Maxwellian. What is the magnetic field distribution? Show that at large positive and negative values of x the magnetic field goes over into $+B_0$ and $-B_0$, respectively.

6-15. Use the concept of the effective potential to derive (6-109).

6-16. Prove using (6-104) that the boundary between filled and empty regions on the $P_x P_y$ phase plane is a parabola, in the problem considered in Sec. 6-4. What is the equation of this parabola?

6-17. Use the $(\mathbf{r,v})$ phase space to derive the results of Sec. 6-4. What are the constants of motion in this phase space and what are the corresponding surfaces?

6-18. Derive (6-114) from (6-108).

6-19. Prove directly that the second term in (6-114) is P_{xx}.

6-20. Use the self-consistent equation (6-101) to directly prove the pressure-balance equation, which reads for this case,

$$P_{xx} + \frac{1}{2\mu_0}\left(\frac{dA}{dx}\right)^2 = P_0$$

6-21. Prove that the addition of trapped particles to the boundary layer, discussed in Sec. 6-4, makes this boundary thicker. (*Hint*: Use the pressure-balance equation from Exercise 6-18; express all quantities in terms of the vector potential, e.g., the thickness $\int dx = \int dA/(dA/dx))$.

6-22. A cold plasma is incident on a uniform magnetic field with $n = n_0$, $v_x = v_0$, $v_y = v_z = 0$, $B = 0$ at $x \to -\infty$, $B_z = B_0$ at $x \to +\infty$. Determine and integrate the self-consistent equations for the boundary layer. Assume that both kinds of particles have equal mass.

6-23. Set up and integrate the self-consistent equations for a plasma slab if the distribution function at $x = 0$ has the form $f(v_x, v_y, v_z) = \lambda \, \delta(E - E_0)$ $\delta(P_y)$ and at $x \to \pm\infty$, $B_z = $ const, $B_x = B_y = 0$. There are no trapped particles.

(a) Both particle species have the same mass.

(b) The ions are heavier than the electrons. Assume that the net charge of the slab is zero.

6-24. A plasma slab with a distribution function as described in the previous problem separates regions of opposing magnetic fields, i.e., $B_z = B_0$ when $x \to \infty$, $B_z = -B_0$ when $x \to -\infty$. Calculate the field distribution assuming $m^+ = m^-$ for simplicity.

6-25. Carry out the calculation of Sec. 6-5 (solenoidal-particle current layer) for relativistic electrons.

(a) Prove that the effective potential in this case is

$$\psi = m_0 c^2 \left[1 + \left(\frac{P_\theta - qrA_\theta}{m_0 cr} \right)^2 \right]^{1/2}$$

(b) Show that the radius of the solenoid becomes

$$R = -\frac{m_0}{q[(B_0 + B_{in})/2]} \frac{v}{[1 - (v^2/c^2)]^{1/2}}$$

6-26. A cold, cylindrical, uniform positronium plasma is situated in a uniform magnetic field B_0. By reversing the direction of the current in the coil, the field outside the plasma becomes $-B_0$. Calculate the field distribution in the plasma assuming that each particle gyrates at the minimum of the potential well after field reversal. Prove that the analytic solution breaks down as the number of particles per unit length in the column reaches a critical value. At this point a surface singularity is formed with diverging current density.

6-27. A plasma is "confined" in the region $-d < x < +d$ by plane electromagnetic standing waves of frequency ω. Each electron has the same velocity, $v_x = v_0$, in the plane $x = 0$. Derive the steady-state self-consistent equations using the equations of motion, assuming charge neutrality for the two plasma components.

6-28. Prove that by slowly varying the external field in the E layer
(a) the various quantities vary in the following way:

$$1/R \propto B_0^{1/2}, \quad N \propto B_0^{1/2}, \quad v \propto B_0^{1/2}, \quad B^s \propto B_0.$$

(b) the enclosed flux remains a constant.

Here $B_0(t)$ is the imposed external magnetic field.

VII

Oscillations and Waves in Uniform
Unmagnetized Plasmas

7-1. Electrostatic Oscillations

Since both electromagnetic fields in vacuum as well as gases in the absence of fields are known to exhibit oscillatory and wave-like behavior, an even richer variety of these phenomena can be expected in plasmas in which particle motion is coupled with the electromagnetic fields. Here, particle motion gives rise to fields and fields result in particle motion. The method applied is the usual self-consistent treatment.

We start with the simple case of a plasma with infinite extent, uniform and cold (no particle moves), without any external field present. This is obviously an equilibrium arrangement. Assume now that one takes a few electrons and displaces them slightly from their equilibrium position. As a result an electro-static field arises, pulling electrons from regions of higher density back to the partly depopulated regions. These forces would restore the original charge neutrality, were it not for the electron inertia, which keeps them moving, resulting in an oscillatory motion around the equilibrium. The ions, because of their much larger mass, can be assumed to remain nearly stationary during the process.

Since we are dealing with small deviations from equilibrium, the equations will be linearized, viz., all nonequilibrium quantities are assumed small (of the first order) and their product will be neglected. The equilibrium electron density n_0 is a zero-order quantity, while the current density

$$\mathbf{J} = q n_0 \mathbf{v} \tag{7-1}$$

is of the first order. The electron "fluid" with velocity distribution $\mathbf{v}(\mathbf{r},t)$ is accelerated by the electric field $\mathbf{E}$, also a first-order quantity. The equation of motion

$$\frac{q}{m}\mathbf{E} = \frac{d\mathbf{v}}{dt} = \frac{\partial \mathbf{v}}{\partial t} + (\mathbf{v} \cdot \nabla)\mathbf{v} \tag{7-2a}$$

becomes in the first order

$$\frac{q}{m} \mathbf{E} = \frac{\partial \mathbf{v}}{\partial t} = \dot{\mathbf{v}} \tag{7-2b}$$

The time derivative of Maxwell's equation

$$\nabla \times \dot{\mathbf{H}} = \dot{\mathbf{J}} + \ddot{\mathbf{D}} \tag{7-3}$$

becomes, using (7-1) and (7-2b) and neglecting higher-order terms,

$$\nabla \times \dot{\mathbf{H}} = q n_0 \dot{\mathbf{v}} + \ddot{\mathbf{D}} = \frac{q^2 n_0}{\varepsilon_0 m} \mathbf{D} + \ddot{\mathbf{D}} \tag{7-4}$$

Taking the divergence of this equation and remembering that $\nabla \cdot \mathbf{D} = q \delta n$, where δn is the first-order electron-density perturbation, one obtains

$$\frac{\partial^2}{\partial t^2} \delta n + \omega_p{}^2 \, \delta n = 0 \tag{7-5}$$

where

$$\omega_p{}^2 = \frac{q^2 n_0}{\varepsilon_0 m} \tag{7-6}$$

is the plasma frequency. Equation (7-5) describes an oscillatory motion of the electrons with this characteristic frequency. The density fluctuations are strictly localized in this approximation; they do not propagate. This fact is expressed by the absence of spatial derivatives in (7-5) and also by the vanishing group velocity $\partial \omega / \partial k$. If oscillations are initiated by displacing particles, those parts of the plasma left with either $\delta n \neq 0$ or $(\partial / \partial t) \, \delta n \neq 0$ will oscillate, while other regions remain unaffected.

The motion of ions can be easily included. Now the current, as described in (7-1), is composed of an electron and ion current with an equation of motion (7-2b) for both species. One finds (see Exercise 7-1) that the charge density oscillates with the frequency

$$\omega_p = \left[\left(\frac{q^2 n_0}{m \varepsilon_0} \right)^+ + \left(\frac{q^2 n_0}{m \varepsilon_0} \right)^- \right]^{1/2} = [(\omega_p{}^+)^2 + (\omega_p{}^-)^2]^{1/2} \tag{7-7}$$

Clearly, the electron term largely exceeds the ionic one. The electron plasma frequency, which lies in the microwave region for gaseous discharges ($n_0 = 10^{10} \text{ cm}^{-3} = 10^{16} \text{ m}^{-3}$, corresponds to $\omega_p \approx 10^9 \text{ sec}^{-1}$), increases with the square root of the density.

Consider now an infinite uniform electron plasma—with a fixed neutralizing ion background—in which the electrons have some kind of equilibrium velocity distribution $f_0(\mathbf{v})$, e.g., a Maxwellian distribution corresponding to a nonzero electron temperature. A small space-dependent perturbation of the electron plasma can again be expected to result in some sort of oscillatory or wave phenomenon as a result of the electric fields produced by the charge separation. To describe this process one replaces the equation of motion by the collisionless Boltzmann equation. To describe small deviations from equilibrium one writes

$$f(\mathbf{r},\mathbf{v},t) = f_0(\mathbf{v}) + f_1(\mathbf{r},\mathbf{v},t) \tag{7-8}$$

where the second term is a small perturbation quantity. Since electrostatic oscillations will be investigated, the $\mathbf{v} \times \mathbf{B}$ term does not appear, and our equation becomes, after neglecting second-order terms,

$$\frac{\partial f_1}{\partial t} + \mathbf{v} \cdot \frac{\partial f_1}{\partial \mathbf{r}} + \frac{q}{m} \mathbf{E} \cdot \frac{\partial f_0}{\partial \mathbf{v}} = 0 \tag{7-9}$$

$\mathbf{E}$ is, of course, again a first-order quantity, and it can be calculated from f_1 with the help of the Poisson equation

$$\varepsilon_0 \nabla \cdot \mathbf{E} = q \int f_1 \, d^3v \tag{7-10}$$

and

$$\nabla \times \mathbf{E} = 0 \tag{7-11}$$

since we are looking for electrostatic oscillations. Since the equilibrium distribution f_0 is given, the self-consistent equations (7-9) to (7-11) suffice to determine the unknown functions f_1 and $\mathbf{E}$ if the initial and boundary conditions are given.

We look now for oscillatory solutions in time for both f_1 and $\mathbf{E}$, assuming that they are proportional to $\exp(-i\omega t)$. The Boltzmann equation now becomes

$$i\omega f_1 = \mathbf{v} \cdot \frac{\partial f_1}{\partial \mathbf{r}} + \frac{q}{m} \mathbf{E} \cdot \frac{\partial f_0}{\partial \mathbf{v}} \tag{7-12}$$

Equation (7-10) can be integrated to obtain $\mathbf{E}$ and the result inserted into (7-12) to obtain a single equation in f_1. Because of the presence of the spatial derivatives, localized oscillations no longer exist. This can also be expected on physical grounds; electrons moving with their equilibrium velocity distribution will carry the disturbances with them.

The next step is to assume a spatial dependence of the form $\exp(i\mathbf{k} \cdot \mathbf{r})$ (see also Exercise 7-2). This means we look for traveling plane-wave solutions of the usual form: $\exp[i(\mathbf{k} \cdot \mathbf{r} - \omega t)]$. Equation (7-10) implies that if f_1 has this form, so does $\mathbf{E}$. Equations (7-9) to (7-11) now become

$$f_1 = \frac{i}{\mathbf{k} \cdot \mathbf{v} - \omega} \frac{q}{m} \mathbf{E} \cdot \frac{\partial f_0}{\partial \mathbf{v}} \tag{7-13}$$

$$i\mathbf{k} \cdot \mathbf{E} = \frac{q}{\varepsilon_0} \int f_1 \, d^3v \tag{7-14a}$$

and

$$\mathbf{k} \times \mathbf{E} = 0 \tag{7-15}$$

Integrating (7-13) over velocity space and writing $\mathbf{E} = (\mathbf{k}/k)E$,

$$\int f_1 \, d^3v = i \frac{q}{m} E \frac{\mathbf{k}}{k} \cdot \int \frac{1}{\mathbf{k} \cdot \mathbf{v} - \omega} \frac{\partial f_0}{\partial \mathbf{v}} \, d^3v \tag{7-16}$$

Using (7-14a) in (7-16), one obtains

$$1 = \frac{q^2}{m\varepsilon_0} \frac{\mathbf{k}}{k^2} \cdot \int \frac{\partial f_0}{\partial \mathbf{v}} \frac{1}{\mathbf{k} \cdot \mathbf{v} - \omega} \, d^3v \tag{7-17a}$$

The right side of this equation is a function of $\mathbf{k}$ and ω only, and according to (7-17a) each pair of $\mathbf{k}$ and ω should be given values to render this expression equal to 1. In other words, (7-17a) is a dispersion relation relating the wave numbers to the frequencies of our plane waves.

Unfortunately, this equation suffers from a serious flaw: Since the denominator of the integrand vanishes for certain values of $\mathbf{v}$, the integral is not defined. Choosing, for instance, a (real) $\mathbf{k}$, there will be different values of ω corresponding to it, unless a prescription is given as to how to integrate around the pole. Unfortunately, (7-17a) does not contain such a prescription and hence seems to be useless.

Physically, this difficulty is caused by electrons moving with exactly the phase velocity of the wave, since, for those, $\mathbf{k} \cdot \mathbf{v} = \omega$. Indeed, if there are no electrons of this kind, if f_0 and $\partial f_0/\partial \mathbf{v}$ vanish for these values of $\mathbf{v}$, the integral can be carried out and the dispersion relation becomes meaningful.

In this case, one can write (7-17a) in a different form by integrating by parts. Since $f_0(\mathbf{v})$ falls off rapidly with large values of the velocity, the surface integral in velocity space vanishes and (7-17a) becomes

$$1 = \frac{q^2}{m\varepsilon_0} \int \frac{f_0(\mathbf{v})}{(\mathbf{k} \cdot \mathbf{v} - \omega)^2} \, d^3v \tag{7-18}$$

The difficulties associated with the dispersion relation can be illuminated from a different angle by looking at the simpler case of a finite number of electron streams, each one traveling with a well-defined velocity V_σ in the same, say x, direction. In this case

$$f_0(\mathbf{v}) = \sum_\sigma n_\sigma \, \delta(\mathbf{v} - \mathbf{V}_\sigma) \tag{7-19}$$

If one considers only waves propagating in the x direction, the vector signs can be dropped. The integral in (7-18) can now be carried out easily, and it yields the dispersion relation

$$F(\omega, k) = \frac{q^2}{m\varepsilon_0} \sum_\sigma \frac{n_\sigma}{(kV_\sigma - \omega)^2} = 1 \tag{7-20}$$

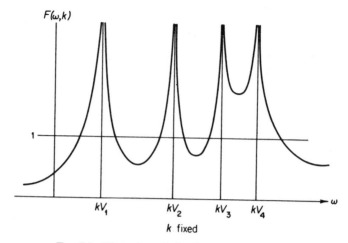

FIG. 7-1. Dispersion relation for electron streams.

Inserting an arbitrary value for k, one can solve for ω and obtain twice as many solutions for each k as the number of streams (Fig. 7-1), and the real roots can be read off the diagram directly. Some pairs of solutions may be complex. There are, in general, two "modes" of wave propagation between two streams adjacent in velocity space, with both of these waves traveling with phase velocities faster than the slower, and slower than the faster, of these streams. Two additional roots fall "outside" the streams.

If one thinks now of the continuous velocity distribution made up of an infinite number of streams, densely packed in velocity space, one expects an infinite number of roots of the dispersion relation to appear. For each k there will be an infinite number of waves whose phase velocities will cover the same region in velocity space as the particle velocities themselves. For an $f_0(\mathbf{v})$, like the Maxwellian distribution, all velocity space is covered and *for each* $\mathbf{k}$ *any* ω *is possible*. In this respect, no dispersion relation exists.

This is in no contradiction with (7-16) or (7-17). One can show that by choosing different ways to integrate around the singularity, one can indeed recover all these roots. From both the mathematical and the physical arguments, it is clear that this multiplicity of roots of the dispersion relation is due to electrons moving with (or nearly with) the phase velocity of a wave. There are two exceptions: the two roots which in Fig. 7-1 fall outside the streams. We shall see later that they also exist in the case of the continuous distribution (although now they do not necessarily fall "outside"), and they represent the real plasma oscillations.

Before separating the real plasma modes from the false ones, let us have a closer look at the latter. Consider a collisionless ideal gas consisting of neutral particles. If one selects a "stream" of particles with equal velocity in this gas and produces a density perturbation in this stream, leaving all other particles unperturbed, this perturbation will propagate with the stream. In case of a periodic perturbation with the wave vector **k**, in a stream of velocity **u**, a traveling wave of frequency $\omega = \mathbf{k} \cdot \mathbf{u}$ will arise, with the phase velocity **u**. Selecting different streams, any kind of wave with arbitrary **k** and ω can be produced, provided streams with any value of **u** are present. If the particles are charged, the situation is somewhat more involved, since the streams are coupled by electrostatic fields, but these modes, where a perturbation is carried " bodily " by a particle stream, are still present. They are our "false" plasma waves.

The way these modes can be produced is highly artificial and unnatural. One usually does not select particles of exactly equal velocities and displace them in space. One rather displaces particles in a region of space irrespective of their velocities. In this case, however, since different " streams" carry the perturbation with different velocities, it is easy to see that in a neutral gas the density perturbation will be smeared out and decay rapidly. In a plasma, however, one expects to find an organized oscillation (with a well-defined frequency) which corresponds to the plasma oscillation found in a cold plasma.

Hence, we reformulate our problem. Instead of looking at normal modes, which include too many "uninteresting" oscillations, we look at the initial-value problem, and ask: What kind of oscillations are sustained in a plasma launched with a "reasonable" initial perturbation?

The spatial variation of the perturbed distribution function and electric field can again be Fourier-decomposed into terms of the form

$$f_{\mathbf{k}}(\mathbf{v},t)e^{i\mathbf{k}\cdot\mathbf{r}} \quad \text{and} \quad \mathbf{E}_{\mathbf{k}}e^{i\mathbf{k}\cdot\mathbf{r}} \tag{7-21}$$

Without loss of generality we choose the x axis along the direction of the vector **k**. Dropping the index **k** for convenience, (7-9) becomes

$$\frac{\partial f_1}{\partial t} + ikv_x f_1 + \frac{q}{m} E_x \frac{\partial f_0}{\partial v_x} = 0 \tag{7-22}$$

while (7-10) again yields

$$ikE_x = \frac{q}{\varepsilon_0} \int f_1 \, d^3 v \tag{7-14b}$$

From (7-15) one finds $E_y = E_z = 0$. To treat the initial-value problem we introduce the Laplace transform of the functions $f_1(\mathbf{v},t)$ and $E(t) = E_x(t)$:

$$F(p,\mathbf{v}) = \int_0^\infty f_1(\mathbf{v},t)e^{-pt} \, dt \tag{7-23}$$

with the inverse transformation

$$f_1(\mathbf{v},t) = \frac{1}{2\pi i} \int\limits_{-i\infty+\sigma}^{+i\infty+\sigma} F(p,\mathbf{v})e^{pt}\, dp \tag{7-24}$$

and similar expressions for $E(t)$. The integration in (7-24) is to be performed along a straight line parallel to the imaginary axis, leaving all singularities of $F(p)$ to the left on the complex p plane. Taking the Laplace transform (multiplying by e^{-pt} and integrating over t) of (7-22) yields

$$(p + ikv_x)F(\mathbf{v},p) + \frac{q}{m} E(p) \frac{\partial f_0}{\partial v_x} = f_1(\mathbf{v},0) \tag{7-25}$$

where $f_1(\mathbf{v},0)$ is the initial perturbation. Equation (7-14b) becomes

$$ikE(p) = \frac{q}{\varepsilon_0} \int F(\mathbf{v},p)\, d^3v \tag{7-26}$$

$F(\mathbf{v},p)$ can be expressed from (7-25) and inserted into (7-26):

$$ikE(p) = \frac{q}{\varepsilon_0} \int \frac{f_1(\mathbf{v},0)}{p + ikv_x}\, d^3v - \frac{q^2}{m\varepsilon_0} E(p) \int \frac{\partial f_0/\partial v_x}{p + ikv_x}\, d^3v \tag{7-27}$$

Solving for $E(p)$ one obtains

$$E(p) = \frac{q}{\varepsilon_0 ki} \left[\int \frac{f_1(\mathbf{v},0)}{p + ikv_x}\, d^3v \right] \Big/ \left[1 + \frac{q^2}{im\varepsilon_0 k} \int \frac{\partial f_0/\partial v_x}{p + ikv_x}\, d^3v \right] \tag{7-28}$$

For any given equilibrium distribution f_0, and initial perturbation $f_1(\mathbf{v},0)$, (7-28) yields the electric field for all time (if the inverse transform is carried out). Of the integrals over velocity space one can formally carry out the integration over v_y and v_z. Introducing the notations $u = v_x$,

$$g(u) = \int f_1(\mathbf{v},0)\, dv_y\, dv_z \tag{7-29}$$

and

$$f_0(u) = \int f_0(\mathbf{v})\, dv_y\, dv_z \tag{7-30}$$

(7-28) becomes, finally,

$$E(p) = \frac{q}{i\varepsilon_0 k} \left[\int\limits_{-\infty}^{+\infty} \frac{g(u)}{p + iku}\, du \right] \Big/ \left[1 - \frac{q^2}{m\varepsilon_0 k} \int\limits_{-\infty}^{+\infty} \frac{\partial f_0}{\partial u} \frac{du}{ku - pi} \right] \tag{7-31}$$

Let us now carry out the inverse transformation,

$$E(t) = \frac{1}{2\pi i} \int\limits_{\sigma-i\infty}^{\sigma+i\infty} E(p)e^{pt}\, dp \tag{7-32}$$

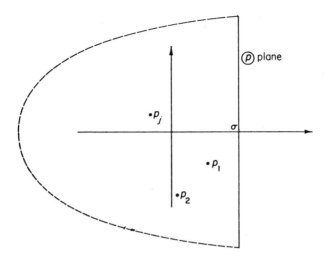

FIG. 7-2. Contour of integration to carry out inverse Laplace transformation.

This can be done by closing the contour in the manner shown in Fig. 7-2. The integral (as well as the integrand) vanishes on the dotted path. From (7-31), $E(p)$ vanishes as $1/p$ for large values of p, while e^{pt} vanishes exponentially for large negative real parts of p. The integral around the closed contour, however, can be expressed with the residua of the integrand at the poles within the contour,

$$E(t) = \sum_{j=0}^{N} \exp(p_j t) \operatorname{Res} E(p_j) \qquad (7\text{-}33)$$

The time dependence of the electric field has really the expected exponential form, except that beside the oscillations represented by the imaginary parts of p_j, there will, in general, also be a damping (if p_j is to the left of the imaginary axis) or amplification (if p_j falls to the right).

To evaluate the frequencies and damping (or amplification) one has to find the p_j's, the poles of (7-31). They fall into two categories: poles of the numerator and zeros of the denominator. The former depends on the precise form of the initial perturbation $g(u)$. For "reasonable" $g(u)$'s it can be shown that the numerator of $E(p)$ has no poles for finite values of p. For some artificially chosen initial perturbations, however, poles do appear, and they correspond to the "false" modes discussed before.

The denominator of $E(p)$, however, does not depend on the initial conditions, and its zeros represent clearly the oscillations we are looking for. These modes oscillate with the values of p_j determined as solutions of

$$H(p_j) = \frac{q^2}{m\varepsilon_0 k} \int_{-\infty}^{+\infty} \frac{df_0/du}{ku - ip_j} \, du = 1 \qquad (7\text{-}17b)$$

This is, however, formally identical with our previous dispersion relation (7-17a), with ip_j playing the role of ω [inserting $p_j = - i\omega$ into (7-33) leads really to the "normal mode" $e^{-i\omega t}$ time dependence]. The former difficulty, the uncertainty as to how to integrate around the poles, no longer arises. $H(p)$ is now an analytic function of p. It is actually part of the analytic function $E(p)$, defined originally only to the *right* of the poles. To define $H(p)$ and $E(p)$ everywhere is only a matter of finding their analytic continuation. This can be performed easily. For positive real parts of p, the integral in $H(p)$ can be carried out without difficulty (see Fig. 7-3). As the real part of p

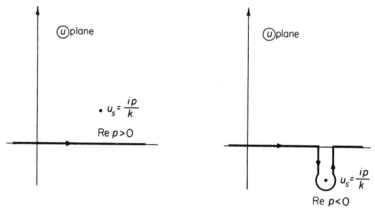

FIG. 7-3. Path of integration on the complex   FIG. 7-4. Path of integration on the com-
u plane for Re $p > 0$.   plex u plane for Re $p < 0$.

changes sign, $H(p)$ would be discontinuous if we retained the same integration path along the real u axis. This can be prevented only by keeping the pole always *above* the path of integration (see Fig. 7-4), leading to the desired analytic continuation of $H(p)$.

The values of p, where $H(p) = 1$, with the now unambiguously defined $H(p)$, yield the complex oscillation frequencies of the electric field.

The transform of the perturbed distribution function $F(\mathbf{v},p)$ can be expressed from (7-25). Unlike $E(p)$, this has singularities on the imaginary axis of p independent of the initial conditions (see Exercise 7-3). As a result, undamped oscillations of the distribution function will always be present.

In conclusion: *A cold electron plasma oscillates unattenuated with the plasma frequency ω_p irrespective of the nature of perturbation. A plasma with a continuous velocity distribution can sustain, in principle, oscillations with any frequency for any given wave number. Which one of these frequencies (or their superposition) one actually finds in a given case depends on the initial preparation of the plasma. It turns out, however, that very artificial preparation is* *required to establish these modes, which involves relatively large perturbations on some selected streams of particles. Unless one of these highly improbable initial conditions can be generated, one finds well-defined plasma oscillations whose (usually complex) frequencies are independent of the initial conditions. These can be calculated from the dispersion relation (7-17b) with the path of integration unambiguously given.*

7-2. Evaluation of the Dispersion Relation

As a first check we evaluate the dispersion relation for cold electrons with the equilibrium distribution function

$$f_0(\mathbf{v}) = n_0\,\delta(\mathbf{v}) \qquad (7\text{-}34)$$

This is just a special case of the multistream plasma, and we again use (7-18) as the dispersion relation. One obtains

$$1 = \frac{n_0 q^2}{m\varepsilon_0}\frac{1}{\omega^2} = \frac{\omega_p^2}{\omega^2} \qquad (7\text{-}35)$$

as expected. [The use of the dispersion relation (7-18) is justified *a posteriori*. With $\omega = \pm\omega_p$, the integrand of (7-17a) is not singular at $u = \omega/k$; hence integration by parts leading to (7-18) may be performed.]

As the next step we introduce an electron temperature low enough so that practically no electrons can be found to travel above a certain (finite) speed. In other words, a cutoff is introduced in the function $f_0(\mathbf{v})$ at a maximum speed $v_{\max}$. In this case, for sufficiently small wave numbers, such that $\omega/k > v_{\max}$, the integrand in the dispersion relation is again well behaved. It is convenient to use (7-18) to evaluate $\omega(k)$. Since the approximation used is valid only for small values of k, one can use $\mathbf{k}\cdot\mathbf{v}/\omega$ as an expansion parameter and find

$$1 = \frac{q^2}{m\varepsilon_0\omega^2}\int\left[1 + 2\frac{\mathbf{k}\cdot\mathbf{v}}{\omega} + 3\left(\frac{\mathbf{k}\cdot\mathbf{v}}{\omega}\right)^2 + \cdots\right]f_0(\mathbf{v})\,d^3v \qquad (7\text{-}36)$$

For a given distribution function these terms can be evaluated. The first term yields n_0, and without the higher-order terms (7-35) is recovered. For an isotropic distribution the second integral vanishes. Retaining only the first three terms, one obtains for an isotropic $f_0(\mathbf{v})$,

$$1 = \frac{\omega_p^2}{\omega^2}\left[1 + 3\frac{\langle(\mathbf{k}\cdot\mathbf{v})^2\rangle}{\omega^2}\right] = \frac{\omega_p^2}{\omega^2}\left[1 + \frac{k^2\langle v^2\rangle}{\omega^2}\right] \qquad (7\text{-}37)$$

This is a second-order equation in ω^2. Solving it, while neglecting higher than second powers in kv/ω, leads to the approximate dispersion equation

$$\omega^2 = \omega_p^2 + \langle v^2\rangle k^2 \qquad (7\text{-}38)$$

For a Maxwellian plasma with an electron temperature T,

$$\omega^2 = \omega_p^{\,2} + \frac{3\kappa T}{m} k^2 \tag{7-39}$$

where κ is the Boltzmann constant. This is the Bohm-Gross dispersion relation.

The phase velocity ω/k of these waves is very large, since the result is restricted to small values of k. The group velocity

$$v_g = \left|\frac{d\omega}{d\mathbf{k}}\right| = \frac{3\kappa T}{m}\frac{k}{\omega} = \frac{v_{\text{thermal}}^2}{v_{\text{phase}}} \tag{7-40}$$

has more physical significance, since it describes the velocity with which perturbations propagate. It is much smaller than the thermal speed.

In actuality (e.g., a Maxwellian distribution) every particle velocity is represented and the cutoff procedure is not justified. In such a case the exact dispersion equation (7-17b) is to be used. The most important deviation from the previous results is the appearance of an imaginary part of ω (real part of p). Equation (7-17b) can be evaluated easily for small-enough values of k, so as to shift the singularity to sufficiently high values of u, where $f_0(u)$ is small (but not zero). In this case, one can assume that the deviation from the previous dispersion relation (7-38) is small, hence the imaginary part of ω is small compared to its real part, the latter being approximately equal to the value obtained in (7-38). The singularity of the integrand in the dispersion relation is, therefore, located close to the real u axis, but at large values of u. Drawing a semicircle around the singularity, with a radius much larger than the distance to the real axis, but much smaller than the distance from the imaginary axis (Fig. 7-5), the integral in (7-17b) can be carried out in two steps. One integrates along the real axis from $-\infty$ to the semicircle and then around the

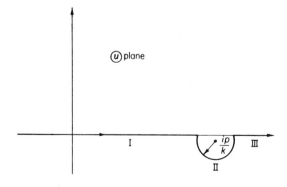

Fig. 7-5. Contour of integration for the evaluation of Landau damping.

semicircle. As stated, both f_0 and df_0/du are small around the singularity and they decrease further to the right; hence the third part (straight section to the right of the semicircle) of the integral is neglected.

The integral on path I has already been evaluated. It is just the integral with the cutoff distribution function, i.e., the right side of (7-37). The integral on path II is the residue of the integrand at the pole multiplied by $i\pi$ (half of the integral all round). The dispersion relation now takes the form

$$\frac{\omega_p^2}{\omega^2}\left[1 + \frac{k^2\langle v^2\rangle}{\omega^2}\right] + \frac{\omega_p^2}{k^2 n} i\pi f_0'\left(\frac{\omega}{k}\right) = 1 \qquad (7-41)$$

where $\omega = ip$ has been substituted.

To evaluate this equation it is convenient to rewrite it as

$$D(\omega, k) + D_1(\omega, k) = 0 \qquad (7-42)$$

where

$$D_1 = \omega_p^2\left\{\frac{k^2\langle v^2\rangle}{\omega^4} + \frac{i\pi}{k^2 n} f_0'\left(\frac{\omega}{k}\right)\right\}$$

is a small correction to $D = (\omega_p^2/\omega^2) - 1$. The roots of the dispersion relation will be in the neighborhood of the roots of D designated ω_0, so that in general $D(\omega_0, k_0) = 0$. Expanding (7-42) about ω_0 yields to lowest order

$$\left.\frac{\partial D}{\partial \omega}\right)_{\omega_0} \delta\omega + D_1(\omega_0, k_0) = 0 \qquad (7-43)$$

We are mostly interested in the imaginary part of $\delta\omega$,

$$\text{Im } \delta\omega = \gamma = -\text{Im } D_1(\omega_0, k_0)\left(\frac{\partial D}{\partial \omega}\right)_{\omega_0}^{-1} = \frac{\pi\omega_p^3 f_0'(\omega_0/k_0)}{2k^2 n} \qquad (7-44)$$

Inserting the distribution function of a Maxwellian plasma one obtains (Exercise 7-6)

$$\gamma_{\text{Maxwell}} = -\left(\frac{\pi}{8}\right)^{1/2} e^{-3/2} \frac{m^{3/2}\omega_p^4}{k^3(\kappa T)^{3/2}} \exp\left(-\frac{\omega_p^2 m}{2k^2\kappa T}\right) \qquad (7-45a)$$

Since γ is negative for such a distribution [because $f'(\omega/k) < 0$], the pole is located on the lower half-plane, as shown in Fig. 7-5, and the oscillation is damped. This phenomenon is known as *Landau damping*. If one introduces the Debye length λ_D from (6-6) into (7-45a), it takes a simpler form:

$$\gamma_{\text{Maxwell}} = -\left(\frac{\pi}{8}\right)^{1/2} e^{-3/2} \frac{\omega_p}{(k\lambda_D)^3} \exp\left[-\frac{1}{2(k\lambda_D)^2}\right] \qquad (7-45b)$$

This formula is valid as long as $\gamma/\omega_p \ll 1$, and this condition is satisfied for $k\lambda_D \ll 1$, namely, for wavelengths much larger than the Debye length. If the wavelength approaches λ_D, the damping factor is of the same order of magnitude as the oscillation frequency. This conclusion—which seems reasonable on the basis of (7-45b)—is confirmed by the numerical analysis of the exact dispersion relation. Therefore, *no organized oscillation with a wavelength shorter than the Debye length persists in a plasma.*

While a Maxwellian plasma supports damped plasma oscillations, this is not true for any kind of distribution function. If $\gamma > 0$, the oscillations grow exponentially and the plasma equilibrium is unstable (or, more precisely, overstable). It is useful to know which velocity distributions $f_0(\mathbf{v})$ are associated with growing modes, which ones with evanescent modes. The question can be formulated this way: For a given distribution function $f_0(u)$, can the integral

$$Z\left(\frac{\omega}{k}\right) = \int_{-\infty}^{+\infty} \frac{f'(u)\, du}{u - (\omega/k)} \tag{7-46}$$

take a real positive value for an ω in the upper half-plane [$\mathrm{Im}(\omega/k) > 0$ for $k > 0$]? If it can, then for some choice of k, (7-17b) will be satisfied with this ω and exponentially growing modes exist. If on the other hand, no such ω can be found, $f_0(u)$ is stable with respect to electrostatic oscillations.

The function Z is analytic and regular in the upper half-plane of its argu-

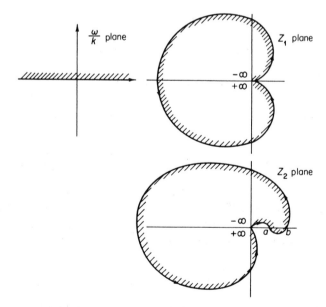

Fig. 7-6. Transformations of the upper half of the ω/k plane into regions of the Z plane. Z_1 corresponds to stable, Z_2 to unstable, configuration.

ment. It can be considered (for $\text{Im}(\omega/k) > 0$) as a transformation, mapping the upper half-plane of ω/k into a region of the Z plane. The mapping also transforms the real axis of the ω/k plane into the boundary of this region in the Z plane in such a way that by moving on the boundary from the "image" point of $\omega/k \to -\infty$ toward the image of $\omega/k \to +\infty$, the region representing the upper half of the ω/k plane is to the left (just as the upper half itself is on the left in the ω/k plane if one moves from $-\infty$ to $+\infty$ on the real axis). Since $Z(-\infty) = Z(+\infty) = 0$, the boundary starts and ends at the origin $Z = 0$. Figure 7-6 shows such mappings. In Z_1 the region $\text{Im}(\omega/k) > 0$ does not contain any point of the positive real axis of Z. Hence the dispersion relation cannot be satisfied for any one of these ω/k's and the configuration is stable. In Z_2 any point between a and b on the real axis corresponds to frequencies with a positive imaginary part which satisfy the dispersion relation; hence unstable oscillations can arise.

The stability criterion can also be formulated in a different way. Unstable oscillations are possible if (and only if) the image of the real axis of ω/k crosses the positive real axis of the Z plane. It is also obvious geometrically that if this happens there must be at least one crossing upward (such as at point b in Z_2).

To put this criterion into a mathematical form, we express $Z(\omega/k)$ on the real axis, moving again on a small semicircle below the singularity and expressing this part of the integral with the residue as we have done before,

$$Z(\xi) = \lim_{\varepsilon \to 0} \left[\int_{-\infty}^{\xi-\varepsilon} + \int_{\xi+\varepsilon}^{+\infty} \right] \frac{f_0'(u)\, du}{u - \xi} + i\pi f_0'(\xi) = P \int_{-\infty}^{+\infty} \frac{f_0'(u)\, du}{u - \xi} + i\pi f_0'(\xi) \quad (7\text{-}47)$$

where the notation ξ has been introduced for $\text{Re}(\omega/k)$ and P denotes the principal value. This $Z(\xi)$ is the equation of the mapping of the real axis ξ. Crossing of the real axis of Z occurs only if there is a ξ where

$$\text{Im } Z(\xi) = \pi f_0'(\xi) = 0 \quad (7\text{-}48)$$

Furthermore, for an upward crossing, $\text{Im } Z(\xi)$ changes from $-$ to $+$; hence, at this point, $f_0(\xi)\ [= f_0(u)]$ has a minimum. It is still possible that the crossing occurs somewhere on the negative real axis of Z, which is of no use. We require, therefore, that at this value of ξ, $\text{Re } Z(\xi) > 0$ for instability. The real part of Z is just the principal value. This becomes, after integrating by parts,

$$\text{Re } Z(\xi) = \lim_{\varepsilon \to 0} \left[\int_{-\infty}^{\xi-\varepsilon} + \int_{\xi+\varepsilon}^{+\infty} \frac{f_0(u)\, du}{(u - \xi)^2} + \frac{f_0(\xi - \varepsilon)}{-\varepsilon} - \frac{f_0(\xi + \varepsilon)}{\varepsilon} \right] \quad (7\text{-}49)$$

To put (7-49) into manageable form, we evaluate the integral

$$f_0(\xi)\lim_{\varepsilon \to 0} \left[\int_{-\infty}^{\xi-\varepsilon} + \int_{\xi+\varepsilon}^{+\infty} \right] \frac{du}{(u - \xi)^2} = 2\frac{f_0(\xi)}{\varepsilon} \quad (7\text{-}50)$$

Combining (7-49) with (7-50) one obtains

$$\mathrm{Re}\, Z(\xi) = P \int \frac{f_0(u) - f_0(\xi)}{(u - \xi)^2}\, du$$

$$+ \lim_{\varepsilon \to 0} \left[\frac{f_0(\xi) - f_0(\xi - \varepsilon)}{\varepsilon} - \frac{f_0(\xi + \varepsilon) - f_0(\xi)}{\varepsilon} \right] \quad (7\text{-}51)$$

The $\lim_{\varepsilon \to 0}$ expression is, however, $f_0'(\xi) - f_0'(\xi) = 0$. Since at $u = \xi$, $f_0(u)$ has a minimum, $\lim_{u \to \xi} [f_0(u) - f_0(\xi)]/(u - \xi)^2$ is finite, hence the principal value sign is unnecessary.

We conclude: *Exponentially growing modes exist if, and only if, there is a minimum of $f_0(u)$ at $u = \xi$, such that $\int_{-\infty}^{+\infty}[f_0(u) - f_0(\xi)](u - \xi)^{-2}\, du > 0$. This is known as the Penrose condition.*

Everything said so far applies also for multiple-component plasmas. The dispersion relation then becomes (see Exercise 7-5)

$$\sum_\sigma \frac{q_\sigma^2}{m_\sigma \varepsilon_0} \frac{\mathbf{k}}{k^2} \cdot \int \frac{\partial f_\sigma}{\partial \mathbf{v}} \frac{1}{\mathbf{k} \cdot \mathbf{v} - \omega}\, d^3 v = 1 \quad (7\text{-}52)$$

Using a reduced distribution function,

$$F(\mathbf{v}) = \frac{m_-}{q_-^2} \sum_\sigma \frac{q_\sigma^2}{m_\sigma} f_\sigma(\mathbf{v}) \quad (7\text{-}53)$$

reduces all multiple-component plasma equations into formulas of electron plasmas with the equivalent electron distribution function $F(\mathbf{v})$. It is clear from (7-53) that the ions appear with a small weighing factor due to their large mass.

It is apparent from the stability criterion above that distribution functions with a single hump, such as the Maxwellian distribution, cannot support growing oscillations, since there exists no minimum of $f_0(u)$. This is in agreement with the negative γ found previously for a Maxwellian plasma.

To illustrate growing modes, we investigate the case of two cold electron plasmas of equal density n moving with the relative velocity $2V$. The distribution function consists of two sharply peaked Gaussians, which go over into δ functions for $T \to 0$ (Fig. 7-7a). There is a minimum at $u = \xi = 0$ and the real part of Z:

$$\int_{-\infty}^{+\infty} \frac{f_0(u)}{u^2}\, du \approx n_0 \left[\frac{1}{(-V)^2} + \frac{1}{V^2} \right] = \frac{2n_0}{V^2} > 0 \quad (7\text{-}54)$$

is indeed positive; hence growing modes can exist. The dispersion relation (7-20) yields for the two streams

$$\frac{q^2 n}{m \varepsilon_0} \left[\frac{1}{(kV - \omega)^2} + \frac{1}{(kV + \omega)^2} \right] = 1 \quad (7\text{-}55)$$

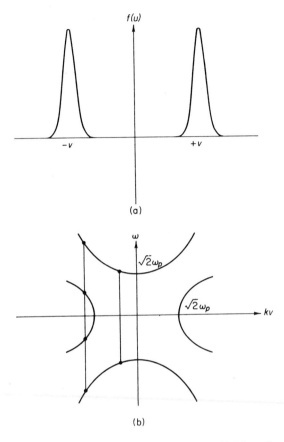

FIG. 7-7. (a) Distribution function for two streams. (b) Dispersion relation.

The $\omega(k)$ curve consists of four branches plotted in Fig. 7-7b. The intersections with the k axis are '

for $\omega = 0$: $$k = \pm\sqrt{2}\,\frac{\omega_p}{V} \qquad (7\text{-}56)$$

and with the ω axis

for $k = 0$: $$\omega = \pm\sqrt{2}\,\omega_p \qquad (7\text{-}57)$$

For short waves, $|k| > \sqrt{2}\,(\omega_p/V)$, there are four real roots; the solutions are purely oscillatory. For long waves $[|k| < \sqrt{2}(\omega_p/V)]$, in addition to two harmonic solutions, two imaginary solutions exist, yielding an exponentially decaying and a growing solution (see Exercise 7-7). It is interesting to note that the range of unstable wave numbers *decreases* with increasing relative velocity. The maximum growth rate is $\omega_p/2$, independent of V (see Exercise 7-8).

The phenomenon of exponentially growing oscillations in two streams in relative motion is known as the two-beam instability. Similar instabilities arise if the electron and ion components of a plasma are in relative motion.

A thermal spread in the streams has a tendency to reduce the two-stream instability, the roots of the dispersion relation become complex, instead of purely real and imaginary, and above a certain temperature the minimum of $f_0(u)$ at $u = 0$ is not "deep enough" and Re $Z(0)$ becomes negative. This change to stable modes takes place when the mean thermal velocity of the streams is roughly equal to V (see Exercise 7-10). Physically, it is a matter of competition between amplification and Landau damping, with the Landau damping winning out when the minimum of the $f_0(u)$ curve is sufficiently filled up.

One can gain a deeper physical insight into the mechanism of Landau damping and instabilities by investigating directly the interaction of the particle motion with the electrostatic wave. We have seen that the Landau damping is caused by particles traveling with a velocity nearly equal to the phase velocity of the wave. Consider an electrostatic mode in the frame moving with the phase velocity ω/k. In the absence of Landau damping and instability one finds a static, sinusoidal electric field and potential in this co-moving frame. For the sake of mathematical convenience, we replace the sinusoidal potential by a periodic triangular function (Fig. 7-8), since this will not alter our qualitative results. This potential is produced mainly by particles *not moving* with the phase velocity, constituting the bulk of the plasma. Let us now intro-

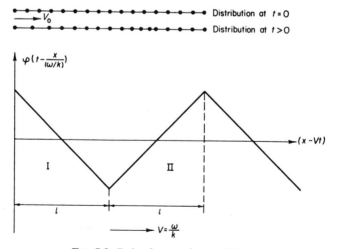

FIG. 7-8. Role of co-moving particles.

duce particles of uniform density, moving with the *small* relative velocity v_0 in this potential field at the time $t = 0$, when the perturbation has been created. Since they move in a potential field, their velocity will change after a short time to

$$v_{1,2} = v_0 + \frac{F_{1,2}}{m} t \tag{7-58}$$

where F_1 is the constant force in region I, F_2 in II ($F_2 = -F_1$). The displacement from the original position of a particle in region I or II is

$$\Delta x_{1,2} = v_0 t + \frac{F_{1,2}}{m} \frac{t^2}{2} \tag{7-59}$$

As a result, the number of particles in each region, which was originally nl in both, has now changed to

$$N_{\text{I}} = n(l - |\Delta x_1| + |\Delta x_2|) = n\left(l + \frac{F_2 - F_1}{m}\frac{t^2}{2}\right) = n\left(l - \frac{F_1}{m}t^2\right) \tag{7-60}$$

and

$$N_{\text{II}} = n(l - |\Delta x_2| + |\Delta x_1|) = n\left(l + \frac{F_1}{m}t^2\right) \tag{7-61}$$

The wave itself moves with $|\omega/k| \gg |v_{1,2}|$, hence the work done per unit time by the moving potential on the (essentially co-moving) particles can be written

$$W = (F_1 N_{\text{I}} + F_2 N_{\text{II}})\left(\frac{\omega}{k}\right) = -2\frac{F_1{}^2}{m}n\frac{\omega}{k}t^2 \tag{7-62}$$

This result has been obtained for the velocities v_0 and V pointing in the same direction. If the direction of v_0 is reversed, N_{I} and N_{II} change roles and W becomes positive, with the same absolute value. This leads to the conclusion that particles moving somewhat faster than the wave loose energy, while those moving slower gain energy from the field. From energy conservation, one concludes that the slow particles picking up energy contribute to (Landau) damping, the fast particles delivering energy to the wave, cause growing modes. Which one wins out depends therefore on the relative number of slower and faster particles in the neighborhood of ω/k—on the sign of $f_0'(\omega/k)$ in agreement with (7-44).

This qualitative reasoning can also be sharpened to a quantitative agreement with our formulas for Landau damping. Counting at the time t all those particle groups which traveled less than half a wavelength $|v_0| < l/t$ [since for the others (7-61) does not hold], one obtains the total power per wavelength,

$$P = -\frac{2\pi q^2 E^2}{m}\frac{\omega}{k^2}\lambda f'\left(\frac{\omega}{k}\right) \tag{7-63}$$

transmitted from the particles to the wave (Exercise 7-9). Here $\lambda = 2l$ and $F = qE$. Equating (7-63) with the rate of change of wave energy [which is according to Exercise 7-12 just twice the electrostatic energy, $\lambda \varepsilon_0(E^2/2)$], one obtains a differential equation for the time variation of E. This contains a damping factor which agrees to within a factor of 2 with γ (see Exercise 7-14).

The instability found in the case of the interpenetrating cold electron streams is of a different character. It is not associated with co-moving particles and, since Re $\omega = 0$, the phase velocity of the wave vanishes. It is actually not a wave, since it does not propagate, and not an oscillation either—rather an exponentially growing disturbance. Starting again with an initial perturbation, one has to investigate the effect produced by the two streams on the potential itself. Since particles moving in either stream are slowed down at the top of the potential hill (they lost kinetic energy "climbing up"), their density there will exceed the particle density at the bottom, where they move faster. The larger electron density, on the other hand, enhances the electrostatic potential. This feedback mechanism leads to the exponentially growing potential wave.

A perturbation in the potential Φ results in a change of particle velocity described by

$$\tfrac{1}{2}m(V + \delta v)^2 + q\Phi = \tfrac{1}{2}mV^2 \tag{7-64}$$

The equation of continuity yields, for an electron stream,

$$(V + \delta v)(n_0 + \delta n) = Vn_0 \tag{7-65}$$

Neglecting second-order terms, (7-64) and (7-65) give for the unbalanced charge density in this potential field due to both streams,

$$\rho_s = 2q\,\delta n = 2\,\frac{q^2 n_0}{mV^2}\,\Phi = 2\,\frac{\omega_p^2}{V^2}\,\varepsilon_0\Phi \tag{7-66}$$

On the other hand, the charge density necessary to maintain a periodic potential with the wave number k is, from Poisson's equation,

$$\rho_\Phi = \varepsilon_0 k^2 \Phi \tag{7-67}$$

Evidently a perturbation becomes unstable when $\rho_s > \rho_\Phi$, or when

$$k^2 < 2(\omega_p^2/V^2) \tag{7-68}$$

which is just the criterion found in (7-56).

7-3. The Plasma as Dielectric Medium

Consider an electron subjected to the electric field $\mathbf{E}_0 \exp[i(\mathbf{k}\cdot\mathbf{r} - \omega t)]$. In the frame moving with the unperturbed electron, the equation of motion is

$$\delta\ddot{\mathbf{r}} = \frac{q}{m}\,\mathbf{E}_0\exp(-i\omega't)\exp[i\mathbf{k}\cdot(\mathbf{r}_0 + \delta\mathbf{r})] \approx \frac{q}{m}\,\mathbf{E}_0\exp[i(\mathbf{k}\cdot\mathbf{r}_0 - \omega't)] \tag{7-69}$$

if the displacement $\delta\mathbf{r}$ is small. Here $\omega' = \omega - \mathbf{k} \cdot \mathbf{v}$ is the Doppler-shifted frequency felt by the particle moving with velocity $\mathbf{v}$ in the absence of perturbation, and $\mathbf{r}_0$ is the unperturbed position in the moving frame. The lowest-order particle displacement is now

$$\delta\mathbf{r} = -\frac{q}{m}\frac{\mathbf{E}_0\exp[i(\mathbf{k}\cdot\mathbf{r}_0 - \omega't)]}{\omega'^2} = -\frac{q}{m}\frac{\mathbf{E}}{(\omega - \mathbf{k}\cdot\mathbf{v})^2} \tag{7-70}$$

where $\mathbf{E} = \mathbf{E}_0\exp[i(\mathbf{k}\cdot\mathbf{r} - \omega t)]$, $\mathbf{r}$ being the unperturbed particle position in the laboratory frame. If this electron is one of the plasma electrons it leaves an unneutralized ion behind when displaced, producing a dipole moment

$$\mathbf{p} = q\delta\mathbf{r} = -\frac{q^2\mathbf{E}}{m(\omega - \mathbf{k}\cdot\mathbf{v})^2} \tag{7-71}$$

while the entire plasma gives rise to a dielectric polarization vector

$$\mathbf{P} = \sum \mathbf{p} = -\frac{q^2}{m}\mathbf{E}\int\frac{f_0\,d^3v}{(\omega - \mathbf{k}\cdot\mathbf{v})^2} \tag{7-72}$$

The dielectric displacement vector in the plasma may be written

$$\mathbf{D} = \varepsilon_0\mathbf{E} + \mathbf{P} = \varepsilon_0\mathbf{E}\left(1 - \frac{q^2}{m\varepsilon_0}\int\frac{f_0\,d^3v}{(\omega - \mathbf{k}\cdot\mathbf{v})^2}\right) = \varepsilon_0\kappa\mathbf{E} \tag{7-73}$$

But $\kappa = 0$ is just the dispersion relation found previously in the absence of resonant particles. Since there is no "real" charge present when the plasma is considered a dielectric, $\mathbf{D} = 0$, $\kappa = 0$ as expected. Resonant particles are not accounted for, since for those $\omega' = 0$ leading to a singularity in (7-70) as indeed they also did in the Boltzmann equation description. Resonant particles constitute the "lossy" part of the dielectric.

It is customary to define the plasma dielectric function

$$\varepsilon(\omega, \mathbf{k}) = 1 - \frac{q^2}{m\varepsilon_0 k^2}\mathbf{k}\cdot\int\frac{\partial f_0/\partial\mathbf{v}}{\mathbf{k}\cdot\mathbf{v} - \omega}d^3v \tag{7-74}$$

with the usual pole integration condition. Conforming to the notation customary in the literature ε rather than κ is used. This may be written

$$\varepsilon = 1 + \chi \tag{7-75}$$

where

$$\chi = -\frac{q^2}{m\varepsilon_0 k^2}\mathbf{k}\cdot\int\frac{\partial f_0/\partial\mathbf{v}}{\mathbf{k}\cdot\mathbf{v} - \omega}d^3v \tag{7-76}$$

is the dielectric susceptibility. It is easy to see from the linearized Boltzmann equation (7-9), also known as the Vlasov equation, that the charge density may be written

$$q \int f_1 \, d^3v = -ik\varepsilon_0 E\chi \tag{7-77}$$

It is of considerable interest to find the momentum and energy associated with plasma oscillations. Suppose that by means of an external grid system an electric wave $E = E(t) \sin(kx - \omega t)$ is slowly switched on the equilibrium plasma. While this field excites plasma oscillations, energy and momentum is being fed into the system, while $E(t)$ changes slowly from $E(-\infty) = 0$ to a finite value. Since it is sufficient to calculate the energy and momentum transmitted per wavelength, we multiply the Boltzmann equation by mv and $\frac{1}{2}mv^2$, respectively, integrate over velocities, and average over a wavelength. This yields

$$\frac{\partial}{\partial t} \left\langle \int mvf \, dv \right\rangle + q \left\langle E \int v \frac{\partial f}{\partial v} \, dv \right\rangle = 0 \tag{7-78}$$

and

$$\frac{\partial}{\partial t} \left\langle \int \tfrac{1}{2}mv^2 f \, dv \right\rangle + \frac{q}{2} \left\langle E \int v^2 \frac{\partial f}{\partial v} \, dv \right\rangle = 0 \tag{7-79}$$

where $\langle \ \rangle = (k/2\pi)\int_0^{2\pi/k} dx$ and the problem can be considered one dimensional along x. Integrating by parts one finds

$$\frac{dP}{dt} = q \left\langle E \int f \, dv \right\rangle \tag{7-80}$$

and

$$\frac{dK}{dt} = q \left\langle E \int vf \, dv \right\rangle \tag{7-81}$$

where P and K are the momentum and energy density, respectively. In order to evaluate the right-hand sides we have to solve for f using the Vlasov equation to lowest order:

$$\frac{\partial f_1}{\partial t} + v \frac{\partial f_1}{\partial x} = \frac{df_1}{dt} = -\frac{q}{m} f_0' E(t) \sin(kx - \omega t) \tag{7-82}$$

where we consider $f_1(t,x,v) = f_1[t,x_0 + vt,v]$, so the left-hand side is just the derivative of f_1 along the unperturbed particle trajectory $(x = x_0 + vt,$ $v = $ const$)$ in phase space. In order to solve (7-82), one merely integrates both sides along the unperturbed trajectories, to find

$$f_1(t,x,v) - f_1(-\infty,x,v) = -\frac{q}{m} \int_{-\infty}^{t} E(t') f_0' \sin(kx_0 + kvt' - \omega t') \, dt' \tag{7-83}$$

Here $f_1(-\infty,x,v) = 0$, and the right-hand side can be integrated by parts twice to yield

$$\frac{m}{q} f_1(t,x,v) = f_0' E(t) \frac{\cos(kx - \omega t)}{kv - \omega} - \int_{-\infty}^{t} \frac{dE}{dt'} f_0' \frac{\cos(kx_0 + kvt' - \omega t')}{kv - \omega} \, dt'$$

$$= f_0' E(t) \frac{\cos(kx - \omega t)}{kv - \omega} - \frac{dE}{dt} f_0' \frac{\sin(kx - \omega t)}{(kv - \omega)^2}$$

$$+ \int \frac{d^2 E}{dt'^2} f_0' \frac{\sin(kx_0 + kvt' - \omega t')}{(kv - \omega)^2} \, dt' \tag{7-84}$$

since $E(-\infty)$, $(dE/dt)(-\infty) = 0$. Now one substitutes $f = f_0 + f_1$ into (7-80) and (7-81). The f_0 term does not contribute since $\langle E \rangle = 0$. Similarly the first term in (7-84) vanishes after averaging over the wavelength since $\langle \sin(kx - \omega t) \cos(kx - \omega t) \rangle = 0$. Furthermore, since E varies very slowly, e.g., $E(t) = E_0 e^{\sigma t}$, where σ is vanishingly small, $d^2 E/dt^2$ can be made arbitrarily small; hence the last integral in (7-84) may be ignored. We do not concern ourselves here with the $\omega = kv$ poles, since we know that they contribute only to momentum and energy *exchange* (e.g., Landau damping), but if there are few resonant particles their contribution to the energy and momentum *content* of the plasma is negligible. For mathematical consistency one may simply take $f_0'(\omega/k) = 0$. Substituting finally (7-84) into (7-80) and (7-81) yields

$$\frac{dP}{dt} = -\frac{q^2}{4m} \int \frac{f_0' \, dv}{(kv - \omega)^2} \frac{dE^2}{dt} \tag{7-85}$$

$$\frac{dK}{dt} = -\frac{q^2}{4m} \int \frac{f_0' v \, dv}{(kv - \omega)^2} \frac{dE^2}{dt} \tag{7-86}$$

or after integration

$$P = -\frac{\omega_p^2}{n} \frac{\varepsilon_0 E^2}{4} \int \frac{f_0' \, dv}{(kv - \omega)^2} \tag{7-87}$$

$$K = -\frac{\omega_p^2}{n} \frac{\varepsilon_0 E^2}{4} \int \frac{f_0' v \, dv}{(kv - \omega)^2} \tag{7-88}$$

The electrostatic field itself does not carry momentum, but it has an energy density $\frac{1}{4}\varepsilon_0 E^2$. So the total energy density is

$$U = \frac{\varepsilon_0 E^2}{4}\left[1 - \frac{\omega_p^2}{n}\int\frac{f_0' v\, dv}{(kv - \omega)^2}\right]$$

$$= \frac{\varepsilon_0 E^2}{4}\left[1 - \frac{\omega_p^2}{kn}\int\frac{f_0'(kv - \omega)\, dv}{(kv - \omega)^2} - \frac{\omega}{k}\frac{\omega_p^2}{n}\int\frac{f_0'\, dv}{(kv - \omega)^2}\right] \quad (7\text{-}89)$$

Upon substitution of the plasma dielectric function (7-74), one finds

$$U = \frac{\varepsilon_0 E^2}{4}\left(\varepsilon + \omega\frac{\partial\varepsilon}{\partial\omega}\right) = \frac{\varepsilon_0 E^2}{4}\frac{\partial}{\partial\omega}(\varepsilon\omega) \quad (7\text{-}90)$$

This result is more general than the proof indicates. It holds for electromagnetic waves as well in a dispersive dielectric, with the modification

$$U = \frac{\varepsilon_0 E^2}{4}\frac{\partial}{\partial\omega}(\varepsilon\omega) + \frac{B^2}{4\mu_0} \quad (7\text{-}91)$$

If the electrostatic wave is an eigenmode of the plasma ($\varepsilon = 0$), the comparison of (7-87) and (7-89) leads to

$$\frac{U}{\omega} = \frac{P}{k} \quad (7\text{-}92)$$

This is again a result of general validity. Quantum mechanically one may associate N quanta with the wave, carrying $N\hbar\omega$ energy and $N\hbar k$ momentum. From this point of view (7-92) is immediately obvious. U/ω is called the "action" of the wave.

Note the energy (and momentum) associated with the wave need not be positive, and depends on the coordinate frame in which it is evaluated. In the wave frame $\omega = 0$; hence $\omega\,\partial\varepsilon/\partial\omega = 0$ and the wave energy vanishes. Or consider, e.g., a wave propagating in a cold beam. The energy of the wave will be positive when viewed from the laboratory frame when $\omega/k > V$ the beam velocity, or negative when $\omega/k < V$ (see Exercise 7-11).

7-4. Further Examples of Electrostatic Modes; Beam-Plasma System and Ion Waves

Obviously depending on the exact distribution function f_0, an infinite variety of dispersion relations can be obtained. Here we study a few of practical or theoretical interest.

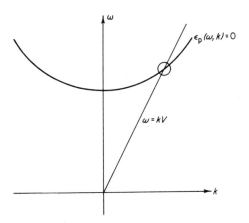

Fig. 7-9. Dispersion curves for the beam-plasma instability.

Suppose a weak electron beam is injected into a plasma, $f_0 = f_{0\,\text{plasma}} + f_{0\,\text{beam}}$. The total dielectric function is from (7-74)

$$\varepsilon(\omega,\mathbf{k}) = \varepsilon_{pl}(\omega,\mathbf{k}) - \frac{q^2\mathbf{k}}{m\varepsilon_0 k^2} \cdot \int \frac{(\partial/\partial\mathbf{v})f_{0\,\text{beam}}}{\mathbf{k}\cdot\mathbf{v} - \omega} d^3v$$

$$= \varepsilon_{pl}(\omega,\mathbf{k}) - \frac{q^2}{m\varepsilon_0} \int \frac{f_{0b}\, d^3v}{(\mathbf{k}\cdot\mathbf{v} - \omega)^2} \qquad (7\text{-}93)$$

Consider a cold beam $f_{0b} = n_b\delta(\mathbf{v} - \mathbf{V})$, where $\mathbf{V}$ is the beam velocity. The dispersion relation is

$$\varepsilon_{pl} - \frac{\omega_b{}^2}{(\mathbf{k}\cdot\mathbf{V} - \omega)^2} = 0 \qquad (7\text{-}94)$$

where $\omega_b = (q^2 n_b/m\varepsilon_0)^{1/2}$, the plasma frequency corresponding to the beam. The weak-beam approximation demands that this quantity be small; hence the second term in (7-94) is small, except of course in the neighborhood of $\omega \approx \mathbf{k}\cdot\mathbf{V}$. One expects the roots of (7-94) to be the roots of $\varepsilon_{pl} = 0$, everywhere except for $\mathbf{k}\cdot\mathbf{V} \approx \omega$, where the presence of the beam introduces a modification. Graphically, if one plots the plasma dispersion curve simultaneously with the $\omega = kV$ line, as in Fig. 7-9, the interesting roots of (7-94) can be expected to be found near the intersection of these lines. This intersection occurs at $\omega_0, \mathbf{k}_0$, where $\varepsilon_p(\omega_0,\mathbf{k}_0) = 0$ and $\omega_0 = \mathbf{k}_0 \cdot \mathbf{V}$. Expanding (7-94) about this point in $\omega = \omega_0 + \delta\omega$, yields for small $\delta\omega$

$$\left.\frac{\partial\varepsilon_{pl}}{\partial\omega}\right)_{\omega_0} \delta\omega - \frac{\omega_b{}^2}{\delta\omega^2} = 0 \qquad (7\text{-}95)$$

or

$$\delta\omega^3 = \frac{\omega_b{}^2}{\partial\varepsilon_{pl}/\partial\omega)_{\omega_0}} \tag{7-96}$$

Note that for simplicity we assumed ω_0 to be real. We have clearly three roots, one real and two complex conjugates. The latter are interesting since one of them has Im $\delta\omega > 0$ leading to instability, with the growth rate

$$\gamma = \frac{\sqrt{3}}{2}\,\omega_b^{2/3}\left(\frac{\partial\varepsilon_{pl}}{\partial\omega}\right)^{-1/3} \tag{7-97}$$

The evaluation of the frequency for a simple case is left for Exercise 7-15. Similarly one may calculate the spatial growth rate for a real frequency and complex k. This is also left as an exercise (Exercise 7-16).

The physical mechanism of the beam-plasma instability is quite simple. In discussing Landau damping we have already seen that a beam propagating somewhat faster than an electrostatic wave delivers energy to the wave. In the presence of the beam the frequency of the unstable wave is downshifted (see Exercise 7-15); consequently the phase velocity of the wave is somewhat below the beam velocity $V = \omega_0/k_0$.

Another kind of wave of considerable interest is the ion wave. We have seen before that adding finite mass ions to a cold electron plasma merely results in a slight frequency shift [see (7-7)] but produces no new wave. If the electrons have a nontrivial (say thermal) velocity distribution a new type of wave emerges. The dielectric function of an electron and ion plasma may be written

$$\varepsilon = 1 - \frac{\omega_{pe}^2}{nk^2}\int\frac{f'_{0e}\,dv}{v - \omega/k} - \frac{\omega_{pi}^2}{nk^2}\int\frac{f'_{0i}\,dv}{v - \omega/k} = 1 + \chi_e + \chi_i \tag{7-98}$$

where the index e refers to the electron, i to the ion component. The velocity v is in the direction of the wave vector and the f_0's are reduced distribution functions (viz., they have been integrated over the other two velocity coordinates). The ions will be considered cold, $f_{0i} = n\delta(v)$, while the electrons have the kinetic temperature T (in energy units), $f_{0e} = \beta e^{-\alpha v^2}$, $\alpha = m/2T$, $\beta = n\sqrt{\alpha/\pi}$. The ion susceptibility is easily evaluated by the usual partial integration to yield

$$\chi_i = -\frac{\omega_{pi}^2}{\omega^2} \tag{7-99}$$

The electronic part is not so easy. If $\omega/k \gg \alpha^{-1/2}$, the thermal velocity, we end up with a slightly modified version of (7-7), a rather uninteresting development. Let us try the opposite approach $\omega/k \ll \alpha^{-1/2}$. This permits a small

ω/k expansion in the principal value integral, while the pole integral is the by now familiar semicircle integral,

$$
\chi_e = \frac{\omega_{pe}^2}{n} \frac{2\alpha\beta}{k^2} P \int \frac{v \exp(-\alpha v^2)\, dv}{v - \omega/k} - i\pi \frac{\omega_{pe}^2}{nk^2} f'_{0e}\left(\frac{\omega}{k}\right)
$$

$$
= \frac{\omega_{pe}^2}{n} \frac{2\alpha\beta}{k^2} \left[\int \exp(-\alpha v^2)\, dv + O\left(\frac{\omega}{k}\alpha^{1/2}\right) \right] + 2i\pi \frac{\omega_{pe}^2}{nk^2}\alpha\beta \frac{\omega}{k} \quad (7\text{-}100)
$$

The imaginary term itself is of order $(\omega/k)\alpha^{1/2}$, so to lowest order

$$
\varepsilon = 1 + \frac{\omega_{pe}^2 m}{k^2 T} - \frac{\omega_{pi}^2}{\omega^2} = 1 + \frac{1}{k^2\lambda_D^2} - \frac{\omega_{pi}^2}{\omega^2} = 0 \quad (7\text{-}101)
$$

where $\lambda_D = \sqrt{T/m}\,\omega_{pe}^{-1}$ is the Debye length, as defined in (6-6). This leads to the frequency of ion waves

$$
\omega_0 = \frac{\omega_{pi} k \lambda_D}{\sqrt{1 + k^2\lambda_D^2}} = \frac{c_s k}{\sqrt{1 + k^2\lambda_D^2}} \quad (7\text{-}102)
$$

where $c_s = \omega_{pi}\lambda_D = \sqrt{T/M}$ is the ion sound velocity (M is the ion mass). The ion sound velocity is approximately the thermal speed the *ions would have* if the electron and ion temperature were equal. In fact, by our assumption the ion thermal speed is zero. Since $\omega_0/k \le c_s \ll \alpha^{-1/2}$, our small ω/k expansion was justified.

Considering now terms of order $(\omega/k)\alpha^{1/2}$, the real term in (7-100) gives an uninteresting small frequency correction, while the imaginary part gives an equally small but more interesting damping term. Writing $\varepsilon = \varepsilon_0 + \varepsilon_1$ and expanding the usual way about ω_0 yields the damping increment

$$
\gamma = \mathrm{Im}\,\delta\omega = -\frac{\mathrm{Im}\,\varepsilon_1(\omega_0,k)}{\partial\varepsilon/\partial\omega)_{\omega_0}} = -\sqrt{\pi}\,\frac{\omega_0^4}{k^3}\left(\frac{m}{2T}\right)^{3/2}\frac{M}{m}
$$

$$
= -\sqrt{\frac{\pi m}{8M}}\,\frac{\omega_0^4}{k^3}c_s^{-3} = -\sqrt{\frac{\pi m}{8M}}\,\frac{k c_s}{(1 + k^2\lambda_D^2)^2} \quad (7\text{-}103)
$$

Indeed $|\gamma/\omega_0| = (\sqrt{\pi m/8M})(1 + k^2\lambda_D^2)^{-3/2} \ll 1$ since $\sqrt{m/M} \ll 1$. This damping is the result of electron Landau damping on ion waves. If the ions are not completely cold, ion waves can persist as long as

$$
\sqrt{\frac{T_i}{M}} \ll \frac{c_s}{\sqrt{1 + k^2\lambda_D^2}} = \sqrt{\frac{T_e}{M}}\,\frac{1}{\sqrt{1 + k^2\lambda_D^2}} \quad (7\text{-}104)
$$

leading to the condition

$$
\frac{T_e}{T_i} \gg 1 + k^2\lambda_D^2 \quad (7\text{-}105)
$$

If this condition is violated, Landau damping on ions imposes severe damping of the waves. Experimentally, ion waves can be observed as long as $T_e/T_i \gtrsim 5$.

For long wavelengths $k\lambda_D \ll 1$, the ion waves are almost dispersionless $\omega/k \approx c_s$, while for $k\lambda_D \gg 1$, $\omega \approx \omega_{pi}$. The dispersion curve is shown on Fig. 7-11.

Consider now a situation where the plasma carries a dc current so that the electrons drift with a velocity u with respect to the ions, and u is small with respect to the electron thermal velocity. The electron distribution function is now $f_{0e} = \beta \exp[-\alpha(v - u)^2]$, where α and β have the values given before. The electronic susceptibility is now to lowest order

$$\chi_{e0} = \frac{\omega_{pe}^2}{n} \frac{2\alpha\beta}{k^2} P \int \frac{(v - u) \exp[-\alpha(v - u)^2]}{v - \omega/k} dv$$

$$= \frac{\omega_{pe}^2}{n} \frac{2\alpha\beta}{k^2} P \int \frac{(v - u) \exp[-\alpha(v - u)^2]}{(v - u) - (\omega/k - u)} dv = \frac{\omega_{pe}^2}{n} \frac{2\beta}{k^2} \sqrt{\alpha\pi} \quad (7\text{-}106)$$

since $(\omega/k - u)\alpha^{1/2} \ll 1$. This is the same result obtained in (7-100) in the absence of a drift; hence ω_0 is unchanged. On the other hand, there is a significant change in Im ε_1

$$\text{Im } \varepsilon_1 = 2\pi \frac{\omega_{pe}^2}{nk^2} \alpha\beta \left(\frac{\omega_0}{k} - u\right) \quad (7\text{-}107)$$

As u approaches ω_0/k the damping decreases to zero. When $u > \omega_0/k$, Im $\varepsilon_1 < 0$ and an instability ensues with the growth rate

$$\gamma = \sqrt{\frac{\pi m}{8M}} \frac{ku - \omega_0}{(1 + k^2\lambda_D^2)^{3/2}} \quad (7\text{-}108)$$

The verification of this formula is easy and it is left for Exercise 7-18. If $\mathbf{k}$ and $\mathbf{u}$ are not collinear $\mathbf{k} \cdot \mathbf{u}$ replaces ku (Exercise 7-19). The situation is reminescent of the beam-plasma problem where the cold drifting ions play the role of the beam. It comes as no surprise that the condition for instability is $u > \omega/k$.

In order to understand the physical mechanism involved in ion waves, consider first the phenomenon of sound waves in an ordinary gas. If a density perturbation is created in a gas, the concomitant pressure perturbation creates a restoring force to redistribute the particles uniformly, while the particle mass represents the inertia necessary for oscillatory behavior. In ion waves a perturbation of the ion density creates a space charge to which electrons respond very rapidly (on the ω_{pe}^{-1} time scale), neutralizing any excess charge created by the ions. While the ions are cold they do not themselves contribute to pressure, but the perturbed warm electron distribution does and creates the restoring force. The electron density distribution is, however, locked to that of the ions; hence any response to the restoring force must move the ions as well, whose inertia is much larger than that of the electrons. Hence the electron pressure (or temperature) appears in the restoring force and the ion mass in the inertial term, as we have found in (7-102) and the definition of c_s. The locking of electron and ion distributions is complete, however, only when $\lambda \gg \lambda_D$ or $k\lambda_D \ll 1$. As $\lambda \sim \lambda_D$ this behavior is modified, and as $k\lambda_D \gg 1$ the ions oscillate in a uniform electron background at the frequency ω_{pi}, in agreement with our dispersion relation.

It is instructive to derive the ion wave equation from the fluid equations. These are essentially the moment equations derived in Chapter III; hence they contain less information then the Boltzmann equation, but are often useful to get quick approximate results. The linearized equation of continuity is for each species

$$\frac{\partial n}{\partial t} + n_0 \frac{\partial v}{\partial x} = 0 \qquad (7\text{-}109)$$

where v is the fluid velocity, n_0 is the equilibrium density, and n is the density perturbation. The linearized momentum conservation equation for electrons, with scalar pressure p_e is

$$n_0 m \frac{\partial v_e}{\partial t} = -eEn_0 - \frac{\partial p_e}{\partial x} \qquad (7\text{-}110)$$

where e is the absolute value of the electron charge. For cold ions $p_i = 0$ so one gets

$$n_0 M \frac{\partial v_i}{\partial t} = eEn_0 \qquad (7\text{-}111)$$

The electron pressure and density are related through the adiabatic law $[(4\text{-}9c)]p_e(n_0 + n_e)^{-\gamma} = $ const to yield after differentiation and linearization

$$\frac{\partial p_e}{\partial x} = \gamma p_0 n_0^{-1} \frac{\partial n_e}{\partial x} \qquad (7\text{-}112)$$

where p_0 is the equilibrium electron pressure. Finally Poisson's equation relates the electric field to the densities

$$\frac{\partial E}{\partial x} = \frac{e}{\varepsilon_0}(n_i - n_e) \qquad (7\text{-}113)$$

Fourier analysis of these equations leads to

$$\omega n_{e,i} = k n_0 v_{e,i} \tag{7-114}$$

$$i n_0 m \omega v_e = e E n_0 + i\gamma \frac{p_0}{n_0} k n_e \tag{7-115}$$

$$-i M \omega v_i = e E \tag{7-116}$$

and

$$i k E = \frac{e}{\varepsilon_0} (n_i - n_e) \tag{7-117}$$

From (7-114) and (7-115)

$$n_e = \frac{-i k e E n_0^2}{n_0 m \omega^2 - \gamma p_0 k^2} \tag{7-118}$$

while from (7-114) and (7-116)

$$n_i = \frac{i k n_0 e E}{M \omega^2} \tag{7-119}$$

Substituting n_e and n_i into Poisson's equation yields the dispersion relation

$$1 = \frac{\omega_{pi}^2}{\omega^2} + \frac{\omega_{pe}^2}{\omega^2 - \lambda k^2} \tag{7-120}$$

where $\lambda = \gamma p_0 / m n_0 = \frac{1}{3} \gamma \langle v^2 \rangle$ since for isotropic pressure $p_0 = \frac{1}{3} n m \langle v^2 \rangle$. If $\omega \gg \omega_{pi}$ the equation becomes

$$\omega^2 = \omega_{pe}^2 + \frac{1}{3} \gamma \langle v^2 \rangle k^2 \tag{7-121a}$$

For one-dimensional compression $\gamma = 3$, and this is just the Bohm-Gross equation for electron plasma oscillations (7-38). If, on the other hand, $\omega \ll \omega_{pe}$ (7-120) becomes

$$1 = \frac{\omega_{pi}^2}{\omega^2} - \frac{3 \omega_{pe}^2}{\gamma \langle v^2 \rangle k^2} \tag{7-121b}$$

Since $m \langle v^2 \rangle = 3T$, this agrees with the ion wave equation (7-101), provided $\gamma = 1$, which holds for an isothermal gas [see (4-9b)]. The requirement that the electron gas behave isothermally in ion waves is not unreasonable, since the wave velocity is much smaller than the thermal velocity the same Maxwellian distribution is maintained throughout the plasma.

Damping depends on the details of the distribution function; hence the moment equations cannot account for it. The same holds for the instability due to slow drifting of electrons through ions. Another instability, however, if the relative drift velocity is large, is of hydrodyanmic origin and follows from the fluid equations as well as the Vlasov equation. It is left for an exercise to show that if the drift velocity exceeds the electron thermal velocity an instability ensues (Exercise 7-17). This is essentially the beam-plasma instability, where the cold drifting ions play the role of the beam.

It is easy to see from (7-120) that no solutions exist for $\omega_{pi} < |\omega| < \sqrt{\omega_{pe}^2 + \omega_{pi}^2}$, hence the only waves described by this equation are the electron and ion waves (Exercise 7-20). This is also clear by inspection since (7-120) is quadratic in ω^2, with only two solutions. If, on the other hand, one solves the dispersion relation for k^2, one finds two conjugate imaginary roots for k in the forbidden ω range. We are accustomed to associating such roots with a spatially growing instability. Since we have just concluded that no waves exist in this region of space, a more careful further investigation is called for.

Complex wave numbers often result in solving wave propagation problems in other areas of physics. The propagation of an electromagnetic wave in a waveguide below the cutoff frequency yields two conjugate imaginary k roots, propagation in a conductive medium yields complex k's etc. In these cases, one simply dismisses the root with a negative imaginary part corresponding to growth in the positive x direction, on valid physical grounds. But what is the physical situation described by these solutions? Suppose a source of radiation is situated at a large positive x ($x \rightarrow \infty$) in an attenuating medium. The evanescent wave generated by this source propagating in the negative x direction looks like an amplifying wave in the positive x direction. Since no boundary condition was set at $x \rightarrow \infty$ these unphysical solutions make their appearance in the solution of the dispersion relation, and are rightfully dismissed.

In plasma physics, however, where genuine spatially amplified waves may exist, a method is required to distinguish these solutions from fake amplification.

Suppose a localized wave source oscillating with a frequency ω and with an exponentially growing amplitude is placed in a medium. If the medium propagates waves at this frequency without attentuation or amplification, the response to the exponentiating source will be a wave train with an amplitude decreasing away from the source. (The wave further from the source has been emitted at an earlier time, when the source amplitude was smaller). If now the medium is an amplifying one for a constant amplitude source, a sufficiently rapidly growing source should turn the wave into a spatially decaying one from the source. This consideration can be used to distinguish real from fake amplification in the following manner: When the solution of the dispersion relation $k = k(\omega)$ presents a root in the lower half of the k plane for real ω, add a positive imaginary part to ω (time growth) and recompute k. If, for a sufficiently large positive imaginary part of ω, the corresponding k moves to

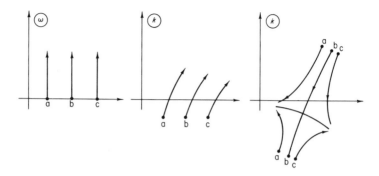

FIG. 7-10. Convective and absolute instability.

the upper half-plane, the wave in question was a genuinely amplifying one. If however, k stays in the lower half-plane, the wave is not really amplifying, but evanescent.

This process which consists essentially of mapping the ω plane into the k plane as shown in Fig. 7-10, encounters an occasional difficulty. At some complex value of ω a double root in k may be encountered arising from the merger of the two different k roots corresponding to the same complex ω. It can be shown, but the proof is not given here, that if the double root arises from the merger of roots coming from different halves of the k plane, the wave exhibits what is called an absolute instability. A wave packet perturbation in some region of space grows *locally* in time. The amplifying wave discussed previously convects and amplifies in space; hence the wave packet perturbation dies out in the region of its original excitation after some time. The amplifying wave is often called a convectively unstable one.

7-5. Transverse Waves in an Infinite Plasma

The phenomenon of electrostatic or plasma oscillations is typical of the plasma state, and it has no parallel in electromagnetic theory. One may ask what happens if electromagnetic waves are propagated in a plasma rather than in free space. At first we investigate again the case of a cold neutral plasma of infinite extent with the uniform particle density n_0. One expects that the electromagnetic fields of the wave set the plasma particles in motion, giving rise to currents. Ions, because of their larger inertia, will contribute less to this current than electrons, and they will, therefore, be neglected in this approximation.

The plasma current density

$$\mathbf{J} = q n_0 \mathbf{v} \tag{7-122}$$

is again inserted into the time derivative of Maxwell's $\nabla \times \mathbf{H}$ equation, to yield after linearization

$$\nabla \times \dot{\mathbf{H}} = qn\dot{\mathbf{v}} + \ddot{\mathbf{D}} \qquad (7\text{-}123)$$

By expressing $\dot{\mathbf{v}}$ from the equation of motion, the $\mathbf{v} \times \mathbf{B}$ term is neglected as a second-order contribution, and (7-123) becomes

$$\nabla \times \dot{\mathbf{H}} = \left(\frac{q^2 n_0}{\varepsilon_0 m} + \frac{\partial^2}{\partial t^2}\right)\mathbf{D} \qquad (7\text{-}124)$$

Looking at phenomena oscillatory in time, $\partial^2/\partial t^2 \to -\omega^2$, and one obtains, using $\nabla \times \mathbf{E} = -\dot{\mathbf{B}}$,

$$-\nabla \times (\nabla \times \mathbf{D}) = \varepsilon_0 \mu_0 (\omega_p{}^2 - \omega^2)\mathbf{D} \qquad (7\text{-}125)$$

This equation can be rewritten

$$-\nabla(\nabla \cdot \mathbf{D}) + \nabla^2 \mathbf{D} = \frac{\omega_p{}^2 - \omega^2}{c^2}\mathbf{D} \qquad (7\text{-}126)$$

Looking for plane-wave solutions with a spatial variation $\exp(i\mathbf{k} \cdot \mathbf{r})$ (7-126) becomes

$$\mathbf{k}(\mathbf{k} \cdot \mathbf{D}) - k^2\mathbf{D} = \frac{\omega_p{}^2 - \omega^2}{c^2}\mathbf{D} \qquad (7\text{-}127)$$

In general the vector $\mathbf{D}$ can be decomposed into a component perpendicular to $\mathbf{k}$ (transverse part) and one parallel to it (longitudinal part). The transverse part yields the electromagnetic plane wave propagating in the plasma. The dispersion relation associated with this wave is

$$k^2 c^2 = \omega^2 - \omega_p{}^2 \qquad (7\text{-}128)$$

while for the longitudinal part the left side of (7-127) vanishes, and one obtains the plasma oscillations with their well-known dispersion relation

$$\omega_p{}^2 - \omega^2 = 0 \qquad (7\text{-}129)$$

For the transverse part (waves), one can see from (7-126) that the presence of the plasma can be described in terms of the index of refraction

$$n = \frac{kc}{\omega} = \left(1 - \frac{\omega_p{}^2}{\omega^2}\right)^{1/2} \qquad (7\text{-}130)$$

The striking feature of the wave propagation in a plasma is the absence of propagation at frequencies $\omega < \omega_p$. At low frequencies n and k become imaginary and the solution chosen in the $\exp(i\mathbf{k} \cdot \mathbf{r})$ form leads to exponential decay. Physically, this cutoff results from the plasma currents becoming equal and opposite to the displacement current.

The phase velocity

$$v_f = \frac{\omega}{k} = [c^2 + (\omega_p^2/k^2)]^{1/2} \tag{7-131}$$

always exceeds, while the group velocity

$$v_g = \frac{\partial \omega}{\partial k} = \frac{k}{\omega} c^2 = \frac{c^2}{v_f} \tag{7-132}$$

is always less than the speed of light.

If the plasma has a nontrivial zero-order velocity distribution (it is "hot"), one uses the collisionless Boltzmann equation in conjunction with Maxwell's equations.

From the latter follows the wave equation

$$\nabla \times (\nabla \times \mathbf{E}) = \nabla (\nabla \cdot \mathbf{E}) - \nabla^2 \mathbf{E} = -\mu_0(\dot{\mathbf{J}} + \varepsilon_0 \ddot{\mathbf{E}}) \tag{7-133}$$

We look for traveling-wave solutions with a space and time dependence of the form $\exp[i(\mathbf{k} \cdot \mathbf{r} - \omega t)]$. In general, quantities will appear with transverse (perpendicular to $\mathbf{k}$) as well as longitudinal (parallel to $\mathbf{k}$) components. Therefore one conveniently writes $\mathbf{E} = \mathbf{E}_{\parallel} + \mathbf{E}_{\perp}$ and $\mathbf{J} = \mathbf{J}_{\parallel} + \mathbf{J}_{\perp}$, where the $\parallel$ and $\perp$ now obviously designate the vector's direction with regard to the wave vector $\mathbf{k}$. Equation (7-133) splits into the two equations

$$-kkE_{\parallel} + kkE_{\parallel} = 0 = -\mu_0(\dot{\mathbf{J}}_{\parallel} + \varepsilon_0 \ddot{\mathbf{E}}_{\parallel}) \tag{7-134}$$

and

$$k^2 \mathbf{E}_{\perp} = -\mu_0(\dot{\mathbf{J}}_{\perp} + \varepsilon_0 \ddot{\mathbf{E}}_{\perp}) \tag{7-135}$$

The first of these equations describes longitudinal electrostatic oscillations.

The electromagnetic wave traveling in the plasma is described by the transverse equation (7-135). Let us look at linearly polarized waves and let $\mathbf{k}$ point in the z and $\mathbf{E}_{\perp}$ in the x direction. With these assumptions and the periodic time dependence, (7-135) beocmes

$$(k^2 - \omega^2/c^2)E = iq\mu_0\omega \int f v_x \, d^3v \tag{7-136}$$

where the $\perp$ subscript has been dropped and the current density expressed with the help of the distribution function in the usual manner. The assumed space and time dependence yields for the collisionless Boltzmann equation —after the usual linearization—the form

$$i(\mathbf{k} \cdot \mathbf{v} - \omega)f_1 + \frac{q}{m} (\mathbf{E} + \mathbf{v} \times \mathbf{B}) \, \partial f_0/\partial \mathbf{v} = 0 \tag{7-137}$$

The transverse electric and magnetic fields are coupled through Maxwell's $\nabla \times \mathbf{E} = -\dot{\mathbf{B}}$ equation, which yields

$$\mathbf{k} \times \mathbf{E} = \omega \mathbf{B} \tag{7-138}$$

Inserting (7-138) into (7-137) and expressing the vectors in their component form one obtains

$$f_1 = i\frac{q}{m}\frac{[\mathbf{E} + \omega^{-1}\mathbf{v} \times (\mathbf{k} \times \mathbf{E})] \cdot \partial f_0/\partial \mathbf{v}}{\mathbf{k} \cdot \mathbf{v} - \omega}$$

$$= i\frac{q}{m}\frac{E\,\partial f_0/\partial v_x + (kE/\omega)[v_x\,\partial f_0/\partial v_z - v_z\,\partial f_0/\partial v_x]}{kv_z - \omega} \tag{7-139}$$

This equation expresses the dependence of the first-order particle-distribution function on the rf electric field. Inserting this into (7-136), which expresses the electric field as a function of the distribution function, closes the circle and yields at once the dispersion relation

$$\omega^2 - k^2 c^2 = \frac{q^2}{m\varepsilon_0}\int\left[\frac{(\omega - kv_z)v_x\,\partial f_0/\partial v_x}{kv_z - \omega} + \frac{kv_x^2\,\partial f_0/\partial v_z}{kv_z - \omega}\right]d^3v \tag{7-140}$$

If the distribution function is such that it contains particles traveling with the velocity $v_z = \omega/k$, the integrand diverges at this point and one encounters the same difficulty we have already dealt with in connection with the electrostatic plasma oscillations. The solution again follows the same line. Instead of following a single Fourier mode in space and time, one looks at the initial-value problem in time (using the Laplace transform technique) of a Fourier component in space. As a result one finds that unless special initial conditions are created, the dispersion relation (7-140) holds, but integrating along v_z in the complex v_z plane the poles must be avoided by integrating *below* them. As a result ω may again be complex. The proof is left to the reader (see Exercise 7-22).

The problem of "resonant" particles traveling with the wave and causing damping or amplification is, however, of far less importance here than it was in the case of electrostatic plasma oscillations. A glance at (7-131) shows that in a cold plasma $\omega/k > c$, and the same holds for most practical zero-order distributions in (7-140). The theory of relativity, however, provides a cutoff at $v_z = c$, even for extremely hot plasmas, but the usual laboratory plasmas do not have an appreciable number of particles even approaching the speed of light. In the case of relativistic plasmas (sometimes of interest in astrophysical applications) the relativistic form of the Boltzmann equation is to be used. Since the $v_z < c$ cutoff is already "built into" this equation, one obtains no resonant particles as expected.

On the other hand, the "resonant" particles are not the only possible source of the frequency becoming complex. One may recall the case of the electrostatic two-stream instability, an exponentially growing disturbance, where the directed motion of the streams feeds energy into the instability. Similar instabilities of the transverse mode can arise for a large class of nonthermal distribution functions.

To see this, one integrates (7-140) by parts in the absence of "resonant" particles and obtains

$$\omega^2 - k^2 c^2 = \omega_p^2 + \frac{q^2 k^2}{m \varepsilon_0} \int \frac{v_x^2 f_0 \, d^3 v}{(k v_z - \omega)^2} \qquad (7\text{-}141)$$

Let us choose, for example, the anisotropic distribution function,

$$f_0(\mathbf{v}) = \delta(v_z) \varphi(v_x, v_y) \qquad (7\text{-}142)$$

where φ is an arbitrary function of the velocity vector perpendicular to the direction of propagation. Inserting (7-142) into (7-141) leads to the dispersion relation

$$\omega^2 - k^2 c^2 - \omega_p^2 \left(1 + \frac{k^2 \langle v_x^2 \rangle}{\omega^2} \right) = 0 \qquad (7\text{-}143)$$

or

$$\omega^4 - (k^2 c^2 + \omega_p^2) \omega^2 - k^2 \omega_p^2 \langle v_x^2 \rangle = 0 \qquad (7\text{-}144)$$

which can be solved for ω^2 at once:

$$\omega^2 = \tfrac{1}{2}\{ k^2 c^2 + \omega_p^2 \pm [(k^2 c^2 + \omega_p^2)^2 + 4 k^2 \omega_p^2 \langle v_x^2 \rangle]^{1/2} \} \qquad (7\text{-}145)$$

One of the roots is always negative for all values of k, and the corresponding values of ω represent an exponentially growing solution.

These *unstable light waves* occur in general when the velocity distribution is anisotropic. Introducing a "smeared out" velocity distribution in v_z reduces the instability in the sense that now only a range of the k values leads to unstable ω's. As soon as the square-average velocity in the z direction approaches the square average in the plane perpendicular to k, the instability disappears.

It is of practical interest to estimate the rate of growth. One may easily see from (7-145) that $-\omega^2$ increases with increasing values of k (short waves grow fastest). For large k's one can expand the square root in (7-145) to obtain

$$\omega^2 = -\frac{k^2 \langle v_x^2 \rangle \omega_p^2}{(k^2 c^2 + \omega_p^2)} \to -\frac{\langle v_x^2 \rangle}{c^2} \omega_p^2 \qquad \text{for} \quad k^2 \to \infty \qquad (7\text{-}146)$$

Hence the maximum growth rate is $(\langle v_x^2 \rangle^{1/2}/c) \omega_p$. It is easy to see that electrostatic instabilities yield growth rates of the order of ω_p (see, e.g., Exercise 7-8). One concludes, therefore, that for nonrelativistic plasmas, the longitudinal (electrostatic) instabilities are as a rule much more dangerous than the transverse ones.

Finally we plot the dispersion curves for a thermal plasma (hot electrons, cold ions) for the two longitudinal (electrostatic) and transverse (electromagnetic) mode in Fig. 7-11. Since $\omega_{pi} = \sqrt{m/M}\,\omega_{pe}$, the scale for ion waves has been distorted for visibility. Only the real part of ω is shown, as obtained from the fluid equations.

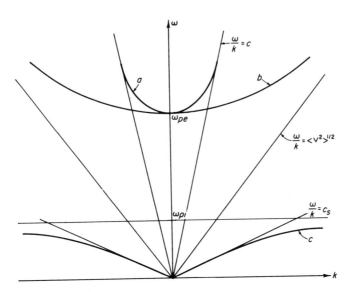

FIG. 7-11. Dispersion curves: (a) e.m. wave; (b) electrostatic electron wave; (c) ion wave.

7-6. Summary

Momentary local deviations from charge neutrality in a plasma give rise to electrostatic plasma oscillations. In a cold, uniform, magnetic-field-free plasma these oscillations take place at the characteristic plasma frequency ω_p and are localized (group velocity $= 0$). In a hot plasma the frequency is complex and obeys the dispersion relation (7-17b), where the integration is carried out in such a way that the pole is kept above the integration path. For a thermal plasma the imaginary part of ω gives rise to Landau damping. This is due to particles that travel with approximately the phase velocity of the wave, interact with it strongly, and in the average extract energy from the wave. For distributions other than thermal, energy may be fed into the wave by the co-moving particles leading to unstable (growing) oscillations. Such runaway processes may also occur in the absence of co-moving particles, such as, for example, in the two-stream instability. The existence of unstable oscillations arising from the perturbation of a distribution function f_0 may be determined by analyzing the dispersion equation (7-17b) on the complex plane. In general, if the function $f_0(u) = \iint f_0 \, dv_y \, dv_z$ has no minimum (single-humped distribution function) there is no instability, while the presence of a deep-enough minimum in $f_0(u)$ leads to electrostatic instabilities (Penrose conditions).

The plasma may be described as a dielectric with a dielectric function $\varepsilon(\omega, k)$, whose zeros for electrostatic waves are the solutions of the dispersion relation, and hence yield the frequencies and wave numbers of the eigenmodes. The energy of an electrostatic wave is just $(\partial/\partial\omega)(\varepsilon\omega)$ times its electrostatic energy, and the ratio of energy and momentum of an eigenmode is ω/k, consistent with the quantum description of waves.

If the plasma consists of hot electrons and cold ions, it can also support ion waves. These waves propagate almost dispersionlessly for $\lambda \gg \lambda_D$, with the ion sound speed $c_s = \sqrt{T/M}$, while for $\lambda \ll \lambda_D$ the frequency approaches ω_{pi} asymptotically. When the plasma supports a current due to a drift of electrons relative to the ions, an instability develops if the drift velocity exceeds c_s. When the drift velocity exceeds the electron thermal velocity, an even stronger hydrodynamic instability, similar to the beam-plasma instability, ensues.

While a root of the dispersion relation with real k and Im $\omega > 0$ corresponds to temporally growing waves, the spatial counterpart of real ω and Im $k < 0$ is more subtle. A closer investigation reveals that depending on the property of mapping of the ω plane to the k plane one might have spatial amplification (convection) or decline (evanescence). In some cases an absolute instability may be present where spatial and temporal growth are coupled, and an initially excited wave packet grows locally in addition to propagating in space.

The plasma can also support electromagnetic (transverse) waves. These can propagate only when the frequency exceeds the plasma frequency. The dispersion curves for the three types of plasma modes are summarized in Fig. 7-11. When the plasma distribution function is sufficiently anisotropic, transverse modes can grow exponentially (unstable light waves).

EXERCISES

7-1. Prove that the charge density oscillates in the first-order approximation with the combination frequency $\omega_p^2 = (\omega_p^+)^2 + (\omega_p^-)^2$ in a cold plasma.

7-2. Write (7-12) combined with (7-10) in the form $\Omega f_1 = \omega f_1$, where Ω is a linear operator. Prove that Ω commutes with the operator $i(\partial/\partial\mathbf{r})$; hence the functions $\exp(i\mathbf{k} \cdot \mathbf{r})$ are also eigenfunctions of Ω.

7-3. Express $F(\mathbf{v},p)$ from (7-25) and find the purely imaginary poles. What kind of time dependence would you predict for $f_1(\mathbf{v},t)$?

7-4. Prove that if the initial distribution function $g(u)$ in (7-31) contains a δ function of the form $\delta(u - u_s)$, undamped oscillations of the electric field with the frequency $\omega = ku_s$ arise.

7-5. Find the dispersion relation for electrostatic plasma oscillations, taking into account the contribution of the ion component with the equilibrium distribution $f_0^+(\mathbf{v})$.

7-6. Derive the Landau damping factor for a Maxwellian plasma (7-45).

7-7. Solve the dispersion relation for electrostatic oscillations for two cold interpenetrating electron streams (7-55) for $\omega^2(kV)$ and show that the roots for ω are either real or imaginary, both coming in $\pm$ pairs.

7-8. At which value of the wave number does the electrostatic two-stream instability grow fastest? Prove that the growth rate of the fastest instability is $\omega_p/2$. [Use the dispersion relation (7-55).]

7-9. Derive (7-63).

7-10. Investigate the dispersion relation for plasma oscillations in the case of two interpenetrating electron streams with an equal thermal spread. Prove that the condition for the disappearance of unstable modes is that the thermal velocity spread (approximately) exceeds the relative stream velocity. (*Hint*: Use a power-series expansion for the distribution functions.)

7-11. Find the energy and momentum densities of an electrostatic wave propagating in a cold electron beam $f_0 \sim \delta(v - V)$. Determine the region of phase velocities for which the energy density is negative.

7-12. The dielectric function corresponding to the Bohm-Gross dispersion relation for electron waves is $\varepsilon = 1 - (\omega_p^2 + k^2 \langle v^2 \rangle) \omega^{-2}$. Prove that the wave energy is twice the electrostatic energy.

7-13. Prove that a plasma with an isotropic velocity distribution cannot support growing plasma oscillations.

7-14. Equate the rate of change of wave energy (twice the electrostatic energy) in a Bohm-Gross wave with the approximate energy absorption by resonant particles (7-63) to find the damping rate γ.

7-15. Prove that the three frequencies of a cold plasma–weak beam system are

$$\omega_{1,2} = \omega_p \left[1 - \left(\frac{\omega_b}{4\omega_p} \right)^{2/3} \pm i\sqrt{3} \left(\frac{\omega_b}{4\omega_p} \right)^{2/3} \right]$$

and

$$\omega_s = \omega_p \left[1 + \left(\frac{\omega_b}{\sqrt{2}\omega_p} \right)^{2/3} \right].$$

7-16. Calculate the spacial growth rate for a beam-plasma system, and show that

$$\text{Im } k = \frac{\sqrt{3}}{2} \left(\frac{\omega_b}{V} \right)^{2/3} \left(\frac{\partial \varepsilon}{\partial k} \right)^{-1/3}$$

7-17. Derive the dispersion relation using fluid equations for cold ions drifting relative to a thermal distribution of electrons. Find the critical drift velocity beyond which the system is unstable. Estimate the growth rate when the drift velocity largely exceeds the thermal velocity.

7-18. Derive (7-108).

7-19. Treat the electron-ion drift instability for small drift velocities **u**, when **u** and **k** are not parallel. Map the region of unstable **k** space.

7-20. Show that (7-120) leads to solutions for real k only for $\omega_{pi} < |\omega| < \sqrt{\omega_{pe}^2 + \omega_{pi}^2}$.

7-21. Prove that the complex k numbers obtained from the Bohm-Gross dispersion relation for $|\omega| < \omega_{pe}$ correspond to evanescent rather than spatially growing modes.

7-22. Treat the problem of transverse waves in a plasma as an initial-value problem. Prove that the dispersion relation is (7-140), and obtain a prescription of integration around the poles.

7-23. Derive the dispersion relation for transverse waves for an electron-ion plasma.

7-24. Prove that transverse waves with frequencies $\omega < \omega_p$ are evanescent in a cold plasma.

7-25. An electromagnetic plane wave of frequency ω impinges on a half-space filled with plasma with plasma frequency $\omega_p > \omega$. What is the characteristic depth of penetration? Prove that for $\omega \ll \omega_p$ this distance is c/ω_p, called the *collisionless skin depth*.

VIII

Waves and Instabilities in Uniform Magnetoplasmas

8-I. Waves in a Cold Magnetoplasma

We are now going to consider a cold plasma immersed in a uniform magnetic field $\mathbf{B}_0$ pointing in the z direction. As a further generalization, ion motion will be taken into account.

While the electromagnetic field equations remain the same, the equations of motion of the particles are now complicated by the presence of the zero-order magnetic field. We have treated this problem in Sec. 2-8, and have found that the velocity of a particle in a static magnetic and oscillating electric field can be expressed as

$$\mathbf{v}_L = \frac{q}{m} \frac{i}{\omega_c - \omega} \mathbf{E}_L \qquad (2\text{-}179)$$

$$\mathbf{v}_R = -\frac{q}{m} \frac{i}{\omega_c + \omega} \mathbf{E}_R \qquad (2\text{-}180)$$

and

$$\mathbf{v}_\parallel = -\frac{q}{m} \frac{i}{\omega} \mathbf{E}_\parallel \qquad (2\text{-}174)$$

where the indices L and R refer to coordinate axes rotating to the left and right about $\mathbf{B}_0$ and parallel to it, respectively. These formulas were derived for a uniform oscillating electric field and no oscillating magnetic field. In the case under consideration, both assumptions are justified. The plasma is cold; hence the particles suffer only small (first-order) displacements from their equilibrium position and acquire small velocities in the first-order wave field. Hence the particle sees the wave field as a uniform oscillating electric field, while the influence of the oscillating magnetic field $\mathbf{v} \times \mathbf{B}$ is only a second-order effect.

The wave produces a plasma current of the density

$$\mathbf{J} = \sum q n_0 \mathbf{v} = \sum q n_0 \overset{\leftrightarrow}{v} \mathbf{E} \qquad (8\text{-}1)$$

where the summation is carried out over the electronic and ionic plasma component, and $\overset{\leftrightarrow}{v}$ is the mobility tensor defined in (2-181a). Equation (8-1) can be inserted into Maxwell's equation, with the assumed periodic time variation $e^{i\omega t}$:

$$\nabla \times \mathbf{H} = \sum qn_0 \overset{\leftrightarrow}{v}\mathbf{E} + i\omega\varepsilon_0\mathbf{E} = i\omega\varepsilon_0\overset{\leftrightarrow}{\kappa}\mathbf{E} \tag{8-2}$$

where the dielectric tensor

$$\overset{\leftrightarrow}{\kappa} = 1 + \frac{\sum qn_0\overset{\leftrightarrow}{v}}{i\omega\varepsilon_0} \tag{8-3}$$

has been defined. In the rotating system $\overset{\leftrightarrow}{\kappa}$ is diagonal with the components

$$\kappa_L = 1 - \frac{q^2n_0}{\omega\varepsilon_0}\left(\frac{1}{m^+}\frac{1}{\omega - \omega_c{}^+} + \frac{1}{m^-}\frac{1}{\omega - \omega_c{}^-}\right)$$

$$= 1 - \frac{q^2n_0(m^+ + m^-)}{\omega^2\varepsilon_0 m^+m^-}\frac{1}{[1 - (\omega_c{}^+/\omega)][1 - (\omega_c{}^-/\omega)]}$$

$$= 1 - \frac{\Pi_p{}^2}{(1 - \beta^+)(1 + \beta^-)} \tag{8-4}$$

$$\kappa_R = 1 - \frac{\Pi_p{}^2}{(1 + \beta^+)(1 - \beta^-)} \tag{8-5}$$

and

$$\kappa_\parallel = 1 - \Pi_p{}^2 \tag{8-6}$$

We introduced the following convenient dimensionless quantities:

$$\Pi_p{}^2 = \frac{(\omega_p{}^+)^2 + (\omega_p{}^-)^2}{\omega^2} \approx \frac{(\omega_p{}^-)^2}{\omega^2} \tag{8-7}$$

$$\beta^+ = |\omega_c{}^+/\omega| = \Omega_i/\omega \tag{8-8}$$

and

$$\beta^- = |\omega_c{}^-/\omega| = \Omega_e/\omega \tag{8-9}$$

where Ω_i and Ω_e are the absolute values of the ion and electron cyclotron frequencies, respectively.

In the nonrotating Cartesian system,

$$\overset{\leftrightarrow}{\kappa} = \begin{pmatrix} \kappa_T & i\kappa_H & 0 \\ -i\kappa_H & \kappa_T & 0 \\ 0 & 0 & \kappa_\parallel \end{pmatrix} \tag{8-10}$$

where it can be easily shown (see Exercise 8-1) that

$$\kappa_T = \frac{\kappa_R + \kappa_L}{2} \tag{8-11}$$

and

$$\kappa_H = \frac{\kappa_R - \kappa_L}{2} \tag{8-12}$$

To arrive at a dispersion relation one combines Maxwell's

$$\nabla \times \mathbf{E} = -i\omega\mu_0\mathbf{H} \tag{8-13}$$

with (8-2) to obtain

$$\nabla \times (\nabla \times \mathbf{E}) = \frac{\omega^2}{c^2} \overleftrightarrow{\kappa}\mathbf{E} \tag{8-14}$$

For a plane wave with the wave vector $\mathbf{k}$, (8-14) becomes

$$\mathbf{k} \times (\mathbf{k} \times \mathbf{E}) + \frac{\omega^2}{c^2} \overleftrightarrow{\kappa}\mathbf{E} = 0 \tag{8-15}$$

This is a system of three homogeneous linear equations for the components of $\mathbf{E}$. The first term in (8-15),

$$\mathbf{k} \times (\mathbf{k} \times \mathbf{E}) = \mathbf{k}(\mathbf{k} \cdot \mathbf{E}) - k^2\mathbf{E} = k_i k_j E_j - k^2 \, \delta_{ij} E_j \tag{8-16}$$

reads in a matrix formulation

$$\mathbf{k} \times (\mathbf{k} \times \mathbf{E}) = \begin{pmatrix} k_x^2 - k^2 & k_x k_y & k_x k_z \\ k_x k_y & k_y^2 - k^2 & k_y k_z \\ k_x k_z & k_y k_z & k_z^2 - k^2 \end{pmatrix} \begin{pmatrix} E_x \\ E_y \\ E_z \end{pmatrix} = \overleftrightarrow{\lambda}\mathbf{E} \tag{8-17}$$

Equation (8-15) now reduces to

$$\left(\overleftrightarrow{\lambda} + \frac{\omega^2}{c^2} \overleftrightarrow{\kappa}\right)\mathbf{E} = 0 \tag{8-18}$$

This system of equations has nontrivial solutions only if

$$\text{Det}\left(\overleftrightarrow{\lambda} + \frac{\omega^2}{c^2} \overleftrightarrow{\kappa}\right) = 0 \tag{8-19}$$

This equation establishes a relationship between ω and $\mathbf{k}$; hence it serves as a dispersion relation. Without loss of generality, one can choose one of the perpendicular components of $\mathbf{k}$, say, $k_x = 0$. Introducing θ, the angle between $\mathbf{B}_0$ and $\mathbf{k}$, the evaluation of the determinant yields after a few algebraic steps (see Exercise 8-2),

$$Ak^4 - B\frac{\omega^2}{c^2} k^2 + C\frac{\omega^4}{c^4} = 0 \tag{8-20}$$

where

$$A = \kappa_T \sin^2\theta + \kappa_\| \cos^2\theta \tag{8-21}$$

$$B = \kappa_R\kappa_L \sin^2\theta + \kappa_\|\kappa_T(1 + \cos^2\theta) \tag{8-22}$$

and

$$C = \kappa_{\parallel}\kappa_R\kappa_L \qquad (8\text{-}23)$$

One finds by investigating (8-20), a quadratic expression in ω^2, that ω^2 is always real; hence ω is real or purely imaginary. This is in agreement with what one expects for a cold plasma, where neither Landau-type damping, nor instabilities feeding on the thermal energy of the plasma, could exist.

The dispersion relation (8-20) can be brought into a more convenient form by dividing by $\cos^2\theta$ and some manipulation (see Exercise 8-2),

$$\tan^2\theta = -\frac{\kappa_{\parallel}(n^2 - \kappa_R)(n^2 - \kappa_L)}{(n^2 - \kappa_{\parallel})(\kappa_T n^2 - \kappa_R \kappa_L)} \qquad (8\text{-}24)$$

where

$$n = \frac{c}{\omega}k = \frac{c}{v_f} \qquad (8\text{-}25)$$

is the index of refraction.

Evaluating the dispersion relation, one finds that after singling out a direction of propagation (θ) the $\omega(k)$ relationship yields not one, but several, roots. There are different *modes of propagation*. The difference between these modes lies not only in the different $\omega(k)$ functions but also in the different polarization of the electromagnetic field. The latter can be evaluated by substituting the ω and $\mathbf{k}$ values obtained from the dispersion relation into (8-18) and calculating the direction of $\mathbf{E}$. The direction and magnitude of $\mathbf{H}$ follows from one of Maxwell's equations [e.g., (8-13)].

We investigate first wave propagation in the two principal directions, along the magnetic field ($\theta = 0$) and perpendicular to it ($\theta = \pi/2$).

For $\theta = 0$ (8-24) yields the following solutions:

$$\kappa_{\parallel} = 0 \qquad (8\text{-}26)$$

$$\kappa_R = n^2 \qquad (8\text{-}27)$$

and

$$\kappa_L = n^2 \qquad (8\text{-}28)$$

Using (8-6) and (8-7), (8-26) becomes

$$\omega^2 = (\omega_p{}^+)^2 + (\omega_p{}^-)^2 \qquad (8\text{-}29)$$

This is the "dispersion relation" for longitudinal plasma oscillations in the absence of a magnetic field. It is physically reasonable that plasma oscillations excited along the direction of the magnetic field are unaffected by the presence of the field.

The wave described by (8-27) is the *electron cyclotron wave*. This wave propagates at all frequencies below the electron cyclotron frequency ($\beta^- > 1$, hence $n^2 > 0$); it has a resonance at $\beta^- = 1$, and attenuates at higher frequencies where κ_R becomes negative. At still higher frequencies propagation again becomes possible. Note that while in the absence of a magnetic field no signal whose frequency is less than the plasma frequency can propagate, here waves of arbitrarily low frequencies can pass. In radio communication this mode is also referred to as the *whistler mode*, since it was observed that audio-frequency noise (initiated by lightning) is also propagated by this mode through the ionosphere along the magnetic field lines of the earth. This is an electron mode (it exists also for infinitely heavy ions), with the electric field circularly polarized and rotating to the right, in the same direction as the electrons.

Equation (8-28) describes the *ion cyclotron wave*. This wave resonates at the ion cyclotron frequency and transmits below that frequency. The electric field vector rotates to the left, with the ions. This mode can be used to heat the ion component of a plasma for thermonuclear applications. Consider feeding a microwave signal into a plasma along the magnetic field lines. If the frequency is somewhat below the ion cyclotron frequency and the incoming wave is linearly polarized, the wave splits up into an ion and an electron cyclotron wave, both circularly polarized. If along the machine the magnetic field is gradually reduced so that somewhere $\Omega_i = \omega$, there is a resonance-type energy feeding into the ion component of the plasma. This arrangement is the *magnetic beach*.

It is interesting to see what happens at very low frequencies. In this case $\Pi_p^2, \beta^+, \beta^- \gg 1$; consequently $\kappa_R = \kappa_L$, and the electron and ion cyclotron wave merge into a single mode:

$$\kappa_R = \kappa_L = 1 + \frac{\Pi_p^2}{\beta^+\beta^-} = 1 + \frac{n_0 q^2 (m^+ + m^-)}{\varepsilon_0 m^+ m^- \omega^2} \frac{m^+ m^- \omega^2}{q^2 B^2} = 1 + \frac{n_0 (m^+ + m^-)}{\varepsilon_0 B^2}$$

(8-30)

This is just the same dielectric constant κ we obtained in (6-14) for the quasi-steady processes. Since κ_H in (8-12) vanishes, the dielectric tensor becomes diagonal with both perpendicular components assuming the value κ, as expected for slow processes. Unless the plasma is very tenuous the second term in κ is much larger than 1, and the dispersion relation of the low-frequency wave becomes

$$n^2 = \frac{c^2}{v_f^2} = \frac{n_0 (m^+ + m^-)}{\varepsilon_0 B^2}$$

(8-31)

and the phase velocity is just the Alfven velocity defined in (4-79),

$$v_f = B/(\mu_0 \rho)^{1/2}$$

(8-32)

where $\rho = n_0 (m^+ + m^-)$ is the mass density. Hence in the low-frequency limit, both electron and ion cyclotron waves degenerate into Alfven waves.

In the other principal direction, $\theta = \pi/2$, one obtains from (8-24) the dispersion relations

$$\kappa_{\|} = n^2 \tag{8-33}$$

and

$$\kappa_R \kappa_L / \kappa_T = n^2 \tag{8-34}$$

Equation (8-33) is simply

$$k^2 c^2 = \omega^2 - [(\omega_p^+)^2 + (\omega_p^-)^2] \tag{8-35}$$

the same as the dispersion relation for transverse waves in a plasma in the absence of a magnetic field [(7-128) including the ion component]. This is a purely transverse linearly polarized electromagnetic wave, where the oscillating electric field vector, as well as the oscillating velocity vector, points in the direction of the static magnetic field (see Exercise 8-3). It is obvious that the magnetic field has no influence on this mode. Owing to its insensitivity to the magnetic field, this mode can be used to measure plasma densities. If one shines microwaves at a frequency exceeding the plasma frequency, at a plasma immersed in a uniform magnetic field, the waves will suffer a decrease in k and a corresponding increase in wavelength, as described by (8-35), compared to their value in free space. This change in wavelength can be measured as a phase shift with microwave interferometric methods and the density computed from ω_p^2. The cutoff at $\omega \to \omega_p$ can also be very sensitively measured.

The other mode described by (8-34) is a more complicated "hybrid" mode, in which the longitudinal (electrostatic) and transverse (electromagnetic) wave characteristics are mixed.

In the absence of $\mathbf{B}_0$ a transverse wave traveling in the y direction gives rise to transverse particle motion in the xz plane. The presence of a $\mathbf{B}_0$ pointing in the z direction couples the particle motion in the x direction with motion in the y direction. This longitudinal oscillatory particle motion in turn results in electrostatic effects along $\mathbf{k}$. In this hybrid mode the electric wave vectors are elliptically polarized in the plane perpendicular to $\mathbf{B}_0$ rotating right- or left-handed.

The dispersion relation of the different modes in the intermediate directions can be conveniently characterized by the wave normal surfaces. These are polar plots of $\mathbf{v}_f$ in the plane containing $\mathbf{B}_0$. At fixed values of the parameters Π_p, β^+, and β^-, one obtains in general two nonintersecting wave-normal surfaces. Slowly changing the values of these parameters results in a deformation of the wave-normal surfaces. This deformation will be continuous; hence it will preserve the topological characters of the surface, unless one passes through a point where $n^2 \to \infty$ or $n^2 = 0$. As we have already seen, at these points wave propagation $(n^2 > 0)$ can change into attenuation $(n^2 < 0)$.

The parameter values resulting in $n^2 \rightarrow \infty$ will be called *resonances* (like the electron and ion cyclotron resonance), and the ones with $n^2 = 0$ the *cutoffs* [like the plasma cutoff in (8-35)].

Choosing a fixed ion-electron mass ratio eliminates one of the parameters β^+, β^-. The entire picture can be summarized in the Allis diagram, as seen in Fig. 8-1. The "parameter space" defined by the axes Π_p^2 and $(\beta^-)^2$ is

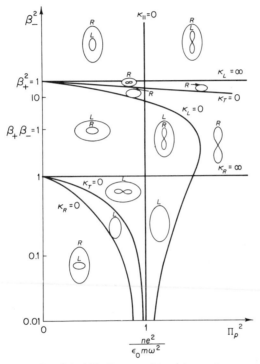

FIG. 8-1. Allis diagram with $m^+/m^- = 4$.

divided by the principal resonance and cutoff lines into separate regions. Each of these regions has its wave-normal surfaces, which may change their shape but not their topology inside the region. Some modes appear, disappear, or change topology by crossing the region boundaries. In Fig. 8-1 the ion-electron mass ratio was chosen as 4 for better visualization and the letters R and L designate electric vectors rotating to the right or left respectively, at $\theta = 0$. (Note, however, that the sense of rotation may change at other values of θ).

The resonances for $\theta = 0$ are, from (8-27) and (8-28),

$$\kappa_R \rightarrow \infty, \quad \beta^- = 1 \qquad \text{electron cyclotron resonance} \qquad (8\text{-}36)$$

and

$$\kappa_L \to \infty, \beta^+ = 1 \qquad \text{ion cyclotron resonance} \qquad (8\text{-}37)$$

For $\theta = \pi/2$ there is no resonance from (8-33), while (8-34) yields the resonance

$$\kappa_T = \frac{\kappa_R + \kappa_L}{2} = 0 = 1 - \frac{\Pi_p{}^2}{2(1 - \beta^+)(1 + \beta^-)} - \frac{\Pi_p{}^2}{2(1 + \beta^+)(1 - \beta^-)} \qquad (8\text{-}38)$$

or

$$(1 - \beta^{+2})(1 - \beta^{-2}) = \Pi_p{}^2(1 - \beta^+\beta^-) \qquad (8\text{-}39)$$

Multiply by ω^4 to get

$$(\omega^2 - \Omega_i{}^2)(\omega^2 - \Omega_e{}^2) = \omega_p{}^2(\omega^2 - \Omega_e\Omega_i) \qquad (8\text{-}40)$$

where $\omega_p = \omega_p{}^-$. The biquadratic equation for ω

$$\omega^4 - \omega^2(\Omega_i{}^2 + \Omega_e{}^2 + \omega_p{}^2) + \Omega_i{}^2\Omega_e{}^2 + \omega_p{}^2\Omega_i\Omega_e = 0 \qquad (8\text{-}41)$$

has two solutions

$$\omega_{1,2}^2 = \tfrac{1}{2}[\Omega_i{}^2 + \Omega_e{}^2 + \omega_p{}^2 \pm \sqrt{(\Omega_i{}^2 + \Omega_e{}^2 + \omega_p{}^2)^2 - 4(\Omega_i{}^2\Omega_e{}^2 + \omega_p{}^2\Omega_i\Omega_e)}]$$

$$\approx \frac{1}{2}\left[\Omega_e{}^2 + \omega_p{}^2 \pm \left(\Omega_e{}^2 + \omega_p{}^2 - \frac{2(\Omega_i{}^2\Omega_e{}^2 + \omega_p{}^2\Omega_i\Omega_e)}{\Omega_e{}^2 + \omega_p{}^2}\right)\right] \qquad (8\text{-}42)$$

since $\Omega_e{}^2 \gg \Omega_i{}^2$.

The plus sign gives the *upper hybrid resonance*

$$\omega_{\text{UH}}^2 = \Omega_e{}^2 + \omega_p{}^2 \qquad (8\text{-}43)$$

and the minus sign the *lower hybrid resonance*

$$\omega_{\text{LH}}^2 = \frac{\Omega_i{}^2\Omega_e{}^2 + \omega_p{}^2\Omega_i\Omega_e}{\Omega_e{}^2 + \omega_p{}^2} = \frac{\Omega_i{}^2 + \omega_{pi}^2}{1 + (\omega_p/\Omega_e)^2} \qquad (8\text{-}44)$$

For a dense plasma in a weak magnetic field $\omega_{pi}^2 \gg \Omega_i{}^2$, and

$$\omega_{\text{LH}}^2 = \frac{\omega_{pi}^2}{1 + (\omega_p/\Omega_e)^2} \qquad (8\text{-}45)$$

or when, in addition $\omega_p{}^2 \gg \Omega_e{}^2$

$$\omega_{\text{LH}}^2 = \left(\frac{\omega_{pi}}{\omega_p}\right)^2 \Omega_e{}^2 = \Omega_i\Omega_e \qquad (8\text{-}46)$$

In general, $\kappa_T = 0$ is represented by a hyperbola in parameter space. One branch goes through the points $\Pi_p = 0$, $\beta^- = 1$ and $\beta^- = 0$, $\Pi_p{}^2 = 1 - (\beta^+)^2$, the other branch through $\Pi_p = 0$, $\beta^+ = 1$ and $\Pi_p \to \infty$, $\beta^+\beta^- = 1$ (see Exercise 8-4).

Resonances can also occur in directions other than principal directions. They can be found by setting $n \to \infty$ in (8-24). One obtains

$$\tan^2\theta_{res} = -\kappa_{\parallel}/\kappa_T \tag{8-47}$$

This resonance occurs at the angle θ_{res} wherever the signs of $\kappa_{\parallel}$ and κ_T are different. In a region where this occurs for a mode, the corresponding wave-normal surface assumes the form of a lemniscate, as seen in Fig. 8-1.

The cutoffs are, for $\theta = 0$,

$$\kappa_R = 0 \tag{8-48}$$

and

$$\kappa_L = 0 \tag{8-49}$$

They are called the cyclotron cutoffs and may be combined into a single equation, $\kappa_R\kappa_L = 0$, or

$$\Pi_p^4 = (1 - \beta^{+2})(1 - \beta^{-2}) \tag{8-50}$$

For $\theta = \pi/2$ one finds the plasma cutoff

$$\kappa_{\parallel} = 0, \qquad \omega^2 = (\omega_p^+)^2 + (\omega_p^-)^2 \tag{8-51}$$

from (8-33), while (8-34) does not yield any new cutoffs. It can be shown that there are no cutoffs in other than the principal directions (see Exercise 8-5).

While our general dispersion relation (8-19) includes electrostatic waves, it is instructive to study those in greater detail. Forming the dot product of $\mathbf{k}$ with (8-15) and recognizing that $\mathbf{E} \parallel \mathbf{k}$ for electrostatic modes yields immediately the electrostatic dispersion relation

$$\mathbf{k} \cdot \boldsymbol{\kappa} \cdot \mathbf{k} = 0 \tag{8-52}$$

which one could have obtained directly from $\nabla \cdot \mathbf{D} = \nabla \cdot \varepsilon_0\boldsymbol{\kappa} \cdot \mathbf{E} = 0$. Choosing again say $k_x = 0, k_y = k_{\perp}$, one finds

$$\kappa_T + \frac{k_{\parallel}^2}{k_{\perp}^2}\kappa_{\parallel} = 0 \tag{8-53}$$

As an interesting example we evaluate this dispersion relation for frequencies much larger than the ion cyclotron frequency but much smaller than the electron cyclotron frequency, so $\beta^+ \ll 1$ and $\beta^- \gg 1$. In this case

$$\kappa_T = 1 - \Pi_p^2 \frac{1 - \beta^+\beta^-}{(1 - \beta^{+2})(1 - \beta^{-2})} \approx 1 + \Pi_p^2 \frac{1 - \beta^+\beta^-}{\beta^{-2}}$$

$$= 1 + \frac{\omega_p^2}{\Omega_e^2}\left(1 - \frac{\Omega_e\Omega_i}{\omega^2}\right) \tag{8-54}$$

and (8-53) becomes

$$1 + \frac{\omega_p{}^2}{\Omega_e{}^2} + \frac{k_{\parallel}{}^2}{k_{\perp}{}^2} - \frac{\omega_p{}^2}{\omega^2}\left(\frac{k_{\parallel}{}^2}{k_{\perp}{}^2} + \frac{\Omega_i}{\Omega_e}\right) = 0 \qquad (8\text{-}55)$$

or

$$\omega^2 = \frac{1}{1 + (\omega_p/\Omega_e)^2 + (k_{\parallel}/k_{\perp})^2}\left(\omega_{pi}^2 + \frac{k_{\parallel}{}^2}{k_{\perp}{}^2}\,\omega_p{}^2\right) \qquad (8\text{-}56)$$

where we recognized that $\Omega_i/\Omega_e = m/M$ and ω_{pi} is the ion plasma frequency. Unless $k_{\parallel}/k_{\perp}$ is very small the ω_{pi}^2 term may be ignored and one is left with electrons oscillating along magnetic field lines driven by the parallel component of the electric field $E_{\parallel} = Ek_{\parallel}/k$; hence the frequency is reduced to essentially $\omega_p k_{\parallel}/k$. The origin of the $(\omega_p/\Omega_e)^2$ term in the denominator is as follows. Due to the polarization drift of electrons in the perpendicular direction, the electrons act as a dielectric and shield out part of the field. The details are left for Exercises 8-7 and 8-8.

When $(k_{\parallel}/k_{\perp})^2$ is small, of order of m/M, ion motion becomes significant. Since $\beta^+ \ll 1$, the ions do not feel the magnetic field. Because the electrons are tied to magnetic field lines, they are inhibited from neutralizing ion space charge. If the electrons were completely stopped, the ions would oscillate at their natural frequency ω_{pi}, just as electrons oscillate at ω_p if the ions are motionless. When $k_{\parallel} = 0$, the electrons can provide some shielding due to their inertial drift. This is just the lower hybrid resonance as described in (8-45). If a small $k_{\parallel}$ is also present, electron motion along field lines can make its contribution. These waves are known as lower hydrid waves.

8-2. Transverse Waves in Hot Magnetoplasmas

By adding a uniform magnetic field to a cold plasma we obtained a rich variety of wave forms. One expects that the introduction of a zero-order velocity distribution will further enhance the complexities of these phenomena.

First of all, the appearance of complex frequencies, signifying damping or growing modes, can be anticipated, as was the case in the absence of a magnetic field. This means there will be energy exchange between the electromagnetic field and the plasma. In the case of damping, energy flows from the field into the plasma, while growing modes (instabilities) extract energy from the particles and feed it into the field. We shall distinguish two types of energy-exchange processes.

The first type may be characterized as *resonance processes*. We have already seen that in electrostatic oscillations particles traveling with a velocity close to ω/k, the phase velocity of the wave, interact strongly with the wave. Whether this interaction results in Landau damping or instability depends on the shape of the distribution function around $v = \omega/k$. In the presence of a magnetic field, this phenomenon appears for any mode where particles travel with the wave for an extended period of time and the wave electric field has a component in the direction of this particle motion. Since the presence of the zero-order magnetic field permits particle motion only along the field (for an extended period of time), only modes with a z-component electric field can produce this kind of interaction and only for $\omega/k < c$.

Another form of resonant interaction that can exist in the presence of $\mathbf{B}_0$ only occurs if some particles "feel" a force, varying with their gyration frequency ω_c acting in the plane perpendicular to $\mathbf{B}_0$ (the xy plane). A particle traveling with the velocity $v_{\parallel}$ along $\mathbf{B}_0$ feels the Doppler-shifted frequency $\omega' = \omega - v_{\parallel}k$. Resonance effects occur for $\omega' \approx \omega_c$ and they are produced by the perpendicular component of the electric field. Again either damping or instability can occur, depending on the distribution function.

The other type of interaction leading to energy exchange may be called *electro-magneto-hydrodynamic*. Here resonance processes do not take part; therefore a detailed knowledge of the distribution function is not essential for the determination of the interaction. We know that by taking moments of the Boltzmann equation in velocity space, macroscopic equations can be obtained. By assuming that some moment of the distribution function (e.g., the heat-conduction tensor) is zero, a closed set of macroscopic equations results. By combining these with Maxwell's equations, one arrives at the equations of electro-magneto-hydrodynamics. These involve $\mathbf{E}$, $\mathbf{B}$, particle and charge density, fluid velocity (average particle velocity), pressure (tensor or scalar), and fluid energy density (temperature). In Chapter III we considered a special case of these equations, the magneto-hydrodynamic set, obtained by assuming the dominance of magnetic interactions.

In treating the general problem of waves in a hot plasma, one again writes the wave equation (8-15), but now the dielectric tensor has to be determined from the Boltzmann equation.

Consider a transverse wave propagating along the uniform magnetic field. The linearized Vlasov equation for electrons may be written

$$\frac{\partial f_1}{\partial t} + \mathbf{v} \cdot \frac{\partial f_1}{\partial \mathbf{x}} + \frac{q}{m}(\mathbf{v} \times \mathbf{B}_0) \cdot \frac{\partial f_1}{\partial \mathbf{v}} = \frac{df_1}{dt} = -\frac{q}{m}(\mathbf{E}_1 + \mathbf{v} \times \mathbf{B}_1) \cdot \frac{\partial f_0}{\partial \mathbf{v}} \quad (8\text{-}57)$$

where $\mathbf{E}_1$, $\mathbf{B}_1$ are the small wave fields. The fact that the left-hand side can be written as a total time derivative, namely, a derivative taken by following a particle along its unperturbed trajectory in phase space, facilitates finding f_1 formally, by carrying out an integral along an unperturbed trajectory

$$f_1(t,\mathbf{x},\mathbf{v}) - f_1(t_0,\mathbf{x}_0,\mathbf{v}_0) = -\frac{q}{m}\int_{t_0}^{t}\left[(\mathbf{E}_1 + \mathbf{v} \times \mathbf{B}_1) \cdot \frac{\partial f_0}{\partial \mathbf{v}}\right]_{t'} dt' \quad (8\text{-}58)$$

where $\mathbf{x}_0$, $\mathbf{v}_0$ are the phase space coordinates of a particle at t_0, that is, located at $\mathbf{x}$, $\mathbf{v}$, at time t. Similarly in the integral all quantities are to be evaluated along the unperturbed phase space trajectory at t'. The unperturbed particle trajectory is a helix where at $t = t'$

$$v_z' = v_z = v_\parallel \tag{8-59}$$

$$v_y' = v_\perp \cos[\omega_c(t' - t) + \psi] \tag{8-60}$$

$$v_x' = v_\perp \sin[\omega_c(t' - t) + \psi] \tag{8-61}$$

Here ψ is the phase angle of the rotating velocity vector in the xy plane at t.

In order to avoid the cumbersome Laplace transform in time, we Fourier transform in space and time, remembering, however, that ω stands for a complex quantity, and one has to integrate around poles with proper care, in velocity space integrals. We choose for space and time dependence of the perturbed quantities $\exp[i(\omega t - \mathbf{k} \cdot \mathbf{x})]$. The Fourier components of $\mathbf{E}_1$ and $\mathbf{B}_1$ are related through Maxwell's equation

$$\mathbf{k} \times \mathbf{E}_1 = \omega \mathbf{B}_1 \tag{8-62}$$

so

$$\mathbf{v} \times \mathbf{B}_1 = \frac{1}{\omega}[(\mathbf{v} \cdot \mathbf{E}_1)\mathbf{k} - (\mathbf{v} \cdot \mathbf{k})\mathbf{E}_1] \tag{8-63}$$

Since in our case $\mathbf{k} \| \mathbf{B}_0$ one has

$$f_1(t,\mathbf{x},\mathbf{v}) - f_1(t_0,\mathbf{x}_0,\mathbf{v}_0)$$

$$= -\frac{q}{m} \int_{t_0}^t \left[\left(1 - \frac{v_\parallel k}{\omega}\right)\mathbf{E} \cdot \frac{\partial f_0}{\partial \mathbf{v}} + \frac{\mathbf{v} \cdot \mathbf{E}}{\omega} k \frac{\partial f_0}{\partial v_\parallel}\right] \exp[i(\omega t' - kz')] \, dt'$$

$$= -\frac{q}{m} \exp[i(\omega t - kz)] \int_{t_0}^t \left[\left(1 - \frac{v_\parallel k}{\omega}\right)\mathbf{E} \cdot \frac{\partial f_0}{\partial \mathbf{v}} + \frac{\mathbf{v} \cdot \mathbf{E}}{\omega} k \frac{\partial f_0}{\partial v_\parallel}\right]$$

$$\times \exp[i(\omega - kv_\parallel)(t' - t)] \, dt' \tag{8-64}$$

where $\mathbf{E}_1 = \mathbf{E} \exp[i(\omega t - kz)]$ and $z' - z = v_\parallel(t' - t)$. Since $f_0 = f_0(v_\perp, v_\parallel)$, $\partial f_0/\partial \mathbf{v}_\perp = (\mathbf{v}_\perp/v_\perp) \partial f_0/\partial v_\perp$. As $\mathbf{E} \perp \mathbf{B}_0$ (the wave is transverse), using (8-60) and (8-61)

$$\mathbf{E} \cdot \frac{\partial f_0}{\partial \mathbf{v}} = [E_x \sin(\omega_c \tau + \psi) + E_y \cos(\omega_c \tau + \psi)] \frac{\partial f_0}{\partial v_\perp} \tag{8-65}$$

where the new variable $\tau = t' - t$ has been introduced. One may now extract from the integral the quantities not dependent on t', to find

$$\int dt' = \left[\left(1 - \frac{v_\parallel k}{\omega}\right)\frac{\partial f_0}{\partial v_\perp} + \frac{kv_\perp}{\omega}\frac{\partial f_0}{\partial v_\parallel}\right]$$

$$\int_{t_0-t}^{0} [E_x \sin(\omega_c\tau + \psi) + E_y \cos(\omega_c\tau + \psi)] \exp[i(\omega - kv_\parallel)\tau] \, d\tau$$

(8-66)

If the wave has a left-hand circular polarization, $E_y = iE_x$ and the integral in (8-66) becomes

$$iE_x \int_{t_0-t}^{0} \exp[i(\omega - kv_\parallel - \omega_c)\tau] \exp(-i\psi) \, d\tau$$

$$= E_x \exp(-i\psi) \frac{1 - \exp[i(\omega - kv_\parallel - \omega_c)(t_0 - t)]}{\omega - kv_\parallel - \omega_c} \qquad (8\text{-}67)$$

Substituting this result in (8-66) and (8-64) yields finally

$$f_1(t,\mathbf{x},\mathbf{v}) - f_1(t_0,\mathbf{x}_0,\mathbf{v}_0)$$

$$= -\frac{q}{m}\frac{E_x \exp(-i\psi)}{\omega - kv_\parallel - \omega_c} \lambda\{\exp[i(\omega t - kz)]$$

$$- \exp[i(\omega t_0 - kz_0)] \exp[-i\omega_c(t_0 - t)]\} \qquad (8\text{-}68)$$

where

$$\lambda = \left(1 - \frac{v_\parallel k}{\omega}\right)\frac{\partial f_0}{\partial v_\perp} + \frac{kv_\perp}{\omega}\frac{\partial f_0}{\partial v_\parallel}.$$

Consider an unstable wave with $\text{Im}\,\omega < 0$. If one chooses $t_0 \to -\infty$, the initial-value terms on both sides disappear since $f_1(-\infty)$ can be chosen arbitrarily small and $\exp(-i\omega\infty) \to 0$. So

$$f_1(t,\mathbf{x},\mathbf{v}) = -\frac{q}{m}\frac{E_x \exp(-i\psi)}{\omega - kv_\parallel - \omega_c}\left[\left(1 - \frac{v_\parallel k}{\omega}\right)\frac{\partial f_0}{\partial v_\perp} + \frac{kv_\perp}{\omega}\frac{\partial f_0}{\partial v_\parallel}\right]\exp[i(\omega t - kz)]$$

(8-69)

If the wave is damped ($\text{Im}\,\omega > 0$) the opposite procedure may be employed. Now one integrates in t' from zero to infinity or in τ from zero to infinity, to end up again with (8-69).

If the wave is polarized to the right, $E_y = -iE_x$ and the integral (8-66) becomes

$$-iE_x \int \exp[i(\omega - k_{\parallel}v + \omega_c)\tau] \exp(i\psi)\, d\tau$$

$$= -\frac{E_x \exp(i\psi)}{\omega - k_{\parallel}v + \omega_c} \{1 - \exp[i(\omega - kv_{\parallel} - \omega_c)(t_0 - t)]\} \quad (8\text{-}70)$$

So the Fourier amplitude of the perturbed distribution function for a right or left polarized wave $f_{R,L}$ can be cast in the form

$$f_{R,L} = \pm \frac{q}{m} \lambda \frac{E_x e^{\pm i\psi}}{\omega - kv_{\parallel} \pm \omega_c} \quad (8\text{-}71)$$

where the upper sign corresponds to right- and the lower to left-handed polarization.

Now one may compute the current density produced by this distribution function

$$J_x = q \int f_{R,L} v_x\, dv^3 = \pm \frac{q^2}{m} E_x \int \lambda \frac{e^{\pm i\psi}}{\omega - kv_{\parallel} \pm \omega_c} \sin \psi v_{\perp}^2\, d\psi\, dv_{\perp}\, dv_{\parallel}$$

$$(8\text{-}72)$$

The ψ integration goes from 0 to 2π to yield

$$\frac{1}{2i} \int_0^{2\pi} (e^{i\psi} - e^{-i\psi}) e^{\pm i\psi}\, d\psi = \pm i\pi \quad (8\text{-}73)$$

which results in the current density

$$J_x = \frac{q^2\pi}{m\omega} iE_x \int \frac{(\omega - kv_{\parallel})\,(\partial f_0/\partial v_{\perp}) + kv_{\perp}\,(\partial f_0/\partial v_{\parallel})}{\omega - kv_{\parallel} \pm \omega_c} v_{\perp}^2\, dv_{\perp}\, dv_{\parallel} \quad (8\text{-}74)$$

A similar calculation can be carried out for J_y leading to a similar expression so $J_x/E_x = J_y/E_y$; hence $\mathbf{J}$ and $\mathbf{E}$ are parallel. The plasma dielectric constant can be calculated from $\mathbf{J} + i\omega\varepsilon_0\mathbf{E} = i\omega\kappa\varepsilon_0\mathbf{E}$ as in (8-2)

$$\kappa_{R,L} = 1 + \frac{J/E}{i\omega\varepsilon_0} = 1 + \frac{q^2\pi}{m\varepsilon_0\omega^2} \int_{-\infty}^{+\infty} dv_{\parallel} \int_0^{\infty} dv_{\perp}\, v_{\perp}^2$$

$$\times \frac{(\omega - kv_{\parallel})\,(\partial f_0/\partial v_{\perp}) + kv_{\perp}\,(\partial f_0/\partial v_{\parallel})}{\omega - kv_{\parallel} \pm \omega_c} \quad (8\text{-}75)$$

The dispersion relation follows from setting $n^2 = \kappa_{R,L}$ [see (8-27) and (8-28)]

$$c^2k^2 = \omega^2 + \frac{q^2\pi}{m\varepsilon_0} \int_{-\infty}^{+\infty} dv_\parallel \int_0^\infty dv_\perp\, v_\perp{}^2 \frac{(\omega - kv_\parallel)\,(\partial f_0/\partial v_\perp) + kv_\perp\,(\partial f_0/\partial v_\parallel)}{\omega - kv_\parallel \pm \omega_c}$$

(8-76)

In case of more than one particle species (electrons and ions), the κ's have to be added, and a summation sign over particle species should be put in front of the integral expressions in (8-75) and (8-76). If there are "resonant" particles that feel the Doppler-shifted frequency as the cyclotron frequency in their coordinate system, one reverts to the usual "trick"; the problem is considered as an initial-value problem and the same dispersion equation (8-76) is regained, but now one has a prescription for integration around the pole on the $v_\parallel$ complex plane. Since the time dependence is now $e^{+i\omega t}$, $p = +i\omega$, one integrates above, not below, the pole.

Let us investigate first the behavior of plasmas with an isotropic velocity distribution. An important special case is a plasma in thermal equilibrium. Since for these the vector $\partial f_0/\partial \mathbf{v}$ is parallel to $\mathbf{v}$, the "magnetic term" in (8-57) is

$$\mathbf{v} \times \mathbf{B}_1 \cdot \frac{\partial f_0}{\partial \mathbf{v}} = 0$$

(8-77)

Consequently, the magnetic field of the wave itself has no influence on such a plasma in the linear approximation. It is easy to see that as a result all factors in the numerator of the integrand in (8-76) that contain k vanish. It can be shown from the dispersion equation that such distributions cannot support growing waves. In the absence of resonant particles ω is always real; in their presence there is damping.

For such a plasma the dispersion relation is from (8-76)

$$1 + \frac{q^2\pi}{m\varepsilon_0\omega} \int_{-\infty}^{+\infty} dv_\parallel \int_0^\infty \frac{\partial f_0/\partial v_\perp}{\omega - kv_\parallel \pm \omega_c} v_\perp{}^2\, dv_\perp - \frac{c^2k^2}{\omega^2} = 0 \qquad (8\text{-}78)$$

The integration over $v_\perp$ can be done by parts, and one may introduce the reduced distribution function $f_0(v_\parallel) = \int f_0 2\pi v_\perp\, dv_\perp$ to find

$$1 - \frac{q^2}{m\varepsilon_0\omega} P \int \frac{f_0(v_\parallel)\, dv_\parallel}{\omega - kv_\parallel \pm \omega_c} - \frac{i\pi}{k}\frac{q^2}{m\varepsilon_0\omega} f_0\!\left(\frac{\omega \pm \omega_c}{k}\right) - \frac{c^2k^2}{\omega^2} = 0 \quad (8\text{-}79)$$

Note that we integrated on a semicircle above the pole. If the imaginary part is sufficiently small, one may expand in the usual manner by writing

$$\kappa_0 - \frac{c^2k^2}{\omega^2} - i\delta\kappa = 0 \qquad (8\text{-}80)$$

with κ_0 designating the first two terms in (8-79). For some real ω_0

$$\kappa_0(\omega_0) - \frac{c^2 k^2}{\omega_0{}^2} = 0 \qquad (8\text{-}81)$$

Searching for a solution to (8-80) in the neighborhood of ω_0

$$\left[\left(\frac{\partial \kappa_0}{\partial \omega} \right)_{\omega_0} + \frac{2c^2 k^2}{\omega_0{}^3} \right] \delta\omega = i\,\delta\kappa)_{\omega_0} = \frac{i\pi}{k} \frac{q^2}{m\varepsilon_0 \omega_0} f\left(\frac{\omega \pm \omega_c}{k} \right) \qquad (8\text{-}82)$$

to yield for the damping increment

$$\gamma = \operatorname{Im} \delta\omega = \frac{\pi q^2}{m\varepsilon_0 k} \left(\omega_0 \frac{\partial \kappa_0}{\partial \omega} + 2\kappa_0 \right)^{-1} f\left(\frac{\omega_0 \pm \omega_c}{k} \right) \qquad (8\text{-}83)$$

It has already been pointed out that this damping has its physical origin in cyclotron acceleration of resonant particles for which $\omega' \approx \omega_c$. We have seen in Sec. 2-8 that a particle in a static magnetic field and oscillating electric field moves with the perpendicular velocity

$$\mathbf{V} = \mathbf{V}_M + \mathbf{V}_E e^{i\omega' t} \qquad (2\text{-}169)$$

where $\mathbf{V}_M$ gyrates at the cyclotron frequency and the other term represents the forced oscillation. As long as $\omega' \neq \omega_c$, $\mathbf{V}_E$ is finite and the average particle energy associated with this motion is constant. In the absence of resonant particles there is no energy exchange between the field and the particle. Near resonance one has to solve the initial-value problem as was done in Exercise 2-28. Switching on, say, a left-hand circularly polarized wave at $t = 0$, such that

$$\begin{aligned} E_x &= E_0 \cos \omega' t \\ E_y &= -E_0 \sin \omega' t \end{aligned} \quad (t \geq 0) \qquad (8\text{-}84)$$

and $\mathbf{E} = 0$ for $t < 0$, and writing $\mathbf{V} = \mathbf{V}_M + \mathbf{V}_{\text{res}}$, one has

$$v_{x\,\text{res}} = \frac{q}{m} \frac{E_0}{\omega' - \omega_c} (\sin \omega' t - \sin \omega_c t)$$

$$v_{y\,\text{res}} = \frac{q}{m} \frac{E_0}{\omega' - \omega_c} (\cos \omega' t - \cos \omega_c t) \qquad (8\text{-}85)$$

for $t > 0$. The power delivered by the field to all particles is

$$P = q\mathbf{E} \cdot \sum (\dot{\mathbf{V}}_M + \mathbf{V}_{\text{res}}) \qquad (8\text{-}86)$$

with the summation extended over all particles. In the absence of the electric field the particle distribution is isotropic so $\sum \mathbf{V}_M = 0$ and the only contribution to P comes from the resonant term. This can be easily calculated from (8-84) and (8-85) to yield,

$$P = \sum \frac{q^2}{m} \frac{E_0^2}{\omega' - \omega_c} \sin(\omega' - \omega_c)t \tag{8-87}$$

Note an important difference with Landau damping, where close to resonance some particles gained and others lost energy. Here, whether ω' is somewhat larger or smaller than ω_c, P is always positive and the wave is damped. This is why the damping factor (8-83) contains the distribution function at resonance and not its derivative. The summation in (8-87) means integration over velocity space

$$P = \frac{q^2}{m} E^2 \int \frac{\sin(\omega - kv_{\parallel} - \omega_c)t}{\omega - kv_{\parallel} - \omega_c} f_0(v_{\parallel}) \, dv_{\parallel} \tag{8-88}$$

dropping the index of E. It looks as if all particles participated in energy exchange. The function $(\sin yx)/x$ is, however, a strongly peaked function around $x = 0$, which approaches $\pi\delta(x)$ for large values of y. So for large values of t, (8-88) becomes

$$P = \frac{q^2}{m} \frac{E^2}{k} \pi f\left(\frac{\omega - \omega_c}{k}\right) \tag{8-89}$$

This power is extracted from the wave whose energy density is

$$U = \frac{\partial}{\partial \omega} (\omega\kappa) \frac{\varepsilon_0 E^2}{2} + \frac{B^2}{2\mu_0} = \frac{\varepsilon_0 E^2}{2} \left(\frac{\partial}{\partial \omega} (\omega\kappa) + \frac{c^2 k^2}{\omega^2}\right) \tag{8-90}$$

since $E/B = \omega/k$ (from $\nabla \times \mathbf{E} = -\dot{\mathbf{B}}$). The electric field is damped, $E \sim e^{-\gamma t}$, so

$$-\frac{dU}{dt} = 2\gamma U = P \tag{8-91}$$

where we have equated the energy loss to the wave with the energy gain of the particles. Since $\kappa = c^2 k^2/\omega^2$

$$\gamma = \frac{P}{2U} = \frac{\pi q^2}{m\varepsilon_0 k} \frac{f_0[(\omega - \omega_c)/k]}{\omega (\partial\kappa/\partial\omega) + 2\kappa}$$

the same result as (8-83).

It still remains to evaluate the real part of the dispersion relation. If the resonant velocity is far out on the tail of the Maxwellian distribution function $[|(\omega \pm \omega_c)/k| \gg V_T]$, one may expand the integral in (8-79) to get

$$\int_{-\infty}^{+\infty} \frac{f(v_{\parallel}) \, dv_{\parallel}}{(\omega \pm \omega_c) - kv_{\parallel}} = \int_{-\infty}^{+\infty} \frac{f(v_{\parallel})}{\omega \pm \omega_c} \left(1 + \frac{kv_{\parallel}}{\omega \pm \omega_c} + \frac{k^2 v_{\parallel}^2}{(\omega \pm \omega_c)^2} + \cdots \right) dv_{\parallel}$$

$$= \frac{n_0}{\omega \pm \omega_c} \left(1 + \frac{k^2 \langle v_{\parallel}^2 \rangle}{(\omega \pm \omega_c)^2} \right)$$

$$= \frac{n_0}{\omega \pm \omega_c} \left(1 + \frac{1}{3} \frac{k^2 \langle v^2 \rangle}{(\omega \pm \omega_c)^2} \right) \qquad (8\text{-}92)$$

So the dispersion relation for one species becomes, ignoring damping

$$\omega^2 - \omega_p^2 \frac{\omega}{\omega \pm \omega_c} \left(1 + \frac{1}{3} \frac{k^2 \langle v^2 \rangle}{(\omega \pm \omega_c)^2} \right) - c^2 k^2 = 0 \qquad (8\text{-}93)$$

For two species, ignoring the usually unimportant thermal correction term, one has the cold plasma dispersion relation [(8-27) and (8-28)]

$$1 - \frac{\omega_p^2}{\omega(\omega \mp \Omega_e)} - \frac{\omega_{pi}^2}{\omega(\omega \pm \Omega_i)} - \frac{c^2 k^2}{\omega^2} = 0 \qquad (8\text{-}94)$$

where again $\Omega_i = |\omega_c^+|, \Omega_e = |\omega_c^-|$. The upper sign corresponds to electron cyclotron (or whistler) modes, the lower to ion cyclotron modes. We have already seen that for $\omega \ll \Omega_i$ Alfven waves are recovered with a phase velocity $\omega/k = V_A$. For higher frequencies, the phase velocity is even smaller, going to zero at the resonances. Since usually $V_A \ll c$, $c^2 k^2/\omega^2 \gg 1$, and the first term may be ignored, leading to the approximate permeability

$$\kappa_{R,L} = - \frac{\omega_p^2}{\omega(\omega \mp \Omega_e)} - \frac{\omega_{pi}^2}{\omega(\omega \pm \Omega_i)} \qquad (8\text{-}95)$$

For whistler waves, $\omega \gg \Omega_i$, the ion term may be ignored, and $\omega \, \partial\kappa/\partial\omega + 2\kappa$ in (8-83) can be calculated (see Exercise 8-9) to yield

$$\left(\omega \frac{\partial\kappa}{\partial\omega} + 2\kappa \right)_{\omega_0} = \frac{(\omega_p^2 + c^2 k^2)^3}{c^2 k^2 \Omega_e^2 \omega_p^2} \qquad (8\text{-}96)$$

This quantity is positive, confirming our contention that the wave is damped.

8-3. Transverse Wave Instabilities

It would be wrong to conclude that resonance leads always to damping, never to instabilities. If one drops the condition of velocity isotropy, effects arising from the wave magnetic field become important. Not that the magnetic field itself exchanges energy with the particles (we know this cannot happen), but the wave magnetic field exerts a force in the z direction on the particles, hence it destroys the isotropy of the distribution in the perpendicular plane (see Fig. 8-2). As a result, $\sum \mathbf{v}_M$ is no longer zero, and this term can lead to negative values of P, hence instability. Consider an originally uniform distribution of particles on the circle drawn in Fig. 8-2. The particles most strongly interacting with the wave magnetic field are those in regions a and b. The ones in a are pushed up, the ones in b down. Also the particles in a and b interact strongest with the electric field. For those in a, $\mathbf{v}_M \cdot \mathbf{E}$ is positive and the particles contribute to damping, while the ones in b deliver energy to the field. In the absence of $\mathbf{H}$ the two power terms cancel. Now take a wave close to resonance but with ω' somewhat smaller than ω_c. If one rotates with the wave, one finds the particles in slow rotation in the direction of the arrow with an angular frequency $\omega_c - \omega'$. Now the populations of region a which move up and to the right find themselves increasingly out of phase with the electric field, while those in b moving down and to the left remain more in phase. Remembering the role of particles in these regions, it is clear that this process

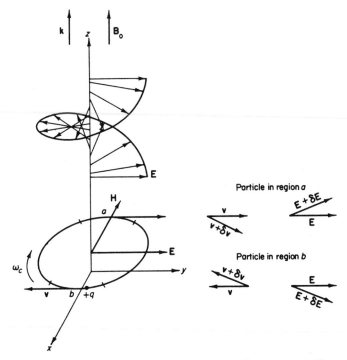

Fig. 8-2. Electromagnetic field in a transverse wave traveling along a uniform magnetic field, and illustration of particle motion in this field.

works in the direction of instability, but it also has to compete with the damping process described before. Which one wins out depends on the distribution function and also on the wavelength.

To demonstrate such an instability we choose the simple distribution function for electrons and ignore ion motion

$$f_0(\mathbf{v}) = \delta(v_\parallel) \, f_0(v_\perp) \tag{8-97}$$

and insert it into the dispersion relation (8-76) to obtain by partial integration

$$\omega^2 = c^2 k^2 + \omega_p^2 \frac{\omega}{\omega - \omega_c} + \omega_p^2 \frac{\frac{1}{2}k^2 \langle v_\perp^2 \rangle}{(\omega - \omega_c)^2} \tag{8-98}$$

where the L mode was chosen for demonstration. Expressing k^2 from (8-98) one finds

$$k^2 = \frac{\omega^2(\omega - \omega_c)^2 - \omega_p^2 \omega(\omega - \omega_c)}{c^2(\omega - \omega_c)^2 + \frac{1}{2}\omega_p^2 \langle v_\perp^2 \rangle} \tag{8-99}$$

For large values of k^2, ω becomes complex. One gets, for example, for $k^2 \to \infty$,

$$(\omega - \omega_c)^2 + \frac{\omega_p^2 \langle v_\perp^2 \rangle}{2c^2} = 0 \tag{8-100}$$

and

$$\omega = \omega_c \pm \frac{\omega_p \langle v_\perp^2 \rangle^{1/2}}{\sqrt{2} c} i \tag{8-101}$$

Note that the growth rate is just as large as that found for the same distribution function in the absence of a magnetic field in (7-146). As expected, growing modes occur when Re $\omega \approx \omega_c$.

If one takes a less artificial distribution function, by allowing some velocity spread along the z axis, one expects this instability to diminish while turning into damping when the velocity distribution becomes isotropic. To see this we derive a criterion for stability from the general dispersion relation (8-76). This can be written in the form

$$k^2 \left(c^2 - \frac{\omega^2}{k^2} \right) = \int_{-\infty}^{+\infty} \frac{F(\omega/k, v_\parallel)}{(\omega/k) - v_\parallel \pm (\omega_c/k)} \, dv_\parallel \tag{8-102}$$

where

$$F\left(\frac{\omega}{k}, v_\parallel \right) = \frac{\pi q^2}{m\varepsilon_0} \int_0^\infty \left[\left(\frac{\omega}{k} - v_\parallel \right) \frac{\partial f_0}{\partial v_\perp} v_\perp^2 + \frac{\partial f_0}{\partial v_\parallel} v_\perp^3 \right] dv_\perp \tag{8-103}$$

If F is small for $v_{||} = (\omega \pm \omega_c)/k$, one may evaluate the integral in (8-102) using again Landau's procedure to find

$$k^2\left(c^2 - \frac{\omega^2}{k^2}\right) = P\int_{-\infty}^{+\infty}\frac{F(\omega/k, v_{||})}{(\omega \pm \omega_c)/k - v_{||}}\, dv_{||} - i\pi F\left(\frac{\omega}{k}, \frac{\omega \pm \omega_c}{k}\right)$$

(8-104)

Choose for instance the "two-temperature distribution" $f_0 = n_0\alpha_\perp^2\alpha_{||}\pi^{-3/2}$ $\times \exp[-(\alpha_\perp^2 v_\perp^2 + \alpha_{||}^2 v_{||}^2)]$, where the distribution functions are Maxwellian in both the parallel and perpendicular direction, but the two temperatures are different. The integrals in (8-103) can be easily evaluated (see Exercise 8-10) to yield

$$F\left(\frac{\omega}{k}, v_{||}\right) = -\frac{\omega_p^2}{\sqrt{\pi}}\alpha_{||}\left[\frac{\omega}{k} - v_{||} + v_{||}\frac{T_\perp}{T_{||}}\right]\exp(-\alpha_{||}^2 v_{||}^2) \quad (8\text{-}105)$$

where the temperature ratio $T_\perp/T_{||} = \alpha_{||}^2/\alpha_\perp^2$.

Following the same procedure as was used for a Maxwellian plasma, the damping rate becomes

$$\gamma = \frac{\pi\, F[\omega/k, (\omega \pm \omega_c)/k]}{\omega\,(\omega\, \partial\kappa/\partial\omega + 2\kappa)_{\omega_0}} \quad (8\text{-}106)$$

where ω_0 again satisfies the $\kappa(\omega_0,k) = c^2k^2/\omega_0^2$ equation. Substituting $v_{||} = (\omega \pm \omega_c)/k$ in (8-105) one finds the point of marginal stability (Exercise 8-11)

$$\omega_m = \mp\omega_c\left(1 - \frac{T_{||}}{T_\perp}\right) \quad (8\text{-}107)$$

For a one-temperature plasma, $T_\perp = T_{||}$, and no point of marginal stability, where damping turns into instability, exists as expected. Using (8-106), (8-96), and (8-105) yields the damping (growth) rate

$$\gamma = \frac{\sqrt{\pi}\alpha_{||}\omega_p^4\Omega_c^2}{(\omega_p^2 + k^2c^2)^3 k}\left[k^2c^2 + \omega_p^2\left(1 - \frac{T_\perp}{T_{||}}\right)\right]\exp\left[-\left(\frac{\alpha_{||}\omega_p^2\Omega_c}{k(\omega_p^2 + k^2c^2)}\right)^2\right]$$

(8-108)

If $(k^2c^2/\omega_p^2) + 1 < T_\perp/T_{||}$, $\gamma < 0$ and the wave is unstable.

As in the case of an unmagnetized plasma, not all instabilities depend on resonant particles. In this case (8-76) may be integrated by parts to read

$$1 = \frac{c^2k^2}{\omega^2} + \frac{q^2}{m\varepsilon_0\omega^2}\int\left(\frac{\omega - kv_{||}}{\omega - kv_{||} \pm \omega_c} + \frac{\frac{1}{2}k^2v_\perp^2}{(\omega - kv_{||} \pm \omega_c)^2}\right)f_0\, d^3v \quad (8\text{-}109)$$

for one species. Since $c^2 k^2/\omega^2 \gg 1$, the left-hand side will again be replaced by zero.

Consider now a current-carrying plasma, where cold electrons drift along the magnetic field with velocity V, so $f_e = n_0 \delta(\mathbf{v} - \mathbf{V})$ and the ions are stationary. For the two species the dispersion relation becomes

$$c^2 k^2 + \frac{\omega_{pi}^2 \omega}{\omega - \Omega_i} + \frac{\omega_{pe}^2(\omega - kV)}{\omega - kV + \Omega_e} = 0 \qquad (8\text{-}110)$$

where the sign of ω_c has been chosen such as to give an ion wave. This is a quadratic equation in ω

$$\omega^2 - \omega\left[kV - \frac{(\Omega_e - \Omega_i)c^2 k^2}{\omega_p^2 + \omega_{pi}^2 + c^2 k^2}\right] + \Omega_i\left[kV - \frac{c^2 k^2 \Omega_e + kV\omega_{pi}^2}{\omega_p^2 + \omega_{pi}^2 + c^2 k^2}\right] = 0$$

$$(8\text{-}111)$$

It is easy to see that this equation admits several unstable regimes. The growth rates are small, less than Ω_i (Exercise 8-12).

We have seen that an anisotropic plasma with $T_\perp > T_\parallel$ can be unstable due to cyclotron resonance of some particles in the distribution function. In the opposite case when $T_\parallel > T_\perp$ a hydromagnetic-type instability exists. This is the so-called *firehose instability* and, while it may be derived from the Vlasov equation (see Exercise 8-13), it is instructive to use the hydromagnetic equations derived in Sec. 3-6 to obtain the same results.

Writing $\mathbf{B} = \mathbf{B}_0 + \mathbf{b}$ and linearizing (3-109) one obtains

$$\nu\left(\frac{\partial \mathbf{v}}{\partial t}\right)_\perp + \nabla_\perp\left(p_\perp + \frac{B^2}{2\mu_0}\right) - \frac{B_0}{\mu_0}\frac{\partial \mathbf{b}}{\partial z}\xi = 0 \qquad (8\text{-}112)$$

where

$$\xi = \frac{p_\perp - p_\parallel}{B^2/\mu_0} + 1 \qquad (8\text{-}113)$$

Since we seek solutions in the form of plane waves we are led to the elimination of the $\nabla_\perp$ term. Equation (3-111) becomes, after linearization,

$$B_0 \frac{\partial \mathbf{v}_\perp}{\partial z} = \frac{\partial \mathbf{b}}{\partial t} \qquad (8\text{-}114)$$

The wave equation follows immediately from (8-112) and (8-114):

$$\frac{\partial^2 \mathbf{v}_\perp}{\partial t^2} - \frac{B_0^2}{\nu\mu_0}\xi\frac{\partial^2 \mathbf{v}_\perp}{\partial z^2} = 0 \qquad (8\text{-}115)$$

This is the same equation obtained for Alfven waves in a perfectly conductive fluid except that the wave velocity is now multiplied by $\xi^{1/2}$. If the pressure is isotropic, $\xi = 1$. If $p_\perp > p_\parallel$, the waves travel faster than in the isotropic case, while $p_\parallel > p_\perp$ leads to slow wave propagation. If, however, $p_\parallel$ is so much larger than $p_\perp$ as to make ξ negative, waves do not propagate but instability develops. This happens if

$$p_\parallel - p_\perp > B_0{}^2/\mu_0 \qquad (8\text{-}116)$$

The rate of growth of this instability is

$$\omega_g = \left[\frac{1}{v_0}\left(p_\parallel - p_\perp - \frac{B_0{}^2}{\mu_0}\right)\right]^{1/2} k \qquad (8\text{-}117)$$

where k is the wave number.

The physical mechanism of this instability is explained on the basis of Fig. 8-3. Consider particles gyrating about the field lines with their $v_\perp$

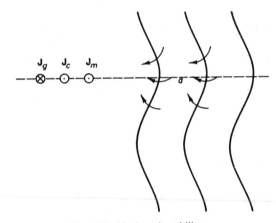

FIG. 8-3. Firehose instability.

velocities, while they move along them with $v_\parallel$. As a result of the curvature of field lines, centrifugal drifts arise, resulting in currents in the direction indicated in the figure. This centrifugal current density $\mathbf{J}_c$ is proportional to the particle density, $v_\parallel{}^2$, particle mass, and inversely proportional to the radius of curvature and the magnetic field. In short, $J_c = p_\parallel/B_0 R$. The other effect caused by the field perturbation is a geometrical one, as seen in Fig. 8-3; there are more positive particles moving into the paper in region a than out of it. The result is a geometrical current $\mathbf{J}_g$ opposing $\mathbf{J}_c$, which is proportional to the magnetic moment μ_m and particle density, and inversely proportional to the radius of curvature. This gives $J_g = p_\perp/B_0 R$. These plasma currents must be compared to the current which is necessary to maintain the given deformation of the uniform magnetic field.

It is easy to show that this current is $J_m = B/\mu_0 R$ (see Exercise 8-14) and it points in the same direction as J_c. If the plasma currents are smaller than the current required to maintain the deformation of field lines, the perturbation swings back and waves result. If, however,

$$J_c - J_g > J_m \qquad (8\text{-}118)$$

the perturbation grows and instability sets in. Equation (8-118) is just the condition for instability obtained in (8-116) in a different way. Since similar instabilities, caused by centrifugal forces, arise in hydrodynamic systems, e.g., in water streaming along a firehose, this instability is also called the *firehose instability*.

Turning now to waves propagating at an angle to the magnetic field, one notes first that transverse and longitudinal modes do not split up as for waves traveling along the field. As a consequence, the dispersion relation, and also the physics of these processes, become rather involved. Some of the better-known instabilities will be discussed briefly.

Consider the superposition of two waves with the space-time dependence $\exp[i(\omega t - k_z z - k_x x)]$ and $\exp[i(\omega t - k_z z + k_x x)]$, respectively. The result is a wave traveling in the z direction (but not a plane wave any more), described by

$$\exp[i(\omega t - k_z z)]\cos k_x x \qquad (8\text{-}119)$$

The magnetic field subjected to such a modulation looks at a certain instant like the one shown in Fig. 8-6, while the pattern moves in the z direction in time.

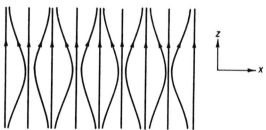

Fig. 8-4 Magnetic field of mirror instability.

If one forgets for the moment about the electric fields associated with such a wave, one finds a magnetic resonance between such waves and particles traveling roughly with the phase velocity ω/k_z along the field lines. They find themselves in a field varying slowly enough in the guiding-center system to preserve the constancy of μ_m. We have seen in conjunction with (2-117) that under such conditions the guiding center moves along the field lines as if subjected to a potential $V = \mu_m B$. The resulting motion of the guiding centers along the periodic condensations and rarefactions of the magnetic field lines is very much like the particle motion in a periodic electrostatic potential. The behavior of such particles moving with approximately the phase velocity of the wave has been discussed in conjunction with Landau damping in Sec. 7-2, and it was found that particles traveling somewhat slower than the wave gain energy, while those traveling somewhat faster lose energy to the wave. A preponderance of the latter may lead to instability. The same arguments leading to the same results also apply here, e.g., no instability exists unless there is a minimum in the distribution function (integrated over $v_\perp$) at some value of $v_\parallel$. For a thermal distribution the so-called *transit time damping* results.

The same magnetic field geometry can also lead to a hydromagnetic instability. Consider a plasma with perpendicular velocities exceeding the longitudinal ones. In the "mirror" geometries of Fig. 8-4 particles are pushed toward regions where the magnetic field is weak. The accumulation of particles at any point leads, on the other hand, to a further local weakening of the field, owing to the diamagnetic effect of gyrating particles. Unless parallel velocities are large enough so that particles can "break out" of the traps, instability develops. One can also see that the most "dangerous" instabilities occur where the wavelength in the z direction is longest. Particles traveling from the weak fields to the constrictions slow down, those moving in the other direction speed up. If the field configuration did not change in the meantime, then after a while one found slower particles at the constrictions than in the rarefactions. Slow particles, however, spend a longer time on a unit length than fast ones; consequently particle density would tend to increase in the constrictions working against instability. Therefore, roughly speaking, if particles can travel a wavelength before the field changes considerably, instability is avoided, while for long waves this competing process can be discounted. Although the derivation of the corresponding dispersion equation is not given here, we mention that the growth rates are comparable to those obtained for $\mathbf{k} \parallel \mathbf{B}_0$ when $T_\perp$ exceeds $T_\parallel$.

8-4. Electrostatic Instabilities

Turning now to electrostatic modes, the perturbed distribution is again evaluated by integrating along particle trajectories, except now $\mathbf{B}_1 = 0$. The equilibrium trajectories are again given by (8-59) to (8-61) and the distribution function is

$$f_1(t,\mathbf{x},\mathbf{v}) = -\frac{q}{m} \int_{-\infty}^{t} \mathbf{E} \cdot \frac{\partial f_0}{\partial \mathbf{v}} \exp[i(\omega t' - \mathbf{k} \cdot \mathbf{x})] \, dt' \qquad (8\text{-}120)$$

where we do not restrict ourselves to propagation parallel to $\mathbf{B}_0$. Since unstable modes are considered, we integrate from $t' = -\infty$ where the initial perturbation vanishes. Introducing again $\tau = t' - t$ the particle positions along the unperturbed trajectory are obtained from integrating (8-59) to (8-51)

$$z' - z = v_{\parallel} \tau \qquad (8\text{-}121)$$

$$y' - y = \frac{v_{\perp}}{\omega_c} [\sin(\omega_c \tau + \psi) - \sin \psi] \qquad (8\text{-}122)$$

$$x' - x = -\frac{v_{\perp}}{\omega_c} [\cos(\omega_c \tau + \psi) - \cos \psi] \qquad (8\text{-}123)$$

Since $\mathbf{E} \parallel \mathbf{k}$, one may write

$$\mathbf{E} \cdot \frac{\partial f_0}{\partial \mathbf{v}} = \frac{E}{k} \mathbf{k} \cdot \frac{\partial f_0}{\partial \mathbf{v}}$$

and

$$\mathbf{k} \cdot \frac{\partial f_0}{\partial \mathbf{v}}(t') = k_{\parallel} \frac{\partial f_0}{\partial v_{\parallel}} + k_{\perp} \frac{v_y'}{v_{\perp}} \frac{\partial f_0}{\partial v_{\perp}} = k_{\parallel} \frac{\partial f_0}{\partial v_{\parallel}} + k_{\perp} \cos(\omega_c \tau + \psi) \frac{\partial f_0}{\partial v_{\perp}} \qquad (8\text{-}124)$$

where $\mathbf{k}_{\perp}$ has been chosen to point in the y direction, without loss of generality. Substituting into (8-120) yields

$$f_1(t,\mathbf{x},\mathbf{v}) = -\frac{q}{m} \frac{E}{k} \int_{-\infty}^{0} \exp\left[i\left(\omega \tau - k_{\parallel} v_{\parallel} \tau - \frac{k_{\perp} v_{\perp}}{\omega_c} [\sin(\omega_c \tau + \psi) - \sin \psi] \right) \right]$$

$$\cdot \left(k_{\parallel} \frac{\partial f_0}{\partial v_{\parallel}} + k_{\perp} \cos(\omega_c \tau + \psi) \frac{\partial f_0}{\partial v_{\perp}} \right) d\tau \, \exp[i(\omega t - \mathbf{k} \cdot \mathbf{x})] \qquad (8\text{-}125)$$

From the identity

$$\exp(i\alpha \sin x) = \sum_{-\infty}^{+\infty} J_n(\alpha) \exp(inx) \qquad (8\text{-}126)$$

where the J_n's are Bessel functions, one may write the integral in the form

$$I = \int_{-\infty}^{0} d\tau \sum_n \sum_m J_n\left(\frac{k_{\perp} v_{\perp}}{\omega_c} \right) J_m\left(\frac{k_{\perp} v_{\perp}}{\omega_c} \right) \exp[i(\omega - k_{\parallel} v_{\parallel} - n\omega_c)\tau]$$

$$\cdot \exp[-i(n - m)\psi]$$

$$\cdot \left(k_{\parallel} \frac{\partial f_0}{\partial v_{\parallel}} + \frac{k_{\perp}}{2} \frac{\partial f_0}{\partial v_{\perp}} \{\exp[i(\omega_c \tau + \psi)] + \exp[-i(\omega_c \tau + \psi)]\} \right) \qquad (8\text{-}127)$$

One may now perform the τ integration, noting that for unstable waves Im $\omega < 0$; hence $e^{i\omega\tau}$ vanishes as $\tau \to -\infty$,

$$I = \sum_{n,m} \exp[-i(n-m)\psi] J_n\left(\frac{k_\perp v_\perp}{\omega_c}\right) J_m\left(\frac{k_\perp v_\perp}{\omega_c}\right) \left[\frac{k_\parallel \, \partial f_0/\partial v_\parallel}{i(\omega - k_\parallel v_\parallel - n\omega_c)}\right.$$
$$\left. + \frac{k_\perp}{2}\frac{\partial f_0}{\partial v_\perp}\left(\frac{\exp(i\psi)}{i[\omega - k_\parallel v_\parallel - (n-1)\omega_c]} + \frac{\exp(-i\psi)}{i[\omega - k_\parallel v_\parallel - (n+1)\omega_c]}\right)\right]$$
$$(8\text{-}128)$$

Rewrite now the $(k_\perp/2)(\partial f_0/\partial v_\perp)$ term in the summation over n

$$\sum_n \left(J_n \frac{\exp[-i(n-1)\psi]}{i[\omega - k_\parallel v_\parallel - (n-1)\omega_c]} + J_n \frac{\exp[-i(n+1)\psi]}{i[\omega - k_\parallel v_\parallel - (n+1)\omega_c]}\right)$$
$$= \sum_v (J_{v+1} + J_{v-1}) \frac{\exp(iv\psi)}{i(\omega - k_\parallel v_\parallel - v\omega_c)} \qquad (8\text{-}129)$$

Making use of the identity

$$J_{v+1}(x) + J_{v-1}(x) = \frac{2v}{x} J_v(x) \qquad (8\text{-}130)$$

and changing the summation index $v \to n$, yields finally a simpler form for I

$$I = \sum_{n,m} \exp[i(m-n)\psi] \frac{J_n(k_\perp v_\perp/\omega_c) J_m(k_\perp v_\perp/\omega_c)}{i(\omega - k_\parallel v_\parallel - n\omega_c)}\left(k_\parallel \frac{\partial f_0}{\partial v_\parallel} + \frac{n\omega_c}{v_\perp}\frac{\partial f_0}{\partial v_\perp}\right) \qquad (8\text{-}131)$$

The $(\omega\mathbf{k})$ Fourier component of f_1 is simply $-(q/m)(E/k)I$ from (8-125). The space charge produced by this distribution function is

$$\rho(\omega,\mathbf{k}) = -\frac{q^2}{m}\frac{E}{k}\int Iv_\perp \, dv_\perp \, dv_\parallel \, d\psi \qquad (8\text{-}132)$$

The ψ integration is easy

$$\int \exp[i(m-n)\psi] d\psi = 2\pi \, \delta_{m,n} \qquad (8\text{-}133)$$

where δ_{mn} is the Kronecker δ ($= 0$ if $m \neq n$, 1 if $m = n$). The dispersion relation follows from Poisson's equation $\rho = -i\varepsilon_0 kE$, to yield

$$1 = -\frac{2\pi}{k^2}\sum_n \frac{q^2}{m\varepsilon_0}\int_{-\infty}^{+\infty} dv_\parallel \int_0^\infty dv_\perp \, v_\perp \frac{J_n^2(k_\perp v_\perp/\omega_c)}{\omega - k_\parallel v_\parallel - n\omega_c}\left(k_\parallel \frac{\partial f_0}{\partial v_\parallel} + \frac{n\omega_c}{v_\perp}\frac{\partial f_0}{\partial v_\perp}\right) \qquad (8\text{-}134)$$

If ions also participate in creating the space charge a similar ion term must be added.

For a cold electron plasma with a distribution function $f_0 = (n/2\pi v_\perp)$ $\delta(v_\perp)\,\delta(v_\parallel)$ this dispersion relation can be evaluated to yield

$$\kappa_e = 1 - \frac{\omega_p^2}{\omega^2}\frac{k_\parallel^2}{k^2} - \frac{\omega_p^2}{\omega^2 - \Omega_e^2}\frac{k_\perp^2}{k^2} = 0 \qquad (8\text{-}135)$$

The details are left for Exercise 8-15. The $k_\parallel$ term represents electron motion parallel to the $\mathbf{B}_0$ field, while the $k_\perp$ term is due to plasma polarization in the perpendicular direction. If $\omega \ll \Omega_e$ one may easily derive this dispersion relation from the electron drift equations (Exercise 8-16).

The label "cold plasma" is somewhat deceptive. All that is needed for this approximation to hold is that the argument of the Bessel functions $k_\perp v_\perp/\Omega_e = k_\perp R_e \ll 1$, and $v_\parallel \ll \omega/k_\parallel$ for most particles. In other words, the perpendicular wavelength must be much larger than the electron gyroradius, and the parallel phase velocity should largely exceed the electron thermal velocity. Even for hot electrons these conditions are easily satisfied for a significant set of wave numbers. Note that for experimental conditions, when the plasma size is finite $\mathbf{k}$ space is limited; $k_\parallel L_\parallel \gtrsim 1$ and $k_\perp L_\perp \gtrsim 1$, where $L_\parallel$, $L_\perp$ are the lengths of the plasma parallel and perpendicular to the field.

In a mirror machine the distribution function by necessity deviates from a Maxwellian. As we have seen before single particles escape if $v_\parallel/v_\perp$ is large, creating a "loss cone" distribution function. Under experimental conditions, electron-electron collisions rapidly wipe out this loss cone for electrons, but this process is considerably slower for ions. The ions then stay confined in the mirror field, while the electrons are kept inside by electrostatic forces due to the ion space charge. We will now show that this distribution function is unstable, due to the so-called *loss cone instability*.

Rewrite (8-134) for the two species as

$$\kappa_e + \frac{2\pi}{k^2}\sum_n \frac{q^2}{M\varepsilon_0}\int_{-\infty}^{+\infty}dv_\parallel\int_0^\infty dv_\perp\, v_\perp \frac{J_n^2(k_\perp v_\perp/\omega_c)}{\omega - k_\parallel v_\parallel - n\Omega_i}\left(k_\parallel\frac{\partial f_0}{\partial v_\parallel} + \frac{n\Omega_i}{v_\perp}\frac{\partial f_0}{\partial v_\perp}\right) = 0 \qquad (8\text{-}136)$$

One expects the ion terms to be relatively small due to the ion mass in the denominator. However, because of the resonant denominator in the ion term, imaginary parts arise from the pole contributions. Recognizing the ion term as χ_i, and integrating above the poles.

$$\mathrm{Im}\,\chi_i = \frac{2\pi^2}{k^2}\sum_n \frac{q^2}{M\varepsilon_0 k_\parallel}\int_0^\infty dv_\perp\, v_\perp J_n^2\!\left(\frac{k_\perp v_\perp}{\Omega_i}\right)\left(k_\parallel\frac{\partial f_0}{\partial v_\parallel} + \frac{n\Omega_i}{v_\perp}\frac{\partial f_0}{\partial v_\perp}\right)_{r_{\parallel n}} \qquad (8\text{-}137)$$

where $v_{\|n} = (\omega - n\Omega_i)/k_{\|}$. Carrying out the usual expansion, ω and $k_{\|}$ satisfy to lowest order the $\kappa_e(\omega_0, k_0) = 0$ equation. Since we want to use κ_e as given in (8-135), we are already committed to choosing large values for $\omega/k_{\|} \gg v_{te}$. Similarly, for ions $f_{0i}(v_{\|} = \omega/k_{\|})$ is exceedingly small. Consequently (8-137) gives a significant contribution only if $v_{\|n} \ll \omega/k_{\|}$; hence $\omega \approx N\Omega_i$, for some $n = N$. The sign of the imaginary part of ω

$$\text{Im } \omega = -\frac{\text{Im } \chi_i}{\partial \kappa_e/\partial \omega} \tag{8-138}$$

depends on the sign of Im χ_i. Figure 8-5 illuminates the situation.

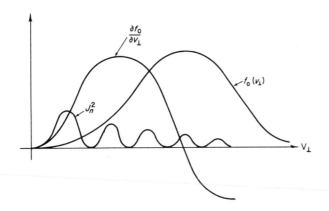

FIG. 8-5. Explanation of the loss cone instability.

For a loss cone distribution, $\partial f_0/\partial v_\perp$ is positive for $v_\perp$ values inside the loss cone. The integral of $\int J_n^2(\partial f_0/\partial v_\perp)\, dv_\perp$ can be positive if the first zero of J_N occurs for sufficiently small values of $v_\perp$ inside the loss cone, since J_n^2 oscillates with a diminishing amplitude as its argument increases. So the second integral in (8-137) can be made positive. Choosing $k_{\|}$ sufficiently small (comparable to the reciprocal machine length), the first integral becomes much smaller than the second. The quantity $\partial \kappa/\partial \omega$ is positive as one may easily see. In fact, unless N is very large (comparable to the mass ratio) $\omega \approx N\Omega_i \ll \Omega_e$ and

$$\kappa_e = 1 + \frac{\omega_p^2}{\Omega_e^2}\frac{k_\perp^2}{k^2} - \frac{\omega_p^2}{\omega^2}\frac{k_{\|}^2}{k^2} \tag{8-139}$$

is a very good approximation. This gives

$$\left.\frac{\partial \kappa_e}{\partial \omega}\right)_{\omega_0} = \frac{2}{\omega_0}\left(1 + \frac{\omega_p^2 k_\perp^2}{\Omega_e^2 k^2}\right) = \frac{2}{N\Omega_i}\left(1 + \frac{\omega_p^2 k_\perp^2}{\Omega_e^2 k^2}\right) \qquad (8\text{-}140)$$

Consequently Im $\omega < 0$ and the plasma is unstable.

From (8-139) $\omega_0 < \omega_p$. Since on the other hand $\omega_0 \approx N\Omega_i$, instability does not occur for very small densities or large fields, when $\omega_p \lesssim N\Omega_i$. Another way to avoid this instability is suggested by the small $k_\parallel$ condition. If the mirror machine is short, instability may again be avoided.

For a long machine, above the density threshold, the dependence of the growth rate on the density may be estimated. Im $\chi_i \sim \omega_p^2/k_\parallel$ for given values of Ω_i, N, $k_\perp$. From (8-139), $k_\parallel \omega_p \approx N\Omega_i k \sqrt{1 + (\omega_p/\Omega_e)^2}$, for $k_\parallel \ll k_\perp$. If $\omega_p/\Omega_e \ll 1$, $k_\parallel \sim \omega_p^{-1}$, and the growth rate is proportional to $\omega_{pi}^2\omega_p$, hence the $\frac{3}{2}$ power of the particle density. A rough estimate gives the order of magnitude for the growth rate in this regime as $\gamma \approx \omega_{pi}^2\omega_p/N^2\Omega_i^2$. Verify this estimate by working out Exercise 8-17.

8-5. Summary

In the presence of a background magnetic field, electrostatic (longitudinal) and electromagnetic (transverse) waves can, in general, no longer be separated. The magnetoplasma is characterized by a tensor dielectric permeability. One finds that, in general, two different modes with different frequencies and polarizations exist for a wave vector **k**. Propagation of electron modes below the plasma frequency is now also possible. Electron and ion cyclotron waves propagate along the magnetic field, and degenerate into transverse Alfven waves for low frequencies and high densities. Hybrid waves propagate transverse to the **B₀** field and produce the upper and lower hybrid resonance.

In a hot thermal plasma a new damping mechanism, cyclotron damping, makes its appearance. In a nonthermal plasma, this resonant damping can turn into an instability, as in the case of a two-temperature plasma, where the average perpendicular particle energy exceeds the parallel one. Nonresonant transverse wave instabilities for parallel propagation were also discussed. The firehose instability destabilizes a plasma with large parallel particle energies, while a current along the magnetic field produced by counter-streaming electrons and ions leads to a transverse beam-plasma instability.

While the equations describing propagation in an arbitrary direction with respect to **B₀** are very complicated, headway can be made by restricting ourselves to electrostatic modes. An important instability arises from this analysis, the loss cone instability. In mirror machines, the distribution function of ions contains a loss cone. An electrostatic electron wave propagating at an angle to the field interacts resonantly with ions leading to an instability once the electron plasma frequency exceeds several times the ion cyclotron frequency. Beyond this (rather low) threshold the instability has a rapidly increasing growth rate, with larger plasma densities.

EXERCISES

8-1. Derive the form of the dielectric tensor $\overleftrightarrow{\kappa}$ in the nonrotating Cartesian system.

8-2. Derive the dispersion relation for a cold plasma in a uniform magnetic field in the form (8-20), as well as in the alternative form (8-24).

8-3. Prove that the dispersion relation (8-35) corresponds to a linearly polarized wave with the electric field vector pointing in the direction of $\mathbf{B_0}$.

8-4. Prove that the plasma resonance $\kappa_T = 0$ curve goes through the following four points in parameter space: $(\Pi_p = 0, \beta^- = 1), (\beta^- = 0, \Pi_p^2 = 1 - \beta^{+2}), (\Pi_p = 0, \beta^+ = 1),$ and $(\Pi_p \to \infty, \beta^+\beta^- = 1)$.

8-5. Prove that for a cold plasma situated in a uniform magnetic field, no cutoffs ($n = 0$) exist outside the principal directions.

8-6. Prove that the boundaries of the regions in the Allis diagram where other than principal resonances occur (8-47) are always principal ($\theta = 0, \pi/2$) resonance and cutoff lines. Consequently the topology of wave normal surfaces is not altered by nonprincipal resonances inside a region bounded by principal resonance and cutoff lines.

8-7. Use the drift equations for electrons in a magnetic field and $\nabla \cdot (\mathbf{J} + \mathbf{D}) = 0$ to derive (8-56). Since $\omega \gg \Omega_i$, the ions can be treated as if they were unmagnetized.

8-8. Use the drift equations to show that the perpendicular dielectric constant for electrons is $\kappa_\perp = (1 + \omega_p^2/\Omega_e^2)$. Prove furthermore that the characteristic frequency of unmagnetized ions oscillating in a dielectric is $\omega = \pm\omega_{pi}/\sqrt{\kappa_\perp}$.

8-9. Derive (8-96).

8-10. Derive (8-105).

8-11. Determine the frequency for marginal stability (8-107).

8-12. Derive the dispersion relation (8-111) and show that unstable regions exist. Investigate for instance long-wavelength perturbations. Can you prove that the growth rate never exceeds Ω_i for any k?

8-13. Use (8-109) to study the nonresonant instability for a two-temperature plasma with $T_\parallel > T_\perp$ for ions. Expand (8-109) treating $|(\omega - kv_\parallel)/\omega_c|$ $\ll 1$ for both species and show that the resulting dispersion relation is identical with the Fourier-analyzed form of (8-118). Assume $\omega/k \ll v_\parallel$.

8-14. Prove that the current density necessary to produce a magnetic field like the one shown in Fig. 8-3 is $B_0/\mu_0 R$, where R is the radius of curvature of the field lines. (The deviation from the uniform magnetic field is small.)

8-15. Use the cold plasma distribution function and (8-134) to derive (8-135).

8-16. Derive the electrostatic dielectric function for a cold electron plasma in a magnetic field, using drift equations, and compare your result with (8-135).

8-17. Estimate the growth rate of the loss cone instability and show that for $\omega_p/\Omega_e \ll 1$, $\gamma \approx \omega_{pi}^2 \omega_p/N^2 \Omega_i^2$.

IX

Nonlinear Waves

9-I. Limits of the Linear Theory

Our treatment of linear waves and instabilities was based on an expansion of the distribution function $f = f_0 + f_1$, where f_1 and the wave fields were treated as small quantities and their products had been neglected.

To appreciate the limits of the linear theory it will suffice to consider plane electrostatic waves (stable or unstable) in a uniform magnetic field free plasma. The equations describing such waves are the Vlasov equation and Poisson's equation. The former is nonlinear in the wave quantities

$$\frac{\partial f_1}{\partial t} + v \frac{\partial f_1}{\partial x} + \frac{q}{m} E \left(\frac{\partial f_0}{\partial v} + \frac{\partial f_1}{\partial v} \right) = 0 \tag{9-1}$$

while the latter is linear. The linear treatment should be a good approximation when $|\partial f_0/\partial v| \gg |\partial f_1/\partial v|$. We have calculated f_1 in the linear approximation in (8-13), and substitute its value into our inequality to find

$$\left| \frac{\partial f_1}{\partial v} \right| = \left| \frac{\partial}{\partial v} \left(\frac{q}{m} E \frac{1}{kv - \omega} \frac{\partial f_0}{\partial v} \right) \right| \ll \left| \frac{\partial f_0}{\partial v} \right| \tag{9-2}$$

as the condition for the validity of the linear approximation. The frequency $\omega(k)$ is, in general, complex and is to be obtained from the dispersion relation. For a given wave both sides of the inequality are functions of v. It is already clear that for ω nearly real, the left-hand side will be largest where $v \approx \omega/k$. Carrying out the differentiation one finds

$$\left| \frac{q}{m} \frac{E}{kv - \omega} \left(f_0'' - \frac{f_0'}{v - \omega/k} \right) \right|_{max} \approx \left| \frac{qkEf_0'}{m(kv - \omega)^2} \right| \ll |f_0'| \tag{9-3}$$

where the prime denotes differentiation with respect to v. Using $\text{Im } \omega = \gamma$ and

269

Re $\omega = kV$ for the growth (damping) rate one finds that the linear theory holds when

$$\gamma \gg \left(\frac{qkE}{m}\right)^{1/2} \equiv \omega_b \qquad (9\text{-}4)$$

If the wave is damped and the initial wave amplitude is small enough to satisfy this inequality, the linear theory is satisfactory. If, however, the wave is unstable, E grows exponentially, and no matter how small its initial value, there will be a time when (9-4) is no longer satisfied and the linear treatment breaks down. Of course, the wave amplitude cannot exponentiate forever; energy conservation will not permit it.

Notice that resonant particles ($v \approx \omega/k$) will be the first to experience nonlinear effects. Other regions of velocity space can behave linearly even for large wave amplitudes. This fact can be usefully exploited.

It would seem from (9-4) that an undamped, stable wave can never be treated linearly. In order to see that this is not so, we have to treat the Vlasov equation as an initial-value problem. Problems involving resonance behavior (like a resonantly driven harmonic oscillator), will yield infinities unless treated more carefully. For ω real the left side of (9-2) leads to infinity for similar reasons.

Consider the problem of a sinusoidal electric field switched on the plasma at $t = 0$ and follow the development in time. The linear evolution of f_1 is given by

$$\frac{\partial f_1}{\partial t} + v\frac{\partial f_1}{\partial x} + \frac{q}{m}E_0 \exp[i(kx - \omega t)]\frac{\partial f_0}{\partial v} = 0 \qquad (9\text{-}5)$$

where $\omega(k)$ satisfies the dispersion relation with ω real. By integrating along the unperturbed trajectory $x = x_0 + vt$, f_1 is easily determined

$$[f_1]_0^t = i\frac{q}{m}E_0 f_0'\left[\frac{\exp\{i[k(x_0 + vt')] - \omega t'\}}{kv - \omega}\right]_0^t \qquad (9\text{-}6)$$

Suppose now that the electric field was switched on at $t = 0$ by external sources (say grids in the plasma) so $f_1(t = 0) = 0$. From (9-6)

$$f_1(t) = i\frac{q}{m}E_0 f_0'\frac{\exp[i(kx - \omega t)] - \exp[ik(x - vt)]}{kv - \omega} \qquad (9\text{-}7)$$

Clearly $f_1(0)$ as well as $\partial f_1/\partial v$ at $t = 0$ vanishes. One expects the linear solution to break down at some later time in the resonant region where $v = \omega/k$. Introduce the new velocity space variable $u = v - \omega/k$, the particle velocity in the wave frame, and expand for small values of kut,

$$f_1 = i\frac{q}{m}Ef_0'\frac{1 - \exp(-ikut)}{ku} \approx i\frac{q}{m}Ef_0'\frac{ikut + \frac{1}{2}k^2u^2t^2}{ku} \qquad (9\text{-}8)$$

where $E = E_0 \exp[i(kx - \omega t)]$ is the wave electric field, and

$$\left(\frac{\partial f_1}{\partial v}\right)_{\omega/k} = \left(\frac{\partial f_1}{\partial u}\right)_0 = i\frac{q}{m} E f_0' \tfrac{1}{2}kt^2 \tag{9-9}$$

where the f_0'' term has again been neglected since it increases only linearly with time, while higher-order terms ignored in (9-8) vanish for $u = 0$. Comparing (9-9) with f_0', one finds that the linear theory holds for times

$$t \ll \sqrt{\frac{m}{qEk}} = \omega_b^{-1}. \tag{9-10}$$

One concludes that even for a very weakly damped wave the linear theory holds for times short compared to ω_b^{-1}. Later, the linear solution again breaks down in the resonant region first [because of the resonant denominator in (9-7)] and the nonlinear effects spread gradually to the nonresonant regions of velocity space, to larger values of u.

The quantity ω_b has a simple physical meaning. Consider particles trapped in the potential wells of the wave. The equation of motion of a particle in the wave frame is

$$\ddot{x} = \frac{q}{m} E \sin kx \approx \frac{q}{m} Ekx \tag{9-11}$$

for particles near the bottom of a potential well ($kx \ll 1$). These particles oscillate with the frequency

$$\omega_b = \left(\frac{qEk}{m}\right)^{1/2} \tag{9-12}$$

called the *bounce frequency*.

Particle trapping is a nonlinear effect not accounted for in the linear treatment. Recall that f_1 is obtained by integration along unperturbed straight line trajectories in phase space. A trapped particle for times short compared to the bounce time does not "know" yet that it is trapped, but around $t \approx \omega_b^{-1}$ it leaves the vicinity of the unperturbed trajectory forever. In Fig. 9-1 the phase space trajectories of particles in the wave frame are plotted. These are lines of constant energy $\tfrac{1}{2}mv^2 + q\phi$, where ϕ is the potential. The unperturbed trajectories are straight horizontal lines $v = $ const. In the linear treatment a wavy deformation of these trajectories develops, but as the separatrix between trapped and untrapped particles is approached, this approximation gets worse and breaks down completely for trapped particles.

When the wave is damped (or grows), the wave potential changes and so does the separatrix, spilling out trapped particles or absorbing new ones. Particles gain or lose energy in the changing potential, and if this happens

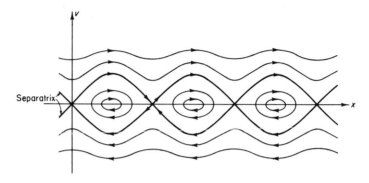

FIG. 9-1. Phase plane trajectories for electrostatic wave in the wave frame.

more rapidly than a bounce time, trapped particle behavior is no longer important. This leads to the condition $\gamma \gg \omega_b$ for linear behavior, as seen earlier.

One may also obtain an estimate of the width in velocity space for the breakdown of linear theory. A particle is trapped in the wave frame as long as $\frac{1}{2}mu^2 < q\phi$, leading to the approximate trapping width $\Delta v \ (\approx u_{\max})$,

$$\Delta v = \sqrt{\frac{q\phi}{m}} = \frac{\omega_b}{k} \qquad (9\text{-}13)$$

In the presence of a magnetic field an other resonance, the cyclotron resonance has been encountered. Consider, e.g., a whistler wave, where particles moving along the magnetic field with the velocity $v_{\parallel} \approx (\omega \pm \omega_c)/k$ gain energy by cyclotron acceleration and thereby damp the wave. The linear theory, however, involves the assumption of $v_{\parallel} = $ const. It is easy to see that due to $\mathbf{v}_{\perp} \times \mathbf{B}_1$ forces ($\mathbf{B}_1$ is the wave field) particles accelerate along $\mathbf{B}_0$ lines, invalidating the linear approximation.

To estimate the breakdown of the linear theory, it is again convenient to work in the wave frame. In such a frame moving with the velocity ω/k, the wave electric field is transformed away, leaving the magnetic field unchanged and constant ($\dot{\mathbf{B}} = -\nabla \times \mathbf{E} = 0$). For the circularly polarized wave under investigation, a helical static magnetic field, composed of $\mathbf{B}_0$ and $\mathbf{B}_1$, is seen in this frame. In fact, particle kinetic energy is conserved; hence particles cannot be indefinitely accelerated in the lab frame as predicted in linear theory. The linearized Vlasov equation may be written down in the wave frame.

$$\frac{\partial f_1}{\partial t} + \mathbf{u} \cdot \frac{\partial f_1}{\partial \mathbf{x}} + (\mathbf{u} \times \mathbf{\Omega}_0) \cdot \frac{\partial f_1}{\partial \mathbf{u}} = \frac{df_1}{dt} = -(\mathbf{u} \times \mathbf{\Omega}_1) \cdot \frac{\partial f_0}{\partial \mathbf{u}} \qquad (9\text{-}14)$$

where $\Omega_0 = (q/m)\mathbf{B}_0$, $\Omega_1 = (q/m)\mathbf{B}_1$, and $\mathbf{u} = \mathbf{v} - (\omega/k)\hat{\mathbf{e}}_3$ is the velocity vector in the wave frame. The unperturbed velocities are $u_\parallel = \text{const}$ and

$$\mathbf{u}_\perp = u_\perp(\hat{\mathbf{e}}_1 \sin \phi + \hat{\mathbf{e}}_2 \cos \phi) \tag{9-15}$$

where the phase angle in velocity space $\phi = \phi_0 \pm \Omega_0 t$. The circularly polarized wave field is

$$\boldsymbol{\Omega}_1 = \Omega_1(\hat{\mathbf{e}}_1 \sin kz + \hat{\mathbf{e}}_2 \cos kz) \tag{9-16}$$

For an isotropic velocity distribution function $f_0(v)$, one finds

$$(\mathbf{u} \times \boldsymbol{\Omega}_1) \cdot \frac{\partial f_0}{\partial \mathbf{u}} = (\mathbf{u} \times \boldsymbol{\Omega}_1) \cdot \frac{\mathbf{u} + (\omega/k)\hat{\mathbf{e}}_3}{v} \frac{\partial f_0}{\partial v} = (\mathbf{u}_\perp \times \boldsymbol{\Omega}_1)_\parallel \frac{\omega}{kv} \frac{\partial f_0}{\partial v} \tag{9-17}$$

and integrate (9-14) to find

$$f_1(t) = -\text{Re}\left(\frac{\omega}{k} \frac{\partial f_0}{\partial v} \frac{v_\perp}{v} \Omega_1 \exp[i(\phi - kz)] \frac{1 - \exp[i(ku_\parallel \mp \Omega_0)t]}{ku_\parallel \mp \Omega_0}\right) \tag{9-18}$$

One now expands (9-18) in the resonance region $ku_\parallel \mp \Omega_0 \approx 0$ for small values of t, and compares $(\mathbf{u} \times \boldsymbol{\Omega}_1) \cdot \partial f_0/\partial \mathbf{v}$ with $(\mathbf{u} \times \boldsymbol{\Omega}_1) \cdot \partial f_1/\partial \mathbf{v}$, to find that the linear theory holds for times

$$t \ll (v_\perp \Omega_1 k)^{-1/2} \tag{9-19}$$

The details are left for Exercise 9-1. As we see, the result is similar to the one found for electrostatic waves, only the electric acceleration $(q/m)E$ is replaced by the magnetic acceleration term $v_\perp\Omega_1$. The breakdown time differs for particles in different parts of the resonant slice in velocity space. For those few particles with $v_\perp$ very large, the breakdown occurs quickly, but most resonant particles in a thermal distribution have $v_\perp \approx v_t$ (the thermal velocity), so one expects the wave to behave nearly linearly for $t \ll (v_t\Omega_1 k)^{-1/2}$. As anticipated, the breakdown occurs due to parallel acceleration of particles due to $\mathbf{v}_\perp \times \mathbf{B}_1$. These particles will no longer resonate with the wave, and feed energy back into the field.

9-2. Nonlinear Effects Associated with Resonant Particles

In Chapter VII, we have seen that plasma oscillations are produced by particles in the bulk of the distribution function, while Landau damping or resonant instabilities are caused by resonant particles exchanging energy with the wave. One may think of the distribution function as consisting of two parts, playing different roles: the bulk particles determine the real part of the frequency, while a few particles in the resonant range in velocity space set the imaginary part. Now we have seen that if the wave becomes weakly nonlinear, only resonant particles are affected by nonlinearity. It is reasonable to conclude that the real part of the wave frequency will only weakly be affected;

the imaginary part, however, could become very different from its linear behavior.

Consider a large amplitude wave that is weakly Landau damped, such that $\gamma \ll \omega_b$. For times short compared to ω_b^{-1}, the wave decays according to linear theory, by the familiar mechanism: resonant particles that travel somewhat faster than the wave feed energy into the wave, while those somewhat slower take energy from it. Since the wave amplitude is large, most resonant particles are trapped, and the stream of particles that traveled forward in the wave frame will be traveling backward after a bounce time and vice versa. While originally more particles traveled slower than the wave than traveled faster, resulting in a net energy loss for the wave, after a bounce time, the situation will be reversed with the particles on the average losing energy to the wave. So one expects Landau damping to stop around $t \approx \omega_b^{-1}$ and wave amplification to commence. After another bounce time, the process starts all over again, leading to amplitude oscillations. This effect has indeed been observed experimentally.

To understand this process in more detail, let us look at the phase space of trapped particles (Fig. 9-2). At $t = 0$, the lower half of phase space is more heavily populated than the upper half. As the phase fluid swirls around, first more particles are lifted up in velocity space than lowered, resulting in kinetic energy increase; hence there is wave energy loss. Later, this process is reversed.

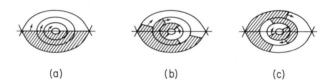

(a)   (b)   (c)

FIG. 9-2. Illustration of phase mixing of trapped particles. Area occupied by a set of particles at (a) $t = 0$, (b) $t < \omega_b^{-1}$, and (c) $t > \omega_b^{-1}$.

The bounce period is not the same, however, for all particles. The approximation used in (9-11) holds only for particles, near the bottom of potential wells. These are particles well inside the separatrix. Closer to the separatrix the period is longer and after a few bounce periods the oscillations in different regions get increasingly out of phase, resulting in the damping of amplitude oscillations, due to this phase mixing process. Figure 9-2a shows the lower half of phase space at $t = 0$, 2b at $t \lesssim \omega_b^{-1}$, and 2c at $t \gg \omega_b^{-1}$.

The time development of the wave amplitude can be calculated solving (9-11) without the $kx \ll 1$ approximation leading to elliptic integrals. The result is what one expects and is sketched in Fig. 9-3. A period of Landau damping is followed by amplitude oscillations with diminishing amplitude

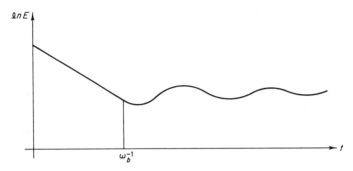

FIG. 9-3. Time development of the amplitude of damped electrostatic wave.

leading to an undamped wave, with the amplitude reduced by a factor of order $\exp(-\gamma/\omega_b)$.

A similar treatment for whistler wave damping leads to qualitatively similar results: linear damping followed by amplitude oscillations and an undamped wave. The characteristic time scale now is $(V_t\Omega_1 k)^{-1/2}$, as expected.

While the problem of an unstable wave exponentiating many times before the linear behavior breaks down is mathematically untractable (except for computer solutions), a qualitative understanding of the principal nonlinear processes is straightforward. As a simple example, consider a beam-plasma system. The growing modes have phase velocities just below the beam velocity. After a few exponentiation times, the fastest wave becomes dominant, and the frequency spectrum narrows down sufficiently so that the single-wave concept is a good approximation. In the linear regime the wave grows, but at one point the beam gets trapped and bouncing back and forth produces amplitude oscillations leading to saturation of the instability. Computer calculations as well as experiments verify this result.

An estimate for the onset of trapping can be obtained, by calculating the velocity of a beam particle in the wave frame from

$$\ddot{x} = \frac{q}{m}\frac{E_0}{2}\exp(\gamma t)\cdot\{\exp[ik(x_0 + Vt)] + \text{c.c.}\} \tag{9-20}$$

where c.c. stands for complex conjugate, V is the initial beam velocity in the wave frame, and the wave was taken to grow according to linear theory. From linear theory $\gamma = \sqrt{3}kV$ (see Exercise 9-2), and one may integrate (9-20) to find the particle velocity. Trapping begins when $\dot{x} = 0$, leading to the trapping potential

$$\phi_{tr} \approx \frac{4}{\sqrt{3}}\frac{mv^2}{q} \tag{9-21}$$

This is also approximately the $\gamma \approx \omega_b$ condition (see Exercise 9-3).

In all these calculations, initial-value problems were posed. A wave is launched at $t = 0$ everywhere, and its time evolution is followed. Experimentally, however, boundary-value problems are often encountered. A wave is launched from a grid with given real ω and its evolution in space is measured. In linear theory, one only converts from the temperal growth (or damping) rate γ to the spatial rate $\Lambda = \gamma/v_g$, where v_g is the wave group velocity. This can easily be verified and is left for Exercise 9-4. Now we must find a conversion from the temporal bounce time ω_b^{-1} to a spatial one k_b^{-1}. Since only resonant particles moving with velocity ω/k are involved, the conversion is clearly $k_b^{-1} = (\omega/k)\omega_b^{-1}$ in the electrostatic case, and $k_b^{-1} = [(\omega \pm \omega_c)/k]\omega_b^{-1}$ for whistlers. Figure 9-3 is still valid for the spacial case by rescaling the time axis $t \to (k/\omega)x$, while the slopes are rescaled to fit spatial damping rates as described.

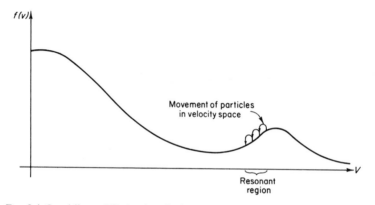

FIG. 9.4 Quasi-linear diffusion in velocity space for the bump on tail distribution.

Now we are turning our attention from single-wave processes to those involving a multitude of waves. As an example, let us look at the *bump on tail* distribution as shown in Fig. 9-4. The unstable region in velocity space produces many waves with different velocities and phases, so resonant particles cannot get trapped.

How is the resonant distribution function then affected? As each wave interchanges energy with its resonant particles, these particles must change their velocity; if they pick up energy they move up, if they lose energy they move down in velocity space. Consequently the distribution function f_0 must vary in time in the resonant region to account for this energy exchange. For the bump on tail distribution, the particles lose energy to the waves, so one expects the well in the distribution function to be filled with particles moving down from the bump.

In order to formulate this concept mathematically, consider the space average (denoted by $\langle\ \rangle$) of the Vlasov equation

$$\frac{\partial}{\partial t}\langle f\rangle + \frac{q}{m}\left\langle E\frac{\partial f}{\partial v}\right\rangle = 0 \tag{9-22}$$

The solution of the linear problem yielded electric fields

$$E = \sum_k E_k \exp[i(kx - \omega_k t)] + \text{c.c.} \tag{9-23}$$

and perturbed distribution functions

$$f_1 = \sum_k f_k \exp[i(kx - \omega_k t)] + \text{c.c.} = \sum_k \frac{if_0'}{kv - \omega_k}\frac{q}{m}E_k \exp[i(kx - \omega_k t)] + \text{c.c.} \tag{9-24}$$

where $\varepsilon(k,\omega_k) = 0$ determines the values of the usually complex ω_k. Note that we have added the complex conjugates to each term. In the linear theory, the space and time variations were put in the complex exponential form for convenience, with the understanding that (say) the real part is physically meaningful. When, however, in the nonlinear treatment, products of these quantities emerge, some care is required since $\text{Re } A \cdot B \neq \text{Re } A \cdot \text{Re } B$, so the real wave fields are explicitly described. In order to avoid carrying factors of $\frac{1}{2}$, E_K and f_K stand for half amplitudes.

Evaluate now the second term in (9-22) using the linear solutions but including their product

$$\left\langle E\frac{\partial f}{\partial v}\right\rangle = \frac{\partial}{\partial v}\left(\sum_k E_k f_k^* + \text{c.c.}\right)$$

$$= \frac{q}{m}\frac{\partial}{\partial v}\sum_k E_k E_k^*\left(\frac{i}{kv - \omega_k} - \frac{i}{kv - \omega_k^*}\right)f_0'$$

$$= \frac{q}{m}\frac{\partial}{\partial v}\sum_k \frac{-2\gamma_k f_0'|E_k|^2}{(kv - \omega_{kr})^2 + \gamma_k^2} \tag{9-25}$$

where $\omega_k = \omega_{kr} + i\gamma_k$. Inserting this into (9-22) leads to the evolution of $\langle f\rangle$. Recall, however, that in the linear theory f_0 played the role of the space-averaged distribution function (the distribution function without the wavy part). For consistency, we must replace f_0' in (9-25) with $\langle f'\rangle$, to take into consideration the slow evolution of the background plasma. Using the notation $\langle f\rangle = F$, (9-22) and (9-25) yield

$$\frac{\partial F}{\partial t} = \frac{\partial}{\partial v}\left(D_v\frac{\partial F}{\partial v}\right) \tag{9-26}$$

where

$$D_v = \frac{q^2}{m^2} \sum_k \frac{2\gamma_k |E_k|^2}{(kv - \omega_{kr})^2 + \gamma_k^2} \tag{9-27}$$

and (9-26) has the appearance of a diffusion equation in velocity space, with the diffusion coefficient D_v. Since D_v is largest where $kv = \omega_{kr}$, the largest diffusion takes place in the resonant region. The process is called *quasi-linear diffusion*.

Carrying out the differentiation in v gives

$$\frac{\partial F}{\partial t} = -4 \sum \omega_{bk}^4 \left[F' \frac{v - \omega_{kr}/k}{[(kv - \omega_{kr})^2 + \gamma_k^2]^2} - \frac{1}{2k^2} \frac{F''}{(kv - \omega_{kr})^2 + \gamma_k^2} \right] \tag{9-28}$$

For γ_k small, the first term dominates, and one may picture the change in the distribution function by plotting the contribution of one of the waves, as shown in Fig. 9-5. The effect is the expected one: it moves particles down the bump.

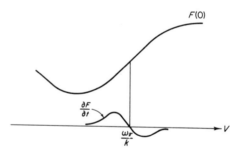

FIG. 9-5. Contribution of a wave to change of F.

In order to complete our quasi-linear equations, one recognizes that F replaces f_0 in all linear equations. Since γ_k depends on F [it is proportional to $(\partial F/\partial v)(\omega/k)$, as we have seen in linear theory] the flattening of F in the resonance region leads to a gradual reduction of the γ_k's. The evolution of $|E_K|^2$ is given by

$$\frac{\partial}{\partial t} |E_k|^2 = 2\gamma_k |E_k|^2 \tag{9-29}$$

just as in linear theory, except γ_k is slowly varying with time. The system now evolves in the following way. The distribution function flattens (forms a plateau) in the resonant region according to (9-26); consequently

$$\gamma_k = \frac{\pi}{2} \frac{\omega_p^3}{k^2 n} F'\left(\frac{\omega_{kr}}{k}\right) \tag{7-44}$$

decreases, reducing the growth of $|E_K|^2$, according to (9-29). Finally, the distribution function becomes stable and there is no further growth of the fluctuations. The final wave amplitude can be estimated, noting that the quasi-linear equations conserve energy (see Exercise 9-6); hence the energy taken out of the particles, while flattening the distribution function, is converted into wave energy. It is interesting to note that trapping in a single-wave field leads to a similar result: the distribution fiunction becomes flat in the resonant region, and the energy difference due to this flattening is balanced by the change in wave energy. The application of quasi-linear theory to a single wave leads to the expected order of magnitude for saturation (see Exercise 9-7).

Quasi-linear theory can be applied when the wave spectrum is broad enough to prevent trapping. One may define the wave correlation time $\tau_c \approx \lambda/\Delta(\omega/k)$, where λ is a characteristic wavelength and $\Delta(\omega/k)$ is the difference between the highest and lowest phase velocities. This is, the time it takes for the fastest wave to move a distance of a wavelength compared to the slowest wave. If $\tau_c \omega_b \ll 1$, particles trapped by a wave are swept out of the wave trough by other waves in a time shorter than the bounce time, and quasi-linear theory is applicable, while in the opposite limit the trapping picture is the correct one.

A different type of phenomenon, involving resonant particles, is the *Landau echo*. We have already seen in Chapter VII that while Landau damping destroys the wave electric field in a few damping times, some perturbation in the distribution function persists. This may also be seen from (9-7). If ω has a negative imaginary part, the term containing $e^{-i\omega t}$ damps out, while the second term containing e^{-ikvt} survives. It is a rapidly oscillating function of v for large values of t, so the integral of f_1 over velocity space vanishes; hence no space charge and no electric field results.

Impose now a second wave on the plasma at some time $T \gg \gamma_1^{-1}$, where γ_1 is the damping increment of the first wave. By this time, the electric field of the first wave has vanished, but the second wave still interacts with the remaining distribution function of the first wave. The Vlasov equation for $t > T$ describes the development of the second wave

$$\frac{\partial f_2}{\partial t} + v\frac{\partial f_2}{\partial x} + \frac{q}{m}(E_2 \exp\{i[k_2 x - \omega_2(t - T)]\} + \text{c.c.})\frac{\partial}{\partial v}(f_0 + f_1 + \text{c.c.}) = 0$$

$$(9\text{-}30)$$

where f_1 is the distribution function of the first wave as given in (9-7). In order to calculate f_2, one has six terms to integrate along particle trajectories. Fortunately, all that work is not necessary. The terms involving f_0 give the linear development of f_2, which combined with Poisson's equation determine $\omega_2(k_2)$ in the usual way. Let us focus our attention on the term involving f_1

and $E_2{}^*$. In order to find its contribution to f_2, the expression

$$\sigma = i \frac{q^2}{m^2} E_1 E_2{}^* \exp\{-i[k_2 x - \omega_2{}^*(t - T)]\} \frac{\partial}{\partial v} \frac{f_0' \exp[ik_1(x - vt)]}{k_1 v - \omega_1}$$

$$(9\text{-}31)$$

has to be integrated along the unperturbed trajectory $x = x_0 + vt$. The index 1 refers to the first wave, and $\exp(-\gamma_1 t) \ll 1$ and has been ignored. One may introduce the function

$$\Phi(v) = \frac{\partial}{\partial v} \frac{f_0'}{k_1 v - \omega_1}$$

to write

$$\sigma = i \frac{q^2}{m^2} E_1 E_2{}^* \exp\{i[(k_1 - k_2)x - k_1 vt + \omega_2{}^*(t - T)]\}$$

$$\times \left(\phi(v) - ik_1 t \frac{f_0'}{k_1 v - \omega_1} \right) \quad (9\text{-}32)$$

In order to find the part of f_2 resulting from this term one carries out the integrals

$$\int_T^t \exp\{i[(k_1 - k_2)(x_0 + vt') - k_1 vt' + \omega_2{}^*(t' - T)]\} \, dt'$$

$$= (\exp\{i[(k_1 - k_2)x - k_1 vt + \omega_2{}^*(t - T)]\}$$
$$- \exp\{i[k_1 - k_2)(x - vt + vT) - k_1 vT]\})[i(\omega_2{}^* - k_2 v)]^{-1} \quad (9\text{-}33)$$

and

$$\int t' \exp\{i[(k_1 - k_2)(x_0 + vt') - k_1 vt' + \omega_2{}^*(t' - T)]\} \, dt'$$

$$= \frac{1}{i(\omega_2{}^* - k_2 v)} \{t \exp[i(k_1 - k_2)x - k_1 vt + \omega_2{}^*(t - T)]$$

$$- T \exp\{i[(k_1 - k_2)(x - vt + vT) - k_1 vT]\} + \text{Eq. (9-33)}\} \quad (9\text{-}34)$$

If the second wave is damped at the rate γ_2, then as $t - T \gg \gamma_2^{-1}$ all terms containing $i\omega_2{}^*(t - T)$ in the exponent vanish. The other terms survive as in the linear solution to give a result of the form

$$f_2 = i \frac{q^2}{m^2} E_1 E_2{}^* \psi(v) \exp[i(k_1 - k_2)x] \exp\{-iv[k_1 T + (k_1 - k_2)(t - T)]\}$$

$$(9\text{-}35)$$

where $\psi(v)$ is a complicated function of v, whose exact form is not significant for our present considerations. The important fact to notice here is that the last exponential is again a rapidly oscillating function of v unless

$$t = \tau = \frac{k_2 T}{k_2 - k_1} \qquad (9\text{-}36)$$

At this time, the integral of f_2 over velocity space gives a finite space charge, and an electric field oscillating with the wave number $k_1 - k_2$. It is clear from the form of (9-32)–(9-34) that $\psi(v)$ contains negative powers of $(\omega_1 - k_1 v)$ and $(\omega_2 - k_2 v)$, so resonant particles are the major contributors to this echo phenomenon. There is, of course, a spatial counterpart to the Landau echo. A grid excites a wave with frequency ω_1, which damps away in a distance short compared to L, where a second grid is placed emitting waves with frequency ω_2. They are also damped away, while further downstream a new

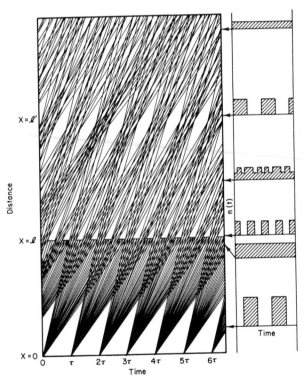

FIG. 9-6. Ballistic space-time trajectories for particles emitted at τ time intervals, and modulated at $x = l$ in time intervals $\tau/2$. [From D. R. Baker, N. R. Ahern, and A. Y. Wong, *Phys. Rev. Lett.* **20**, 318 (1968).]

wave emerges at a distance $X = \omega_2 L/(\omega_2 - \omega_1)$ (see Exercise 9-9). The effect has been observed experimentally and is in good agreement with the theory.

The Landau echo is basically ballistic in nature and can be visualized by considering the simplified picture of Fig. 9-6. The figure shows the ballistic space-time trajectories of particles with different velocities emitted at $x = 0$ at τ time intervals, simulating the periodicity of a wave. Before the particles reach $x = l$, the time periodic structure disappeared due to the velocity scatter. At $x = l$ the particles are modulated at a different frequency (in the figure twice the original). Further downstream, this modulation also disappears, but an echo emerges at $l' = \omega_2 l/(\omega_2 - \omega_1) = 2l$. The particle densities at different positions are also plotted as a function of time.

9-3. Mode Coupling

In the presence of many waves, the linear solution yields

$$f_1(\mathbf{r},\mathbf{v},t) = \sum_{\mathbf{k}} f_{\mathbf{k}}(\mathbf{v},\mathbf{k},\omega_{\mathbf{k}}) \exp[i(\mathbf{k}\cdot\mathbf{r} - \omega_{\mathbf{k}}t)] \qquad (9\text{-}37)$$

$$\mathbf{E}_1(\mathbf{r},t) = \sum_{\mathbf{k}} \mathbf{E}_{\mathbf{k}}(\mathbf{k},\omega_{\mathbf{k}}) \exp[i(\mathbf{k}\cdot\mathbf{r} - \omega_{\mathbf{k}}t)] \qquad (9\text{-}38)$$

where $\omega_{\mathbf{k}}(\mathbf{k})$ is to be determined from the linear dispersion relation. Note that these solutions contain the complex conjugates; for each $\mathbf{k}$ there is a $-\mathbf{k}$ and $\omega_{-\mathbf{k}} = -\omega_{\mathbf{k}}^*$ and $\mathbf{E}_{-\mathbf{k}} = \mathbf{E}_{\mathbf{k}}^*$, $f_{-\mathbf{k}} = f_{\mathbf{k}}^*$. This can be verified from linear theory (see Exercise 9-10).

Carry out a perturbation expansion of the Vlasov equation to second order, writing

$$f = f_0 + f_1 + f_2 \qquad \text{and} \qquad E = E_1 + E_2$$

to find

$$\frac{\partial f_2}{\partial t} + \mathbf{v}\cdot\frac{\partial f_2}{\partial \mathbf{r}} + \frac{q}{m}\mathbf{E}_2\cdot\frac{\partial f_0}{\partial \mathbf{v}} = -\frac{q}{m}\frac{\partial}{\partial \mathbf{v}}\cdot(f_1\mathbf{E}_1)$$

$$= -\frac{q}{m}\frac{\partial}{\partial \mathbf{v}}\cdot\sum f_{\mathbf{k}'}\mathbf{E}_{\mathbf{k}''}\exp\{i[(\mathbf{k}' + \mathbf{k}'')\cdot\mathbf{r} - (\omega' + \omega'')t]\} \quad (9\text{-}39)$$

To keep notations simpler $\omega' = \omega_{\mathbf{k}'}$, etc., has been introduced. The second-order Poisson's equation

$$\nabla\cdot\mathbf{E}_2 = \frac{q}{\varepsilon_0}\int f_2\,dv \qquad (9\text{-}40)$$

links the second-order fields and distribution functions. Equation (9-39) is an inhomogeneous equation, with a given periodic space and time dependence. It is evident that the space and time dependence of f_2 and $\mathbf{E}_2$ must be given by $\exp[i(\mathbf{k} \cdot \mathbf{r} - \omega t)]$ terms, where for each $\mathbf{k}'$, $\mathbf{k}''$ pair $\mathbf{k} = \mathbf{k}' + \mathbf{k}''$ and $\omega = \omega' + \omega''$. In other words, two linear waves couple to a third wave whose wave number and frequency is the sum of those of the generating waves. These matching conditions have a simple interpretation, when viewed quantum mechanically. The two driving waves are made up of quanta, with energies $\hbar\omega'$, $\hbar\omega''$ and momenta $\hbar\mathbf{k}'$, $\hbar\mathbf{k}''$. They can supply energy and momentum to a wave whose quanta contain energy $\hbar(\omega' + \omega'')$, and momentum $\hbar(\mathbf{k}' + \mathbf{k}'')$.

The lowest-order wave-coupling processes are therefore three-wave interactions, and it suffices to single out one of the triplets that satisfies the matching conditions. For such a triplet one has from (9-39) and (9-40)

$$i(\mathbf{k} \cdot \mathbf{v} - \omega)f_2(\mathbf{k}) + \frac{q}{m}\mathbf{E}_2(\mathbf{k}) \cdot \frac{\partial f_0}{\partial \mathbf{v}} = -\frac{q}{m}\frac{\partial}{\partial \mathbf{v}} \cdot [f(\mathbf{k}_i)\mathbf{E}(\mathbf{k}_j) + f(\mathbf{k}_j)\mathbf{E}(\mathbf{k}_i)]$$

(9-41)

and

$$i\mathbf{k} \cdot \mathbf{E}_2 = \frac{q}{m} \int f_2 \, d^3v \tag{9-42}$$

where $\omega = \omega_i + \omega_j$, $\mathbf{k} = \mathbf{k}_i + \mathbf{k}_j$.

The generalization to more than one particle species is trivial, by writing the right-hand side of (9-42) as a sum over species. One may now solve (9-41) for f_2 and substitute into (9-42), to find

$$\mathbf{k} \cdot \mathbf{E}_2 = \frac{q^2}{m\varepsilon_0}\mathbf{E}_2 \cdot \int \frac{(\partial f_0/\partial \mathbf{v}) \, d^3v}{\mathbf{k} \cdot \mathbf{v} - \omega}$$
$$+ \frac{q^2}{m\varepsilon_0} \int \frac{(\partial/\partial \mathbf{v}) \cdot [f(\mathbf{k}_i)\mathbf{E}(\mathbf{k}_j) + f(\mathbf{k}_j)\mathbf{E}(\mathbf{k}_i)]}{\mathbf{k} \cdot \mathbf{v} - \omega} d^3v \tag{9-43}$$

From $\nabla \times \mathbf{E}_2 = 0$, $\mathbf{k} \| \mathbf{E}_2$ and (9-43) becomes

$$E_2\left(1 - \frac{q^2}{m\varepsilon_0 k^2}\mathbf{k} \cdot \int \frac{(\partial f_0/\partial \mathbf{v}) \, d^3v}{\mathbf{k} \cdot \mathbf{v} - \omega}\right)$$
$$= E_2\varepsilon(\omega, k)$$
$$= \frac{q^2}{m\varepsilon_0 k} \int \frac{(\partial/\partial \mathbf{v}) \cdot [f(\mathbf{k}_i)\mathbf{E}(\mathbf{k}_j) + f(\mathbf{k}_j)\mathbf{E}(\mathbf{k}_i)]}{\mathbf{k} \cdot \mathbf{v} - \omega} d^3v \tag{9-44}$$

where ε is just the dielectric function (7-74). Substitute now $f(k_i)$, $f(k_j)$, to find

$$E_2\varepsilon(\omega_1\mathbf{k}) = i\frac{q^3}{m^2\varepsilon_0}\frac{E_iE_j}{|\mathbf{k}_i + \mathbf{k}_j|}\int\frac{1}{(\mathbf{k}_i + \mathbf{k}_j)\cdot\mathbf{v} - (\omega_i + \omega_j)}$$

$$\cdot\left(\frac{\mathbf{k}_i}{k_i}\cdot\frac{\partial}{\partial\mathbf{v}}\frac{1}{\mathbf{k}_j\cdot\mathbf{v} - \omega_j}\frac{\mathbf{k}_j}{k_j}\cdot\frac{\partial}{\partial\mathbf{v}}\right.$$

$$\left. + \frac{\mathbf{k}_j}{k_j}\cdot\frac{\partial}{\partial\mathbf{v}}\frac{1}{\mathbf{k}_i\mathbf{v} - \omega_i}\frac{\mathbf{k}_i}{k_i}\cdot\frac{\partial}{\partial\mathbf{v}}\right)f_0\,d^3v \tag{9-45}$$

where $\mathbf{E}_i$, $\mathbf{E}_j$ stand for $\mathbf{E}(\omega_i)$, $\mathbf{E}(\omega_j)$. One may abbreviate this expression

$$E_2(\omega_i + \omega_j, \mathbf{k}_i + \mathbf{k}_j)\varepsilon^{(1)}(\omega_i + \omega_j, \mathbf{k}_i + \mathbf{k}_j) = iE_iE_j\varepsilon^{(2)}(\omega_i,\omega_j,\mathbf{k}_i,\mathbf{k}_j) \tag{9-46}$$

where $\varepsilon^{(1)} = \varepsilon$ is henceforth called the first-order dielectric function, and the second-order dielectric function $\varepsilon^{(2)}$ has been introduced. (Note that it does not have the dimension of a dielectric function.) Considering $\mathbf{k}_i$ and $\mathbf{k}_j$ parallel permits us to drop the vector sign and simplify the notation somewhat. In that case

$$\varepsilon^{(2)}(\omega_i,\omega_j,k_i,k_j) = \frac{q^3}{m^2\varepsilon_0}(k_i + k_j)^{-1}$$

$$\cdot\int\frac{1}{(k_i + k_j)v - (\omega_i + \omega_j)}\frac{\partial}{\partial v}\left[\left(\frac{1}{k_iv - \omega_i} + \frac{1}{k_jv - \omega_j}\right)f_0'\right]dv \tag{9-47}$$

where f_0 is the reduced one-dimensional distribution function. Integrating by parts and rearranging terms leads to a more symmetrical form of $\varepsilon^{(2)}$

$$\varepsilon^{(2)} = \frac{q^3}{m^2\varepsilon_0}\int\frac{f_0'\,dv}{(k_iv - \omega_i)(k_jv - \omega_j)[(k_i + k_j)v - (\omega_i + \omega_j)]} \tag{9-48}$$

One may also decompose the integral as a sum of three integrals to yield

$$\varepsilon^{(2)} = \frac{q}{m}\frac{1}{(k_i\omega_j - k_j\omega_i)^2}$$

$$\cdot\{(k_i + k_j)^3\chi(\omega_i + \omega_j, k_i + k_j) - k_i^3\chi(\omega_i,k_i) - k_j^3\chi(\omega_j,k_j)\} \tag{9-49}$$

where χ is the dielectric susceptibility defined in (7-76). The mathematical manipulations leading to these alternate forms of the second-order dielectric function are left for Exercise 9-11. The distribution function, at the combination frequency can now be expressed from (9-41), using (9-46)

$$f_2(k_i + k_j, \omega_i + \omega_j)$$

$$= -\frac{q}{m} \frac{E_i E_j}{(k_i + k_j)v - (\omega_i + \omega_j)}$$

$$\cdot \left\{ f_0' \frac{\varepsilon^{(2)}}{\varepsilon^{(1)}} + \frac{q}{m} \frac{\partial}{\partial v} \left[\left(\frac{1}{k_i v - \omega_i} + \frac{1}{k_j v - \omega_j} \right) f_0' \right] \right\} \qquad (9\text{-}50)$$

It is clear from (9-46) and (9-50) that both the electric field and the distribution function of the driven wave become large when $\varepsilon^{(1)}$ is small, namely, the driven wave is close to resonance. The problem is similar to that of a driven harmonic oscillator, where the combination of the first-order waves acts as a driver. When the oscillator is driven near its eigenfrequency (here frequency and wave number) the amplitude gets large. Right at resonance our method breaks down and the initial-value problem should be solved, unless the driven wave is damped.

Let us investigate now the physical processes that give rise to nonlinear mode coupling. To do this consider the motion of a charged particle in the field of two waves $E_1 = e_1 \exp[i(k_1 x - \omega_1 t)] + \text{c.c.}$ and $E_2 = e_2 \exp[i(k_2 x - \omega_2 t)] + \text{c.c.}$ In the unperturbed rest frame of the particle, one may apply the oscillation-center approximation to describe the motion of the particle. The electric field at the instantaneous particle position is approximately

$$E(x) = E(x_0) + \delta x \left. \frac{\partial E}{\partial x} \right)_{x_0} \qquad (9\text{-}51)$$

where x_0 is the equilibrium particle position, and

$$\delta x = -\frac{q}{m(\omega_i')^2} E_i - \frac{q}{m(\omega_j')^2} E_j \qquad (9\text{-}52)$$

with $\omega_i' = \omega_i - k_i v$, $\omega_j' = \omega_j - k_j v$ the Doppler-shifted frequencies in the particle frame. To the lowest order, the electric field felt by the particle is $E(x_0)$, producing oscillations at the frequencies ω_i', ω_j' and giving rise to the first-order dielectric function as discussed in 7-3. At the next order, the combination frequencies of the two waves emerge due to the $\delta x \, \partial E / \partial x$ term producing

$$E(\omega_i' + \omega_j') = -i \frac{q}{m} E_i E_j \left[\frac{k_i}{(\omega_j')^2} + \frac{k_j}{(\omega_i')^2} \right] \qquad (9\text{-}53)$$

giving rise to particle displacements

$$\delta x(\omega_i' + \omega_j') = -\frac{q}{m}\frac{E(\omega_i' + \omega_j')}{(\omega_i' + \omega_j')^2} = i\frac{q^2}{m^2}\frac{E_i E_j}{(\omega_i' + \omega_j')^2}\left[\frac{k_i}{(\omega_j')^2} + \frac{k_j}{(\omega_i')^2}\right]$$

$$(9\text{-}54)$$

These particle motions produce space charge, oscillating at the same frequency in the plasma. To calculate it, consider a stream of particles with (unperturbed) velocity v and uniform density N. Calculate now the perturbed particle density in the stream frame. Particles originally located along the

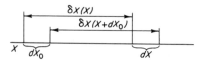

FIG. 9-7. Particle displacements in x.

length dx_0 occupy the range dx, as the result of displacements δx, with the density $n(x + \delta x)$ (Fig. 9-7). From particle number conservation

$$N\, dx_0 = n(x + \delta x)\, dx \approx \left(n(x) + \frac{\partial n}{\partial x}\,\delta x\right)dx \qquad (9\text{-}55)$$

while evidently $dx/dx_0 = 1 + (d/dx)\,\delta x$. The perturbed density becomes up to second order in small quantities

$$\delta n = n - N = -\frac{d}{dx}\,(n\,\delta x) \qquad (9\text{-}56)$$

Selecting now the $\omega_i' + \omega_j'$ Fourier component

$$\delta n(\omega_i' + \omega_j') = -i(k_i + k_j)[N\,\delta x(\omega_i' + \omega_j')$$
$$+ n(\omega_i')\,\delta x(\omega_j') + n(\omega_j')\,\delta x(\omega_i')] \qquad (9\text{-}57)$$

This expression can now be evaluated, using (9-54) and the first-order particle densities and displacements (see Exercise 9-12), to yield

$$\delta n = -\frac{q^2 N(k_i + k_j)E_i E_j}{m^2 \omega_i' \omega_j'(\omega_i' + \omega_j')}\left(\frac{k_i}{\omega_i'} + \frac{k_j}{\omega_j'} + \frac{k_i + k_j}{\omega_i' + \omega_j'}\right) \qquad (9\text{-}58)$$

As we have seen before, the plasma can be decomposed into individual particle streams of density $N = f_0\, dv$, and the charge density produced by all particles becomes in the laboratory frame

$$\rho(\omega_i + \omega_j) = -\frac{q^2}{m^2} E_i E_j (k_i + k_j)$$

$$\cdot \int \frac{f_0}{(\omega_i - k_i v)(\omega_j - k_j v)[(\omega_i + \omega_j) - (k_i + k_j)v]}$$

$$\cdot \left(\frac{k_i}{\omega_i - k_i v} + \frac{k_j}{\omega_j - k_j v} + \frac{k_i + k_j}{(\omega_i + \omega_j) - v(k_i + k_j)} \right) dv \qquad (9\text{-}59)$$

$$= -E_i E_j (k_i + k_j)\varepsilon^{(2)}$$

Integration by parts of (9-48) verifies the last equality. The dielectric displacement corresponding to this space charge is

$$D(\omega_i + \omega_j) = -i\frac{\rho(\omega_i + \omega_j)}{k_i + k_j} = iE_i E_j \varepsilon^{(2)}(\omega_i, \omega_j, k_i, k_j) \qquad (9\text{-}60)$$

which is just the right-hand side of (9-46).

In this calculation only the forces arising from the two driving waves have been considered, leaving out the interaction of the plasma particles with each other. The space charge generated at the frequency $\omega_i + \omega_j$, however, produces an electric field at the same frequency, and the plasma responds to this field. This response is described by the first-order dielectric function at this frequency and

$$E(\omega_i + \omega_j) = \frac{D(\omega_i + \omega_j)}{\varepsilon^{(1)}(\omega_i + \omega_j, k_i + k_j)} \qquad (9\text{-}61)$$

This is the same result obtained in (9-46) by the more formal expansion.

The waves in the plasma can be viewed as a collection of coupled oscillators. In the linear approximation, the coupling vanishes and the linear eigenmodes emerge. In the second order, two excited modes couple to a third oscillator and drive it. If this oscillator is driven near its own resonance, $\varepsilon^{(1)}(\omega_i + \omega_j, k_i + k_j) \approx 0$, the amplitude becomes large. This behavior can be expected, e.g., when the dispersion curve is nearly linear $\varepsilon^{(1)} \approx 0$ for $\omega/k = \text{const}$, as in ion acoustic waves or Alfven waves, since then if ω_i, k_i and ω_j, k_j are eigenmodes, so is $\omega_i \pm \omega_j, k_i \pm k_j$. This problem will be considered later in more detail.

So far we have ignored wave damping. An interesting situation arises when the driving waves are undamped, but the beat wave experiences Landau damping. In this process, called *nonlinear Landau damping*, the energy absorbed by the particles interacting with the beat wave must be supplied by the primary waves. This should result in a change of dispersion relation for these waves, since their frequency must acquire an imaginary part. This effect can be seen by going to third order in the expansion of the Vlasov equation. Here a beat wave, say $\omega_1 - \omega_2$, heats with a primary wave ω_2 to produce a wave at a primary wave frequency, in our example ω_1. This self-interaction of waves with themselves leads to a modification of their dispersion relation.

The third-order Vlasov equation for such a process is

$$i(kv - \omega)f(k,\omega,v) + \frac{q}{m}\left(E(k)\frac{\partial f_0}{\partial v} + \frac{\partial}{\partial v}\sum_{k_i}[E(k + k_i)f(-k_i)\right.$$

$$\left. + E(-k_i)f(k + k_i)]\right) = 0 \quad (9\text{-}62)$$

where $E(k + k_i)$ and $f(k + k_i)$ are second-order quantities resulting from the beating of the k and k_i modes.

Expressing $f(k)$ and substituting into Poisson's equation yields

$$E(k)\left(1 - \frac{q^2}{m\varepsilon_0 k}\int\frac{f_0'\,dv}{kv - \omega}\right) - \frac{q^2}{m\varepsilon_0 k}\sum_{k_i}\left[E(k + k_i)\int\frac{(\partial/\partial v)f(-k_i)}{kv - \omega}\,dv\right.$$

$$\left. + E(-k_i)\int\frac{(\partial/\partial v)f(k + k_i)}{kv - \omega}\,dv\right] = 0 \quad (9\text{-}63)$$

The second-order quantities $E(k + k_i)$ and $f(k + k_i)$ may now be substituted from (9-46) and (9-50). This leads to

$$E(k)\left\{\varepsilon^{(1)}(\omega,k)\right.$$

$$+ \sum |E(k_i)|^2\frac{q^3}{m^2\varepsilon_0 k}\left[\frac{\varepsilon^{(2)}(\omega_i,\omega_j,k_i,k_j)}{\varepsilon^{(1)}(\omega + \omega_i,k + k_i)}\int\frac{1}{kv - \omega}\frac{\partial}{\partial v}\left(\frac{-f_0'}{k_i v - \omega_i}\right.\right.$$

$$+ \frac{f_0'}{(k + k_i)v - (\omega + \omega_i)}\Big)\,dv$$

$$+ \frac{q}{m}\int\frac{1}{kv - \omega}\frac{\partial}{\partial v}\frac{(\partial/\partial v)[(kv - \omega)^{-1}f_0' + (k_i v - \omega_i)^{-1}f_0']}{(k + k_i)v - (\omega + \omega_i)}\,dv\Bigg]\Bigg\} = 0$$

$$(9\text{-}64)$$

From the definition of $\varepsilon^{(2)}$, (9-47), one recognizes that the first integral times $q^3/m^2\varepsilon_0 k$ is just $\varepsilon^{(2)}(-\omega_i,\omega + \omega_i,-k_i,k + k_i)$. The last integral is proportional to what may be called the third-order dielectric function $\varepsilon^{(3)}$. The quantity in the curly brackets is the third-order dispersion relation for the (ω,k) wave

$$\varepsilon^{(1)}(\omega,k) + \sum |E(k_i)|^2\left[\frac{\varepsilon^{(2)}(\omega_i,\omega_j,k_i,k_j)\varepsilon^{(2)}(-\omega_i,\omega + \omega_i,-k_i,k + k_i)}{\varepsilon^{(1)}(\omega + \omega_i,k + k_i))}\right.$$

$$\left. + \frac{q^4}{m^3\varepsilon_0 k}\varepsilon^{(3)}\right] = 0 \quad (9\text{-}65)$$

or briefly

$$\varepsilon^{(1)} + \delta\varepsilon = 0 \tag{9-66}$$

Since $\delta\varepsilon$ is proportional to $|E|^2$, it is small, and one may find the frequency shift in the usual manner

$$\delta\omega = -\delta\varepsilon(\omega_0)\left(\frac{\partial\varepsilon^{(1)}}{\partial\omega}\right)_{\omega_0}^{-1} \tag{9-67}$$

where ω_0 satisfies the linear dispersion relation $\varepsilon^{(1)} = 0$. Consider the interaction of two primary waves $k = k_1$ and $k_i = -k_2$. The resonant denominators occur at $v_R = (\omega_1 - \omega_2)/(k_1 - k_2) = \Delta\omega/\Delta k$. The imaginary part of $\delta\varepsilon$ is

$$\text{Im } \delta\varepsilon = |E(-k_2)|^2 \text{ Im}\left\{-\frac{(\varepsilon^{(2)})^2}{\varepsilon^{(1)}} + \frac{q^4}{m^3\varepsilon_0 k_1}\varepsilon^{(3)}\right\} \tag{9-68}$$

where we recognized from (9-48) that $-\varepsilon^{(2)}(\omega_2,\Delta\omega_1 k_2,\Delta k) = \varepsilon^{(2)}(\omega_1,-\omega_2, k_1,-k_2)$. In order to evaluate the above expression we separate the ε's into real and imaginary parts, by writing $\varepsilon^{(1)} = \alpha_1 + i\beta_1$, $\varepsilon^{(2)} = \alpha_2 + i\beta_2$, to obtain

$$\text{Im } \frac{(\varepsilon^{(2)})^2}{\varepsilon^{(1)}} = \frac{1}{|\varepsilon^{(1)}|^2}\left[-\beta_1(\alpha_2{}^2 + \beta_2{}^2) + 2\beta_2(\alpha_1\alpha_2 + \beta_1\beta_2)\right] \tag{9-69}$$

The imaginary parts of the ε's come entirely from the resonances at $v_R = \Delta\omega/\Delta k$ as assumed, and can be easily evaluated. If $\Delta\omega > 0$

$$\beta_1 = -\frac{q^2\pi}{m\varepsilon_0\Delta k^2}f_0'\left(\frac{\Delta\omega}{\Delta k}\right) \tag{9-70}$$

$$\beta_2 = -\frac{q^3\pi}{m^2\varepsilon_0\Delta k}\frac{1}{\omega^2}f_0'\left(\frac{\Delta\omega}{\Delta k}\right) \tag{9-71}$$

where

$$\omega = \omega_1 - v_R k_1 = \omega_2 - v_R k_2 \tag{9-72}$$

is the Doppler-shifted frequency of both waves in the group velocity frame v_R. The imaginary part of $\varepsilon^{(3)}$ can now be evaluated (see Exercise 9-15)

$$\text{Im } \frac{q^4}{m^3\varepsilon_0 k_1}\varepsilon^{(3)} = -\frac{\pi q^4}{m^3\varepsilon_0\omega^4}f_0'\left(\frac{\Delta\omega}{\Delta k}\right) \tag{9-73}$$

Inserting these results into (9-68) yields finally

$$\text{Im } \delta\varepsilon = -|E_2|^2 \frac{q^2\pi}{m\varepsilon_0 \, \Delta k^2} f_0'\left(\frac{\Delta\omega}{\Delta k}\right)$$

$$\cdot \left[\frac{\alpha_2{}^2 + \beta_2{}^2}{|\varepsilon^{(1)}|^2} - 2\frac{q \, \Delta k}{m \, \omega^2}\frac{\alpha_1\alpha_2 + \beta_1\beta_2}{|\varepsilon^{(1)}|^2} + \frac{q^2 \, \Delta k^2}{m^2\omega^4} \right]$$

$$= -|E_2|^2 \frac{q^2\pi}{m\varepsilon_0 \, \Delta k^2} f_0'\left(\frac{\Delta\omega}{\Delta k}\right) \left| \frac{q \, \Delta k}{m\omega^2} - \frac{\varepsilon^{(2)}}{\varepsilon^{(1)}} \right|^2 \qquad (9\text{-}74)$$

resulting in the growth rate of the k_1 wave

$$\gamma = \text{Im } \delta\omega = \left(\frac{\partial\varepsilon^{(1)}}{\partial\omega}\right)^{-1} |E_2|^2 \frac{q^2\pi}{m\varepsilon_0 \, \Delta k^2} f_0'\left(\frac{\Delta\omega}{\Delta k}\right) \left| \frac{q \, \Delta k}{m\omega^2} - \frac{\varepsilon^{(2)}}{\varepsilon^{(1)}} \right|^2 \qquad (9\text{-}75)$$

For a Maxwellian plasma, $f_0' < 0$, leading to damping. What about the second wave ω_2? The derivation is clearly the same except for one important point: $\Delta\omega$ is now negative; hence the pole integrations are carried out above the pole, resulting in a change in sign for β_1, β_2, and Im $\varepsilon^{(3)}$. Consequently, γ changes sign, and an amplification of the lower-frequency wave results.

This result can also be derived, recognizing that the fundamental mechanism is the exchange of energy and momentum of the two waves with particles, moving with the resonant velocity v_R. Designating the wave energy densities by U and momentum densities by P one writes

$$\frac{dU_1}{dt} + \frac{dU_2}{dt} + \frac{dT}{dt} = 0 \qquad (9\text{-}76)$$

$$\frac{dP_1}{dt} + \frac{dP_2}{dt} + \frac{1}{v_R}\frac{dT}{dt} = 0 \qquad (9\text{-}77)$$

where dT/dt is the time rate of change of particle kinetic energy. Since $U/P = \omega/k$ for both waves and $v_R = \Delta\omega/\Delta k$, one can solve for

$$\frac{dU_1}{dt} = -\frac{\omega_1}{\omega_1 - \omega_2}\frac{dT}{dt}$$

$$\frac{dU_2}{dt} = \frac{\omega_2}{\omega_1 - \omega_2}\frac{dT}{dt} \qquad (9\text{-}78)$$

One can see immediately that if $\omega_1 > \omega_2$, the first wave loses while the second gains energy. It also follows that

$$\frac{d}{dt}\left(\frac{U_1}{\omega_1} + \frac{U_2}{\omega_2}\right) = 0 \qquad (9\text{-}79)$$

The quantity U/ω is the wave action, or number of quanta in a wave; hence (9-79) expresses the conservation of plasmons in the process. We are dealing with the stimulated decay of plasmons with energy $\hbar\omega_1$, decaying into plasmons of energy $\hbar\omega_2$ with the excess energy (and momentum) being absorbed by the particles.

In order to find the decay and growth rates of the waves, one has to evaluate dT/dt. The resonant particles are subjected to a force at the difference frequency, which accelerates them toward minima of the field intensity (the so-called ponderomotive force as described in Sec. 2-9). This second-order wave electric field was derived in (9-53), and gives for this case

$$E_p = -i\frac{q}{m} E_1 E_2 \frac{\Delta k}{\omega^2} \tag{9-80}$$

since $\omega_i' = \omega_j' = \omega$. A similar force also acts on the nonresonant particles, which respond by setting up a space charge wave at the difference frequency, as we have seen, to produce an electric field at this frequency (9-61).

$$E_{s.c.} = iE_1 E_2 \frac{\varepsilon^{(2)}}{\varepsilon^{(1)}} \tag{9-81}$$

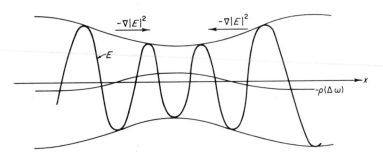

FIG. 9-8. Ponderomotive forces and space charge set up by them in a beat wave field.

where s.c. stands for space charge wave field (Fig. 9-8). This space charge wave tends to act against the ponderomotive field by filling up the minima in $|E|^2$ with particles that repel the resonant particles.

A resonant particle is subjected to the combined effects of these forces, resulting in an effective field

$$E_{eff} = iE_1 E_2 \left(\frac{\varepsilon^{(2)}}{\varepsilon^{(1)}} - \frac{q\,\Delta k}{m\omega^2}\right) \tag{9-82}$$

It is now easy to evaluate the energy exchange dT/dt of the resonant particles in this field; write

$$U = \frac{\varepsilon_0 E^2}{4} \, \omega \, \frac{\partial \varepsilon}{\partial \omega} \, e^{-2\gamma t}$$

and evaluate the growth rate from (9-78). The result is, of course, again (9-75) (see Exercise 9-16).

These physical considerations suggest that the two terms in (9-82) nearly cancel, with the space charge wave field significantly reducing the ponderomotive force. More-detailed calculations show that this is indeed the case, for electron plasma oscillations damping on electrons. There are other instances, however, where nonlinear Landau damping is significant.

Consider, for instance, the case where electron waves damp on ions. The ponderomotive force acting on ions is small due to the large ion mass; hence the $q\,\Delta k/M\omega^2$ term may be neglected. Ions only feel the space charge wave, represented by the $\varepsilon^{(2)}/\varepsilon^{(1)}$ term. In $\varepsilon^{(2)}$ the effect of ions can be ignored ($\varepsilon^{(2)} \sim 1/m^2$), in concert with our notion that the space charge wave is produced by electrons. There are, however, two reasons why the ions play an important role in $\varepsilon^{(1)}$.

First, both electron waves have frequencies only slightly above ω_p; hence the difference frequency $\Delta\omega \ll \omega_p$. Therefore the beat wave is very far from an electron wave resonance, and $\varepsilon^{(1)}(\Delta\omega, \Delta k) \gg 1$ if only electrons are assumed to participate. Since $\Delta\omega$ is in the frequency range of ion waves, the real part of $\varepsilon^{(1)}$ is reduced by ion motion. In other words, while the beat wave is not an eigenmode (sometimes it is called a quasi-mode), it acts more like an ion wave than an electron wave.

Second, the resonant velocity $\Delta\omega/\Delta k$ is of the order of the group velocity of electron waves $v_e^2(\omega/k)^{-1}$, much smaller than the electron thermal velocity, but can be comparable to the ion thermal velocity. So ions can Landau damp on the beat wave, and the ion term Im $\varepsilon^{(1)}$ can be sizable.

The detailed calculation of the damping rate is left for Exercise 9-18.

9-4. Parametric Processes

We have seen that two large waves can drive other waves at the beat frequency and wave number. If the beat wave is an eigenmode of the system, it will be driven to large amplitude. In this case it is not necessary that both driving waves be large to obtain coupling. Qualitatively, if $\varepsilon^{(1)}$ in (9-46) is very small, one of the driving modes, say E_j, can be small itself, and a finite amplitude beat wave E_k may be driven. Now the large amplitude wave E_i starts to beat with E_k. If the frequency of E_i is ω_0 (and $-\omega_0$), and that of E_j is ω_1 (and $-\omega_1$), then the resonant beat wave frequency is $\omega_0 - \omega_1$ (and

$\omega_1 - \omega_0$). When the large amplitude wave E_i, to be called the pump wave, beats with the resonant beat wave E_k, it produces a wave with frequency $\omega_0 + (\omega_1 - \omega_0) = \omega_1$, and $-\omega_0 + (\omega_0 - \omega_1) = -\omega_1$, which is the E_j wave itself. So the E_j wave will increase in the process. When this larger E_j wave beats with E_i again an increased E_k wave results and so on. This process leading to the exponential increase of both E_j and E_k is called the *parametric decay instability*.

One may easily give a mathematical description of this process. From (9-46)

$$E(\omega_1 - \omega_0) = iE(\omega_1)E(-\omega_0) \frac{\varepsilon^{(2)}(\omega_1, -\omega_0)}{\varepsilon^{(1)}(\omega_1 - \omega_0)} \tag{9-83}$$

and

$$E(\omega_1) = iE(\omega_0)E(\omega_1 - \omega_0) \frac{\varepsilon^{(2)}(\omega_0, \omega_1 - \omega_0)}{\varepsilon^{(1)}(\omega_1)} \tag{9-84}$$

As a simplication of notation, the wave numbers have been omitted from the arguments of the E's and ε's. Combining these equations leads at once to

$$\varepsilon^{(1)}(\omega_1) = -|E(\omega_0)|^2 \frac{\varepsilon^{(2)}(\omega_1, -\omega_0)\varepsilon^{(2)}(\omega_0, \omega_1 - \omega_0)}{\varepsilon^{(1)}(\omega_1 - \omega_0)} \tag{9-85}$$

For a given pump wave amplitude and ω_0, this equation may be regarded as a nonlinear dispersion relation for the determination of $\omega_1(k_1)$. It is clear that this relationship is different from the linear one. If ω_1 was the solution of the linear dispersion relation, the left-hand side would be $\varepsilon^{(1)}(\omega_1, k_1) = 0$, in contradiction with (9-85); in particular, since it was assumed that $\omega_1 - \omega_0$, $k_1 - k_0$ corresponds to an eigenmode as well, so the denominator on the right would vanish. This indicates that by solving (9-85) a frequency shift emerges, changing both the real and imaginary parts. The shift in the imaginary part leads to instability.

Instead of pursuing this formalism further, a different, physically more transparent approach will be followed, using the fluid equations. Before embarking on this course, however, a few words concerning the matching conditions are in order.

The matching conditions (say $\omega_0 = \omega_1 + \omega_2$, $k_0 = k_1 + k_2$), require that the three waves, when represented as vectors in (ω, k) space, satisfy the rules of vector addition. The requirement that the three waves be eigenmodes implies that the end points of these vectors be situated on dispersion curves. An example for collinear waves in an unmagnetized plasma is shown in Fig. 9-9. Here the pump wave is an electromagnetic wave, which decays into an ion acoustic wave and an other backward propagating electromagnetic wave. This process is called *stimulated Brillouin backscatter*. A similar construction can be made for an electromagnetic wave decaying into an electron

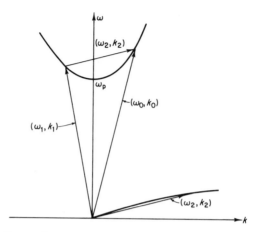

FIG. 9-9. Decay of an e.m. wave into an ion and a backscattered e.m. wave.

plasma wave and a backward propagating electromagnetic wave, called *stimulated Raman backscatter.* From the constructions it is easy to see that an electromagnetic wave can not decay into two other electromagnetic waves, nor can an electron wave decay into two other electron waves. An ion wave, however, if it is on the linear portion of the dispersion curve ($k \ll \omega_{pi}/c_s$), may decay into two ion waves. This process will be investigated later in more detail.

The backscatter instabilities have a simple physical interpretation. Consider a light wave with wavelength λ_0 impinging on a plasma with an electron density varying with a wavelength $\lambda = \frac{1}{2}\lambda_0$. This density perturbation forms a one-dimensional lattice, resulting in a partial reflection of the electromagnetic wave. The reflected wave and incoming wave produce a standing wave component, whose intensity maxima are located at the minima of the electron density wave. We know from the oscillation-center approximation that electrons will be expelled by the ponderomotive force from regions where the field intensity is large. This enhances the density perturbation, leading to more backscattering and larger standing wave amplitudes, and so on. The effect is most pronounced when the density perturbation corresponds to an eigenmode, viz., an electron or ion wave. Since such waves propagate themselves, the frequency of the backscattered wave will be Doppler shifted. This is in agreement with Fig. 9-9, where the ion wave has roughly half the wavelength (twice the wave number) of the pump wave, and the frequency of the backscattered wave $\omega_1 = \omega_0 - c_s k_2$ due to the Doppler shift. An electron plasma wave can backscatter on ion waves in the same manner.

In treating the Brillouin process, considerable simplification results from recognizing that the light wave frequency is significantly higher than that of the ion wave. To distinguish the two types of waves, the index f (fast) will designate the electromagnetic waves, while the index s (slow) is reserved for the ion wave. The electron density will be written as $N + n$, where N is the equilibrium density and n is the density perturbation. The electromagnetic waves obey the wave equation

$$\nabla^2 \mathbf{E}_f - \frac{1}{c^2} \ddot{\mathbf{E}}_f = \mu_0 \dot{\mathbf{J}}_f \approx -\mu_0 e (N + n) \dot{\mathbf{v}}_f \tag{9-86}$$

since $\dot{n}\mathbf{v}_f \ll n\dot{\mathbf{v}}_f$, as n varies slowly. In the electromagnetic wave $\dot{\mathbf{v}}_f = -(e/m)\mathbf{E}_f$, so

$$\left(c^2 \nabla^2 - \frac{\partial^2}{\partial t^2} - \omega_p^2 \right) \mathbf{E}_f = \omega_p^2 \frac{n}{N} \mathbf{E}_f \tag{9-87}$$

where ω_p is the electron plasma frequency corresponding to the equilibrium density N. The left-hand side yields the linear dispersion relation, while the right-hand side couples the density fluctuations to the electromagnetic wave. This equation expresses the propagation of the wave in a plasma with non-uniform density.

Next, a nonlinear equation for the slow density modulation is to be described. The equation of continuity gives

$$\dot{n} + N \nabla \cdot \mathbf{v} + \nabla \cdot (n\mathbf{v}) = 0 \tag{9-88}$$

while the electron equation of motion yields

$$\dot{\mathbf{v}} + (\mathbf{v} \cdot \nabla)\mathbf{v} = -\frac{e}{m}(\mathbf{E} + \mathbf{v} \times \mathbf{B}) - \frac{1}{m(N + n)}\nabla p \tag{9-89}$$

Let us look at the nonlinear terms. The product of two slow quantities leads to harmonic generation in the ion wave, not to be considered here, while the product of two fast quantities can give a slow beat wave. Since $n_f = 0$, $p_f = 0$, $n\mathbf{v}$ and $n \nabla p$ are harmonic terms to be ignored. Since the fast wave is transverse $(\mathbf{v}_f \cdot \nabla)\mathbf{v}_f = 0$ while $(\mathbf{v}_s \cdot \nabla)\mathbf{v}_s$ is an uninteresting harmonic term. The only term to provide coupling from the fast to the slow wave is the $\mathbf{v}_f \times \mathbf{B}_f$ term. From

$$\nabla \frac{E_f^2}{2} = \mathbf{E}_f \times (\nabla \times \mathbf{E}_f) = \frac{m}{e} \dot{\mathbf{v}}_f \times \dot{\mathbf{B}}_f \tag{9-90}$$

and writing

$$\mathbf{v}_f = \tilde{\mathbf{v}}(x,t) \exp(i\omega_f t) + \text{c.c.} \tag{9-91}$$

$$\mathbf{B}_f = \tilde{\mathbf{B}}(x,t) \exp(i\omega_f t) + \text{c.c.} \tag{9-92}$$

where ω_f is a typical electromagnetic wave frequency. Since there are two of these (the pump and backscattered waves), $\tilde{\mathbf{v}}$ and $\tilde{\mathbf{B}}$ are themselves slowly varying quantities in time to describe the frequency modulation resulting from the beating of these waves. Hence the slowly varying component is

$$(\dot{\mathbf{v}}_f \times \dot{\mathbf{B}}_f)_s = \omega_f{}^2(\tilde{\mathbf{v}} \times \tilde{\mathbf{B}}^* + \text{c.c.}) = \omega_f{}^2(\mathbf{v} \times \mathbf{B})_s \qquad (9\text{-}93)$$

where $\dot{\tilde{\mathbf{v}}}, \dot{\tilde{\mathbf{B}}}$ are small; hence they were ignored. The term

$$\frac{e}{m}(\mathbf{v} \times \mathbf{B})_s = \nabla \frac{e^2(E_f{}^2)_s}{2m^2\omega_f{}^2} \qquad (9\text{-}94)$$

is just the ponderomotive acceleration [(2-196) and (2-197)]. We may now rewrite the remaining terms of (9-88) and (9-89),

$$\dot{n}_s + N\,\nabla \cdot \mathbf{v}_s = 0 \qquad (9\text{-}95)$$

and

$$\dot{\mathbf{v}}_s = -\frac{e}{m}\mathbf{E} - \nabla\left(\frac{p}{mN} + \frac{e^2\langle E_f{}^2\rangle}{2m^2\omega_f{}^2}\right) \qquad (9\text{-}96)$$

where $\langle\ \rangle = (\ )_s$, the "fast time scale average," has been introduced, designating the slowly varying envelope of a fast oscillating quantity. The ponderomotive potential plays the same role as the pressure, or (writing $\mathbf{E} = -\nabla\varphi$) the electrostatic potential. The time derivative of (9-95) may now be combined with (9-96) to yield

$$\ddot{n}_s - \frac{e}{m}N\,\nabla \cdot \mathbf{E} - \nabla^2\left(\frac{p}{m} + \frac{Ne^2\langle E_f{}^2\rangle}{2m^2\omega_f{}^2}\right) = 0 \qquad (9\text{-}97)$$

If the light wave frequency greatly exceeds the plasma frequency, the density perturbation can be caused by electron waves, and writing $\nabla \cdot \mathbf{E} = -(e/\varepsilon_0)n_s$ one has a nonlinear equation of electron waves, driven by the light waves

$$\ddot{n}_s + \omega_p{}^2 n_s - v_e{}^2\,\nabla^2 n_s = \frac{\omega_p{}^2}{\omega_f{}^2}\frac{\varepsilon_0}{2m}\nabla^2\langle E_f{}^2\rangle \qquad (9\text{-}98)$$

where $v_e{}^2$ is the mean square thermal velocity previously designated by $\langle v^2\rangle$.

When treating ion waves, an expression similar to (9-97) is obtained for the ion density fluctuations, except that $e \to -e$, $m \to M$, and electron pressure is replaced by ion pressure. Since ion waves are strongly damped unless ions are cold, we will treat the cold ion case. Consequently the ion pressure term is ignored, and so is the ponderomotive force contribution since $M \gg m$. So the ion density varies according to

$$\ddot{n}_i + \frac{e}{M}N\,\nabla \cdot \mathbf{E} = 0 \qquad (9\text{-}99)$$

Substituting $\nabla \cdot \mathbf{E}$ from (9-99) into (9-97) gives

$$\ddot{n}_e + \frac{M}{m}\ddot{n}_i - \nabla^2\left(\frac{p}{m} + \frac{Ne^2\langle E_f^2\rangle}{2m^2\omega_f^2}\right) = 0 \qquad (9\text{-}100)$$

The first term is much smaller than the second one and may be neglected. If one sets $n = n_e \approx n_i$, a good approximation for the linear part of the ion dispersion curve. and substitutes $p = \frac{1}{3}Nmv_e^2 = NT_e$ as in (7-121), one finds

$$\ddot{n} - c_s^2\,\nabla^2 n = \frac{\omega_p^2}{\omega_f^2}\frac{\varepsilon_0}{2M}\nabla^2\langle E_f^2\rangle \qquad (9\text{-}101)$$

The left-hand side is just the linear ion wave equation, while the right-hand side expresses the ponderomotive driving force of the fast wave.

Equations (9-87) and (9-101) contain all the necessary information to study the interaction of light waves with ion acoustic waves.

Very similar expressions may be derived for electron plasma oscillations (fast waves) interacting with ion waves. From the equations of continuity and motion and Poisson's equation, one finds for the fast electron wave

$$\ddot{E}_f + \omega_p^2 E_f - v_e^2\frac{\partial^2}{\partial x^2}E_f = -\frac{\omega_p^2}{N}nE_f + \underbrace{v_e^2\frac{n}{N}\frac{\partial^2}{\partial x^2}E_f}_{\text{small}} \qquad (9\text{-}102)$$

where the last term is usually ignored ($k^2 v_e^2 \ll \omega_p^2$). The details are left for Exercise 9-20. This equation without the last term is similar to the one describing electromagnetic waves (9-87). When deriving the nonlinear ion wave equation from (9-88) and (9-89), the nonlinear driving term comes from $(\mathbf{v}_f \cdot \nabla)\mathbf{v}_f$, to give (see Exercise 9-21)

$$\langle(\mathbf{v}_f \cdot \nabla)\mathbf{v}_f\rangle = \nabla\frac{e^2\langle E_f^2\rangle}{2m^2\omega_f^2} \qquad (9\text{-}103)$$

the same expression we obtained earlier for $(e/m)\langle\mathbf{v}_f \times \mathbf{B}_f\rangle$. This should not be surprising, considering the derivation of the ponderomotive force in Chapter II, where no distinction was made between longitudinal and transverse waves. The nonlinear ion wave equation is again (9-101).

To obtain a dispersion relation we describe the pump wave electric field as $\mathbf{E}_0^+ \exp[i(\mathbf{k}_0 \cdot \mathbf{x} - \omega_0 t)] + \mathbf{E}_0^- \exp[-i(\mathbf{k}_0 \cdot \mathbf{x} - \omega_0 t)]$, with ω_0 real and $\mathbf{E}_0^- = (\mathbf{E}_0^+)^*$, and the density perturbation $n = n^+ \exp[i(\mathbf{k} \cdot \mathbf{x} - \omega t)] + n^- \exp[-i(\mathbf{k} \cdot \mathbf{x} - \omega t)]$, where ω is to be determined, and $n^- = (n^+)^*$. The product of the pump field and n in (9-87) gives rise to a scattered electric field, with the combination frequency and wave number. From (9-87)

$$[(\omega \pm \omega_0)^2 - c^2(\mathbf{k} \pm \mathbf{k}_0)^2 - \omega_p^2]\mathbf{E}^{\pm} = \omega_p^2\frac{n^+}{N}\mathbf{E}_0^{\pm} \qquad (9\text{-}104)$$

where $\mathbf{E}^{\pm} = \mathbf{E}(\omega \pm \omega_0, \mathbf{k} \pm \mathbf{k}_0)$ are so-called *sideband* fields to be considered small compared to the pump field. In (9-101), terms that vary with $\exp[i(\mathbf{k} \cdot \mathbf{x} - \omega t)]$ are collected to yield

$$(\omega^2 - c_s^2 k^2)n^+ = \frac{\omega_p^2}{\omega_f^2} \frac{\varepsilon_0}{2M} k^2 (\mathbf{E}_0^+ \cdot \mathbf{E}^- + \mathbf{E}_0^- \cdot \mathbf{E}^+) \qquad (9\text{-}105)$$

Expressing $\mathbf{E}^+$ from (9-104) and inserting into (9-105) yields the dispersion relation

$$\omega^2 - c_s^2 k^2 = \Lambda \left[\frac{1}{(\omega - \omega_0)^2 - \omega_p^2 - c^2(\mathbf{k} - \mathbf{k}_0)^2} \right.$$
$$\left. + \frac{1}{(\omega + \omega_0)^2 - \omega_p^2 - c^2(\mathbf{k} + \mathbf{k}_0)^2} \right] \qquad (9\text{-}106)$$

where

$$\Lambda = \frac{e^2 \omega_p^2}{2mM\omega_f^2} k^2 |E_0|^2 = \frac{e^2 |E_0|^2}{2m^2 \omega_f^2} k^2 \omega_{pi}^2 \qquad (9\text{-}107)$$

is the so-called strength parameter, while $eE_0/m\omega_f$ is the oscillatory electron velocity in the pump field. The denominators can be simplified by noticing that the pump wave satisfies the electromagnetic dispersion relation, and ignoring ω/ω_0 terms

$$(\omega \pm \omega_0)^2 - \omega_p^2 - c^2(\mathbf{k} \pm \mathbf{k}_0)^2 \approx \pm 2\omega\omega_0 \mp 2\mathbf{k} \cdot \mathbf{k}_0 c^2 - k^2 c^2 \qquad (9\text{-}108)$$

leading to a more manageable form of (9-106)

$$(\omega^2 - c_s^2 k^2) \left[\left(\omega - \frac{c^2 \mathbf{k} \cdot \mathbf{k}_0}{\omega_0} \right)^2 - \frac{c^4 k^4}{4\omega_0^2} \right] = \Lambda \frac{c^2 k^2}{2\omega_0^2} \qquad (9\text{-}109)$$

With the notations $c^2 k^2/2\omega_0 = \delta$ and $c^2 \mathbf{k}_0/\omega_0 = \mathbf{v}_g$, the group velocity of the pump wave, one obtains

$$(\omega^2 - c_s^2 k^2)[(\omega - \mathbf{k} \cdot \mathbf{v}_g) + \delta] = \Lambda \frac{\delta}{\omega_0} \frac{1}{(\omega - \mathbf{k} \cdot \mathbf{v}_g) - \delta}$$
$$\approx -\frac{\Lambda}{\omega_0} \left(1 + 2 \frac{\mathbf{k} \cdot \mathbf{k}_0}{k^2} \right)^{-1} \qquad (9\text{-}110)$$

since $\omega/\delta \ll 1$, unless ω_0 is very close to ω_p.

We look first for instabilities around the point ω_2, k_2 in Fig. 9-9. This point is characterized by $c_s k = kv_g - \delta$ (see Exercise 9-22). Writing $\omega = c_s k + \varepsilon$, one finds

$$\varepsilon^2 = -\frac{\Lambda}{2c_s k} \left(1 + 2 \frac{\mathbf{k} \cdot \mathbf{k}_0}{k^2} \right)^{-1} < 0 \qquad (9\text{-}111)$$

Hence this mode is indeed unstable. If ω_0 is not very close to ω_p, $v_g \gg c_s$, so $kv_g \approx \delta$; therefore $k_0 \approx \frac{1}{2}k$, as expected on the basis of the physical picture given earlier. This gives the growth rate

$$\gamma = \frac{1}{2}\sqrt{\frac{\Lambda}{c_s k}} \tag{9-112}$$

for stimulated Brillouin backscatter. This is a good approximation as long as $\gamma \ll c_s k$. If the pump field is sufficiently strong so that $\gamma \gg c_s k = kv_g - \delta$,

$$\omega^3 \approx -\frac{\Lambda}{2\omega_0} \tag{9-113}$$

and $\gamma = \sqrt{3}\,\Lambda/4\omega_0$. Since Re $\omega \neq c_s k$, the ion wave is not an eigenmode of the system. The large ponderomotive force overwhelms the electron pressure, and serves as the main driving mechanism in this so-called *reactive quasi-mode*.

Another case where the ion mode is driven off resonance is the *filamentation instability*. Consider an ion density perturbation perpendicular to the incoming wave, so the light propagates along constant density lines. The plasma dielectric constant for light waves $\kappa = 1 - \omega_p^2/\omega^2$ is larger in density troughs than along the peaks and so is the index of refraction $\sqrt{\kappa}$. It is known from optics that light propagates preferentially along regions with a high index of refraction. A fiber with large index of refraction forms a light pipe to guide the radiation, which suffers total reflection on the pipe surface. Hence, in a plasma, light will be concentrated along density troughs, giving rise to a ponderomotive force to deepen these troughs further, trapping more light and so on.

To investigate this instability we choose $\mathbf{k} \perp \mathbf{k}_0$ in (9-109), to find

$$(\omega^2 - c_s^2 k^2)(\omega^2 - \delta^2) = \Lambda \frac{\delta}{\omega_0} \tag{9-114}$$

This biquadratic can be easily solved, and one finds $\omega^2 < 0$ (hence instability) if $\Lambda > c_s^2 k^2 \omega_0 \delta$. Note that Λ is proportional to k^2, while the right-hand side of the inequality is proportional to k^4. Instability arises for a given pump strength once k is small (the wavelength of the ion quasi-mode long). This is a purely growing instability since Re $\omega = 0$.

In *laser fusion*, a solid pellet is irradiated for a short time with high-intensity laser light in order to cause the compression and heating of the center to fusion temperatures. The pellet surface turns very early into a plasma corona, and the laser light interacts with this plasma. On the outer regions of this corona the density is low $\omega_0 \gg \omega_p$, and processes like those described here take place. Backscattering reflects some of the light, while filamentation aids its penetration. At some point, as the light propagates into denser regions, it arrives at the cutoff $\omega_0 \approx \omega_p$ region, the so-called critical surface. It is hoped that the light energy is absorbed there.

One of the processes found here is the *modulational instability*. Consider the case when $k \ll k_0$, the ion perturbation wavelength is much longer than the electromagnetic wavelength. In this case $kv_g \gg \delta$, and an instability arises around $\omega = kv_g$. Set $\omega = kv_g + \varepsilon$ in (9-109), while ignoring $\delta \ll kv_g$ gives approximately

$$\varepsilon^2 = \delta^2 - \Lambda \frac{\delta}{\omega_0 k^2} \frac{1}{c_s^2 - v_g^2} \tag{9-115}$$

If $v_g \lesssim c_s$, which is only possible for ω_0 very close to ω_p, $\varepsilon^2 < 0$, indicating instability. Of course our approximations break down for $c_s = v_g$, and Im ω stays finite.

This instability is related to the filamentary one. A train of light waves propagating at the group velocity becomes trapped in a density well, and deepens it by ponderomotive pressure.

Right at the resonance layer $k_0 = 0$, and beyond it k_0 becomes imaginary, leading to an exponential decay of the wave field. In this region the pump field can be represented by a uniform oscillating electric field. Such a field couples to density perturbations along $\mathbf{E}_0$ and produces the *oscillating two-stream instability* (OTSI).

The physical mechanism is illustrated on Fig. 9-10. The pump field causes the electrons to oscillate at the pump frequency, giving rise to an oscillating space charge. This space charge produces an electric field E_k of its own, oscillating at the same frequency ω_0, with the wave number of the density ripples k. The total electric field is $E_0 + E_k$, and it gives rise to a ponderomotive

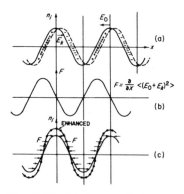

FIG. 9-10. Illustration of the OTSI: (a) electron displacement in oscillating E_0 field; (b) ponderomotive force from E_0 and E_k; (c) enhancement of the initial density wave due to the ponderomotive force.

potential proportional to $\langle (E_0 + E_k)^2 \rangle$. This potential has the wave number k, and can either enhance or diminish the density perturbation, depending on the relative phases of E_0 and E_k. The former case leads to the OTSI, since increased density ripples produce an increase of E_k, enhancing the pondero-motive force F. Equation (9-102) determines E_k, and if $\omega_0^2 < \omega_p^2 + v_e^2 k^2$ (as is indeed the case at the resonant layer), the phases are right to drive the OTSI. The details are left for Exercise (9-23).

So far we have ignored wave damping. Now we are going to discuss briefly two important effects resulting from damping of the linear modes.

First, in the presence of damping, a finite threshold pump intensity is required for the onset of the instabilities. The fluid equations do not naturally include damping, but if collisional or Landau damping rates of these modes are known, a phenomenological way to incorporate them into our equations exists. Replacing $\partial^2/\partial t^2$ by $\partial^2/\partial t^2 + 2\Gamma_i \, \partial/\partial t$ in (9-87), (9-101), and (9-102), where the Γ_i are the damping rates of these three modes, gives an adequate description of the effects of linear damping. In Exercise 9-24 you are asked to calculate some thresholds resulting from linear wave damping.

It is clear that energy loss mechanisms other than damping have a similar effect. Consider, e.g., that the pump wave and the decay products interact only in a restricted region of space. This can occur in an inhomogeneous medium, where resonant coupling is restricted to the region where the coupling conditions are satisfied, or when the pump wave itself occupies only a finite region of space, from which the decay products can escape. One can estimate this convective energy loss in the following way. If L is the length of the interaction region, and v_g is the component of the group velocity in the same direction, the loss of wave energy is described by the equation $L \, dU/dt = -v_g U$, where U is the energy density of the wave. This results in the effective convective damping rate $\gamma_{conv} = v_g/2L$.

The second effect is more subtle. We have seen that if the beat wave of two modes is not an eigenmode, but is Landau damped, nonlinear Landau damping of the higher-frequency wave, and growth of the lower-frequency wave arises. If the higher-frequency wave is the pump wave, and the lower-frequency one a small electron density flunctuation, the damping of the beat mode leads to an exponential increase of the density fluctuation. This mode is the *resistive quasi-mode* or *stimulated Compton scattering*.

In the previous section, the formulas for nonlinear Landau damping for electrostatic waves have been derived. It is clear, however, from the alternative derivation of these formulas, based on energy and momentum conservation, that they apply for electromagnetic waves as well.

When the beat frequency is much less than the primary wave frequencies, the first term in (9-57) originating in the ponderomotive force is much larger than the other two, whose origin is in the nonlinearity of the equation of continuity. This observation permits us to write for the slow waves a simplified second order Vlasov equation for electrons

$$\frac{\partial f_e}{\partial t} + \mathbf{v} \cdot \frac{\partial f_e}{\partial \mathbf{x}} - \frac{e}{m} (\mathbf{E} - \nabla\psi) \cdot \frac{\partial f_0}{\partial \mathbf{v}} = 0 \tag{9-116}$$

where $\psi = -(e/2m)\langle E_f^2\rangle/\omega_f^2$ is the ponderomotive potential resulting from the fast waves. A similar equation may be written for ions, except that the ionic ponderomotive potential can be ignored ($m/M \ll 1$). Combining the two Vlasov equations with Poisson's equation results in the generalization of the low-frequency electron fluctuations to replace (9-105)

$$n_e = -\frac{\varepsilon_0 k^2}{2m\omega_f^2} (\mathbf{E_0}^+ \cdot \mathbf{E}^- + \mathbf{E_0}^- \cdot \mathbf{E}^+) \frac{\chi_e(1 + \chi_i)}{\varepsilon^{(1)}} \tag{9-117}$$

where $\varepsilon^{(1)}$, χ_e, χ_i are the linear complex dielectric function, electronic, and ionic permeabilities of the slow wave, respectively. The proof is left for Exercise 9-25. Combining (9-117) with (9-104) leads to the generalized form of (9-106), which includes stimulated Compton scattering:

$$\frac{1}{\chi_e} + \frac{1}{1 + \chi_i} = -\frac{\varepsilon_0 k^2 \omega_p^2}{2m\omega_f^2} \frac{|E_0|^2}{N} \left(\frac{1}{(\omega + \omega_0)^2 - c^2(\mathbf{k} + \mathbf{k_0})^2 - \omega_p^2} \right.$$

$$\left. + \frac{1}{(\omega - \omega_0)^2 - c^2(\mathbf{k} - \mathbf{k_0})^2 - \omega_p^2} \right) \tag{9-118}$$

In some processes the frequencies of all modes are comparable, so one can no longer divide waves into fast and slow ones. Such a case will be illustrated by treating the decay of an electromagnetic pump wave of frequency $\omega_0 = 2\omega_p$ into two electrostatic waves with frequency ω_p. Here a photon decays into two plasmons. The electron velocity in the combined wave field consists of the velocity in the field of the e.m. wave $\mathbf{v}_0$, and that produced by the electrostatic decay products $\mathbf{v}_1$ and $\mathbf{v}_2$, so that $\mathbf{v} = \mathbf{v}_0 + \mathbf{v}_1 + \mathbf{v}_2$ and $n = n_1 + n_2$ ($n_0 = 0$). So the equation of continuity (9-88) becomes

$$\dot{n} + N \nabla \cdot (\mathbf{v}_1 + \mathbf{v}_2) + \mathbf{v}_0 \cdot \nabla n = 0 \tag{9-119}$$

where products of small quantities have been neglected and $\nabla \cdot \mathbf{v}_0 = 0$. For the pump wave one sets $\mathbf{v}_0 = \mathbf{v}_0^+ \exp[i(\mathbf{k_0} \cdot \mathbf{x} - \omega_0 t)] + \mathbf{v}_0^- \exp[-i(\mathbf{k_0} \cdot \mathbf{x} - \omega_0 t)]$, while $\mathbf{v}_1 = v_1(\mathbf{k}/k) \exp[i(\mathbf{k} \cdot \mathbf{x} - \omega t)]$ and for $\mathbf{v}_2, \mathbf{k} \to \mathbf{k} - \mathbf{k_0}$, $\omega \to \omega - \omega_0$. Since one expects $\omega \approx \omega_p$, the $\omega + \omega_p \approx 3\omega_p$ wave will not be excited. Now the Fourier components, $\mathbf{k}$ and $\mathbf{k} - \mathbf{k_0}$ can be evaluated to yield

$$-\omega n_1 + N k v_1 + \mathbf{v}_0^+ \cdot \mathbf{k} n_2 = 0 \tag{9-120}$$

$$-(\omega - \omega_0)n_2 + N|\mathbf{k} - \mathbf{k_0}|v_2 + \mathbf{v}_0^- \cdot \mathbf{k} n_1 = 0 \tag{9-121}$$

Note that $\mathbf{k_0} \cdot \mathbf{v}_0 = 0$ for the e.m. wave.

For simplicity we will use the cold plasma approximation and neglect the pressure term in the equation of motion (9-89). Using (A-11) one writes

$$(\mathbf{v} \cdot \nabla)\mathbf{v} = \tfrac{1}{2}\nabla(\mathbf{v} \cdot \mathbf{v}) - \mathbf{v} \times (\nabla \times \mathbf{v}_0) \tag{9-122}$$

since $\nabla \times \mathbf{v}_{1,2} = 0$. Since $\nabla \times \dot{\mathbf{v}}_0 = -(e/m)\nabla \times \mathbf{E}_0 = (e/m)\dot{\mathbf{B}}_0$, $\nabla \times \mathbf{v}_0 = (e/m)\mathbf{B}_0$.

In our ordering $\mathbf{v} \cdot \mathbf{v} = v_0{}^2 + 2\mathbf{v}_0 \cdot (\mathbf{v}_1 + \mathbf{v}_2)$, and (9-89) becomes

$$\dot{\mathbf{v}}_1 + \dot{\mathbf{v}}_2 + \frac{e}{m}(\mathbf{E}_1 + \mathbf{E}_2) + \nabla[\mathbf{v}_0 \cdot (\mathbf{v}_1 + \mathbf{v}_2)] = 0 \tag{9-123}$$

Taking the divergence and using Poisson's equation yield for the two Fourier components,

$$\omega k v_1 - \omega_p{}^2 \frac{n_1}{N} - k^2 \mathbf{v}_0{}^+ \cdot \frac{\mathbf{k}}{|\mathbf{k} - \mathbf{k}_0|} v_2 = 0 \tag{9-124}$$

$$(\omega - \omega_0)|\mathbf{k} - \mathbf{k}_0|v_2 - \omega_p{}^2 \frac{n_2}{N} - (\mathbf{k} - \mathbf{k}_0)^2 \mathbf{v}_0{}^- \cdot \frac{\mathbf{k}}{k} v_1 = 0 \tag{9-125}$$

or

$$v_1\left[\omega k - \frac{k(\mathbf{v}_0{}^+ \cdot \mathbf{k})(\mathbf{v}_0{}^- \cdot \mathbf{k})}{\omega - \omega_0}\right]$$

$$\approx v_1 \omega k = \omega_p{}^2 \frac{n_1}{N} + \frac{k^2}{(\mathbf{k} - \mathbf{k}_0)^2} \frac{\omega_p{}^2}{\omega - \omega_0} \mathbf{k} \cdot \mathbf{v}_0{}^+ \frac{n_2}{N} \tag{9-126}$$

where the approximation holds for a relatively weak pump such that $v_0 \ll \omega_p/k$, and k and $|k - k_0|$ are of the same order of magnitude. Similarly,

$$v_2(\omega - \omega_0)|\mathbf{k} - \mathbf{k}_0| = \omega_p{}^2 \frac{n_2}{N} + \frac{(\mathbf{k} - \mathbf{k}_0)^2}{k^2} \frac{\omega_p{}^2}{\omega} \mathbf{k} \cdot \mathbf{v}_0{}^- \frac{n_1}{N} \tag{9-127}$$

Substituting (9-126) and (9-127) into the equations of continuity (9-120) and (9-121) results in two homogeneous equations in n_1 and n_2

$$n_1(\omega_p{}^2 - \omega^2) + n_2 \mathbf{k} \cdot \mathbf{v}_0{}^+ \left(\omega + \frac{\omega_p{}^2}{\omega - \omega_0} \frac{k^2}{(k_0 - k)^2}\right) = 0 \tag{9-128}$$

$$n_2[\omega_p{}^2 - (\omega - \omega_0)^2] + n_1 \mathbf{k} \cdot \mathbf{v}_0{}^- \left[\omega - \omega_0 + \frac{\omega_p{}^2}{\omega} \frac{(k - k_0)^2}{k^2}\right] = 0 \tag{9-129}$$

Substituting $\omega_0 = 2\omega_p$, $\omega = \omega_p + \delta$, where $\delta \ll \omega_p$, yields the dispersion relation

$$4\delta^2 = (\mathbf{k} \cdot \mathbf{v}_0{}^+)(\mathbf{k} \cdot \mathbf{v}_0{}^-)\left(1 - \frac{k^2}{(\mathbf{k} - \mathbf{k}_0)^2}\right)\left(1 - \frac{(\mathbf{k} - \mathbf{k}_0)^2}{k^2}\right) \tag{9-130}$$

Since $(1 - x)(1 - x^{-1})$ is always negative for $x > 0$, δ is imaginary to yield a growing solution.

To find the fastest-growing wave, one takes $\mathbf{v}_0$ in (say) the x and $\mathbf{k}_0$ in the y direction, while $\mathbf{k}$ has both x and y components. Maximizing (9-130) with respect to k_x^2 (see Exercise 9-26), one finds that k_x of the fastest-growing mode is the geometric mean of k_0 and k_y, $(k_x^2)_{max} = k_0 k_y$ and the corresponding growth rate is $\gamma_{max} = \frac{1}{2}k_0 v_0$.

In laser fusion, this process takes place at the quarter critical layer, (the quarter critical layer is where the density is a quarter of that of the critical layer $\omega_0 = \omega_p$). The further beating of the incoming e.m. waves with the half frequency plasma waves produces light waves at $\frac{1}{2}\omega_0$ and $\frac{3}{2}\omega_0$, observed experimentally.

It should be clear that in magnetized plasmas, with the large number of possible eigenmodes, the number of possible parametric processes is enormous. They play an important role in the radio-frequency heating of magnetoplasmas (e.g., in lower hybrid wave heating). These processes can be analyzed in ways similar to those we have carried through for field-free plasmas.

9-5. Large Amplitude Waves

In treating parametric processes, the large amplitude pump wave was held constant and the decay products were considered small. This is basically a linear treatment with the imposed pump wave amplitude given, and linearization performed on the other modes. A truly nonlinear treatment should make no such distinction, and treat all waves on an equal footing.

The nonlinear equations describing coupled electron and ion waves are (9-101) and (9-102), where E_f contains all fast waves including the pump. Without loss of generality one may write $E_f = \frac{1}{2}[\tilde{E}(x, t)\exp(-i\omega_p t) + \text{c.c.}]$, and realize that the temporal variation of $\tilde{E}$ must be slow compared to the plasma frequency time scale, since all fast wave frequencies are close to ω_p. Consequently (9-102) becomes approximately

$$2i\omega_p \dot{\tilde{E}} + v_e^2 \frac{\partial^2}{\partial x^2} \tilde{E} = \omega_p^2 \frac{n}{N} \tilde{E} \qquad (9\text{-}131)$$

as well as the complex conjugate of this equation. In (9-101) the fast time scale average is over the plasma frequency time scale, so one has

$$\ddot{n} - c_s^2 \frac{\partial^2}{\partial x^2} n = \frac{\varepsilon_0}{4M} \frac{\partial^2}{\partial x^2} |\tilde{E}^2| \qquad (9\text{-}132)$$

It would be desirable to solve these two coupled equations exactly for initially specified n and $\tilde{E}$. Since we do not know how to carry out this program (except for computer solutions), we do the next best thing. Conservation laws (see Exercise 9-28) and special solutions can be found. If the ion perturbation is stationary $\tilde{n} = 0$ and $n = -\varepsilon_0(4Mc_s{}^2)^{-1}|\tilde{E}|^2$. Similarly, if the ion perturbation is stationary in some frame moving with the velocity V,

$$n = -\varepsilon_0[4M(c_s{}^2 - V^2)]^{-1}|\tilde{E}|^2$$

and n may be eliminated from (9-131) to yield

$$i\dot{\tilde{E}} + \frac{v_e{}^2}{2\omega_p}\frac{\partial^2}{\partial x^2}\tilde{E} + \frac{e^2}{8\omega_p mM(c_s{}^2 - v^2)}|\tilde{E}|^2\tilde{E} = 0 \qquad (9\text{-}133)$$

For obvious reasons, this is called the *nonlinear Schrödinger equation*, where $\tilde{E}$ plays the role of the ψ function and $|\tilde{E}|^2$ is proportional to a negative potential. Our intuition in solving the linear Schrödinger equation helps us in finding solutions to the nonlinear one. We know that in a sufficiently deep potential well, bounded solutions exist for ψ. Such a bounded solution, however, produces a localized potential well in the nonlinear equation, where, since $|\psi|^2$ is proportional to the negative potential in the Schrödinger equation, the wave function can "dig its own well" and hence confine itself. In our case, the well is a depression of the ion density n, and it is dug by the ponderomotive force exerted by the confined electon wave.

We try a solution of the form $\tilde{E} = A(x - Vt)\exp[i(k_0 x - \omega_0 t)]$, where the propagating amplitude function is real. Inserting into (9-133) and separating real and imaginary parts yields (see Exercise 9-29)

$$k_0 = \frac{\omega_p}{v_e{}^2}V \qquad (9\text{-}134)$$

and

$$A'' - \left(k_0{}^2 - \frac{2\omega_p}{v_e{}^2}\omega_0\right)A + \frac{2\omega_p\lambda}{v_e{}^2}A^3 = 0 \qquad (9\text{-}135)$$

where λ is the coefficient of the $|\tilde{E}|^2\tilde{E}$ term in (9-133). Using the abbreviations a and b for the coefficients of the A and A^3 terms, multiplying by A', and integrating yields

$$\tfrac{1}{2}(A')^2 - \tfrac{1}{2}aA^2 + \frac{b}{4}A^4 = E \qquad (9\text{-}136)$$

where E is the constant of integration. The character of the solutions is revealed by considering the motion of a unit mass in the potential $U = -\tfrac{1}{2}aA^2 + (b/4)A^4$ (think of A as a position x, and the derivative as a time derivative). The energy conservation equation is then (9-136) and by plotting $U(A)$ (Fig. 9-11) we see that for a given value of E, A oscillates in a bounded region. The solution of (9-136) can also be found analytically in terms of

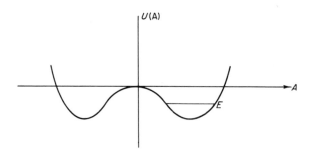

FIG. 9-11. Potential well obtained by integration of nonlinear Schrödinger equation.

elliptic integrals. In general, $A(x - Vt)$ is a periodic function. The special case $E = 0$ yields the most interesting solution. Consider the mass being released at $A = 0$ with zero energy. It will take an infinitely long time for the mass to be accelerated to finite velocity, but then it will move out to its maximum excursion $A_{max}^2 = 2a/b$ move back toward $A = 0$, where it settles down as $t \to \infty$. The solution (which is a sech function) is sketched in Fig. 9-12. This solution shows a solitary envelope pulse propagating with velocity V. The electron wave oscillates and propagates within the envelope. The entire solution is called *envelope soliton*. Note that this solution can only be obtained if $a > 0$ or $\omega_0 < (v_e^2/2\omega_p)k_0^2$. Since the frequency of E_f is $\omega_p + \omega_0$, the restriction on ω_0 implies that E_f cannot oscillate faster than the Bohm-Gross frequency in the envelope soliton. The containment mechanism of the fast wave becomes now clear: in the density depression the local Bohm-Gross frequency is less than the frequency of the electron oscillation; hence the wave can propagate, but it is cut off outside the depression.

While the envelope soliton is just one of the multitude of possible solutions of the nonlinear Schrödinger equation [not to mention the coupled (9-131) and (9-132)] there are reasons to believe that it is one of the physically significant solutions. Computer solutions as well as analytic calculations have

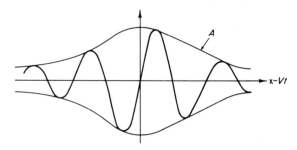

FIG. 9-12. Envelope soliton solution of nonlinear Schrödinger equation.

shown the stability of envelope solitons with respect to small perturbations. Interacting solitons tend to maintain their identity after passing through each other, and have been experimentally observed.

Equations (9-131) and (9-132) can be generalized to three space dimensions, where the $\partial^2/\partial x^2$ in (9-131) becomes replaced by $\nabla\nabla\cdot$ and in (9-132) by ∇^2. A three-dimensional calculation, as well as computer solutions, shows the disintegration of the envelope solitons on a slow (ionic) time scale when subjected to transverse perturbations, showing a filamentary instability. One may also construct solitons with spherical or cylindrical symmetry, but one finds that they collapse toward a point or line singularity. When approaching the singularity, however, several approximations used to derive our equations break down. When deriving the ion wave equation, for instance, $N \gg n$ was assumed. Taking further nonlinearities into account one finds that the spherical and cylindrical solitons no longer collapse.

As the next example for nonlinear wave behavior we study the interaction of ion acoustic waves ($\omega/k \approx c_s$) with each other. These waves interact strongly since for any triplet of eigenmodes $\omega_1 + \omega_2 \approx \omega_3$ and $k_1 + k_2 \approx k_3$ is satisfied. This interaction is described by

$$\varepsilon^{(1)}(\omega,k)E(\omega,k) = i \sum_{\omega',k'} E(\omega',k')E(\omega - \omega',k - k')\varepsilon^{(2)}(\omega - \omega',\omega',k - k',k')$$

$$(9\text{-}137)$$

or in terms of potentials ($E = -ik\varphi$) and replacing the summation by integral

$$k\varepsilon^{(1)}(\omega,k)\varphi(\omega,k) = \iint k'(k - k')\varphi(\omega',k')\varphi(\omega - \omega',k - k')\varepsilon^{(2)} \, d\omega' \, dk'$$

$$(9\text{-}138)$$

One writes now $\omega = c_s k + \delta\omega$ and expands

$$\varepsilon^{(1)}(\omega,k) = \varepsilon^{(1)}(c_s k,k) + \frac{\partial\varepsilon^{(1)}}{\partial\omega}\delta\omega \qquad (9\text{-}139)$$

Since for ion waves

$$\varepsilon^{(1)} = 1 - \frac{\omega_{pi}^2}{\omega^2} + \frac{1}{k^2\lambda_D^2} \qquad (7\text{-}101)$$

one finds $\varepsilon^{(1)}(c_s k) = 1$, and

$$\left.\frac{\partial\varepsilon^{(1)}}{\partial\omega}\right)_{c_s k} = \frac{2\omega_{pi}^2}{c_s^3 k^3} = \frac{2}{\omega_{pi}k^3\lambda_D^3} \qquad (9\text{-}140)$$

In evaluating $\varepsilon^{(2)}$ we use (9-48) and set approximately $\omega/k = c_s$ for each wave, resulting in the approximate expression

$$\varepsilon^{(2)}(\omega',\omega - \omega',k',k - k') = \sum_{+-} \frac{q^3}{m^2\varepsilon_0} \frac{1}{k'k(k - k')} \int \frac{f_0' \, dv}{(v - c_s)^3} \quad (9\text{-}141)$$

where the summation is over the two particle species. Substitution of these results into (9-138) gives

$$\varphi(k,\omega)(\delta\omega + \tfrac{1}{2}k^3 c_s \lambda_D^2) - \sum_{+-} \frac{q^3}{2m^2\varepsilon_0} c_s \lambda_D^2 k$$

$$\cdot \int \frac{f_0' \, dv}{(v - c_s)^3} \int \varphi(\omega',k')\varphi(\omega - \omega',k - k') \, d\omega' \, dk' = 0 \quad (9\text{-}142)$$

In order to transform this equation to the space-time domain we multiply by i and carry out an inverse Fourier transform. Since $\delta\omega = \omega - c_s k$, the first term is $\mathscr{F}[(\partial/\partial t + c_s \, \partial/\partial x)\varphi(x,t)]$, where $\mathscr{F}$ stands for Fourier transform. The second term contains k^3; hence it is the Fourier transform of the third x derivative, while the integral is a convolution, representing the transform of φ^2. This results in a nonlinear differential equation of the form

$$\left(\frac{\partial}{\partial t} + c_s \frac{\partial}{\partial x}\right)\varphi + A \frac{\partial^3 \varphi}{\partial x^3} + B \frac{\partial}{\partial x} \varphi^2 = 0 \quad (9\text{-}143)$$

with the constants

$$A = \tfrac{1}{2}c_s \lambda_D^2 \quad (9\text{-}144)$$

$$B = \sum_{+-} \frac{q^3}{2m^2\varepsilon_0} c_s \lambda_D^2 \int \frac{f_0' \, dv}{(v - c_s)^3} \quad (9\text{-}145)$$

In the frame moving with the velocity c_s (9-143) becomes

$$\frac{\partial\varphi}{\partial t} + A \frac{\partial^3 \varphi}{\partial x^3} + B \frac{\partial}{\partial x} \varphi^2 = 0 \quad (9\text{-}146)$$

the *Korteweg–de Vries* or briefly *KDV* equation. In the absence of the non-linear term this equation is satisfied by a backward traveling wave of the from $\exp[i\sqrt{V/A}(x + Vt)]$. This is not surprising since in the c_s frame the ion wave propagates backward due to dispersion. In order to understand the effect of the nonlinear term, we have plotted in Fig. 9-13 a sine wave describing φ at $t = 0$, the corresponding φ^2 and the deformed wave at a somewhat later time due to the nonlinear term. Since $\delta\varphi = -B(\partial\varphi^2/\partial x) \, \delta t$, φ increases where $\partial\varphi^2/\partial x > 0$ and decreases where $\partial\varphi^2/\partial x < 0$. The result is

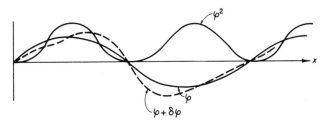

Fig. 9-13. Illustration of the nonlinear deformation of ion wave potential.

a steepening of the descending portions of the φ curve. The characteristic time scale for steepening in the absence of the dispersive term is

$$\tau^{-1} \sim B \frac{\partial \varphi}{\partial x}.$$

One may evaluate B from (9-145) and find $B \approx q/Mc_s$ (see Exercise 9-31). The velocity an ion acquires during a wave period is of the order $qE/M\omega \approx qE/Mc_s k$, so τ is the time it takes an ion to move about a wavelength. The backward wave motion as described by the dispersive term can prevent the disastrous steepening of the wave, and well-defined wave, as well as soliton, solutions can be obtained. The form $\varphi(x,t) = \varphi(x - Vt)$ inserted into (9-146) yields

$$-V\varphi' + A\varphi''' + B(\varphi^2)' = 0 \qquad (9\text{-}147)$$

This equation can be integrated to yield

$$-\frac{V}{2}\varphi^2 + \frac{A}{2}(\varphi')^2 + \frac{B}{3}\varphi^3 = E \qquad (9\text{-}148)$$

Regarding this as the energy conservation equation of mass A in the potential $U = -\frac{1}{2}V\varphi^2 + \frac{1}{3}B\varphi^3$, and plotting the potential in Fig. 9-14, we see that oscillating solutions exist for $V < 0$ and a soliton solution for $V > 0$. The amplitude of the soliton is $\varphi_{\max} = 3V/2B$. Such solitons have been observed in both laboratory experiments and computer solutions. They exhibit remarkable stability properties; two solitons propagating with different velocities interact when they overlap in a nonlinear fashion, then part company, each one preserving its original shape and velocity.

We will undertake now a more precise description of these solitons. As mentioned before, our treatment of ion waves holds only for relatively small density perturbations. In general, the electron density in a slowly varying potential ϕ is described by the isothermal law $n_e = N \exp(e\phi/T)$. It is easy to see (Exercise 9-32) that by expanding this expression for $e\phi/T \ll 1$, the usual expressions for ion waves are recovered. In the frame moving with the velocity

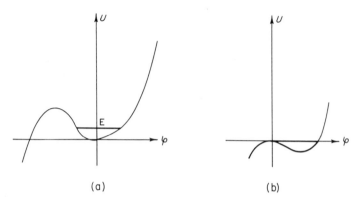

FIG. 9-14. Potential wells for (a) $V < 0$ and (b) $V > 0$, obtained from integration of KDV equation.

V, where the wave or soliton is stationary, from continuity $n_i v_i = NV$ for the ions, and from energy conservation $\frac{1}{2}Mv_i^2 + e\phi = \frac{1}{2}MV^2$. One may express the ion density n_i and substitute into Poisson's equations

$$\varphi'' = -\frac{e}{\varepsilon_0} N\left(\frac{VM}{\sqrt{V^2M^2 - 2e\varphi M}} - e^{e\phi/T}\right) \qquad (9\text{-}149)$$

It is clear that this equation includes nonlinearities of higher order than the ones discussed before, while the linear expansion in φ leads to the linear ion wave dispersion relation (Exercise 9-33). It is convenient to introduce the dimensionless potential $\psi = 2e\phi/MV^2$ to find

$$\frac{1}{2}\left(\frac{V}{\omega_{pi}}\right)^2 \psi'' = \exp\left(\frac{MV^2}{2T}\psi\right) - \frac{1}{\sqrt{1-\psi}} \qquad (9\text{-}150)$$

After multiplication by ψ' and integration one obtains

$$\frac{1}{2}\left(\frac{V}{2\omega_{pi}}\right)^2 \psi'^2 - \frac{T}{MV^2}\exp\left(\frac{MV^2}{2T}\psi\right) - \sqrt{1-\psi} = C \qquad (9\text{-}151)$$

Plot again the "potential function" $U(\psi) = -\alpha e^{\psi/2\alpha} - \sqrt{1-\psi}$, where $\alpha = T/MV^2 = c_s^2/V^2$ is the reciprocal Mach number squared. Note that $U(\psi)$ is real only for $-\infty < \psi \leq 1$, and only the positive square root has physical significance (the ion density cannot be negative). $U'(0) = 0$ and $U''(0)$ is positive for $\alpha > 1$, negative for $\alpha < 1$. The curves corresponding to $\alpha > 1$ and $\alpha < 1$ are shown in Figures 9-15a,b. For $\alpha > 1$ (the wave velocity is less than c_s) no soliton solutions exist in accordance with the KDV result. Nonlinear wave solutions do exist as seen from Fig. 9-15a. Meaningful solutions, satisfying the boundary conditions, exist only when ψ is oscillatory.

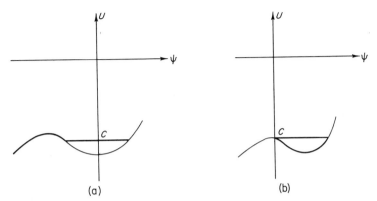

FIG. 9-15. Potentials from integration of the nonlinear ion wave equation: (a) $\alpha > 1$, (b) $\alpha < 1$.

If $\alpha > 4$, i.e., the wave velocity is less than half of the ion acoustic velocity, the singular point at $\psi = 1$ is lower than the maximum on the left and an oscillatory solution can reach $\psi = 1$. As seen from (9-150), at $\psi = 1$ the ion density becomes infinite. This corresponds to a strong steepening of the wave, with ions originally in different positions catching up with each other. Here one reached the threshold of wave breaking, as in water waves moving into a beach. In plasmas, this process corresponds to the overtaking and crossing of particles, leading to a turbulent breakup of the wave pattern, involving particle trapping in potential wells.

For $\alpha < 1$, solitons can be formed, in agreement with our findings from the KDV equation. Now, however, a catastrophic breakdown of solitons becomes possible, once $U(1) \leq U(0)$ or $-(1 + \alpha) \leq -\alpha e^{1/2\alpha}$. The threshold is around $\alpha = 0.4$, so one concludes that soliton solutions are limited to the range $1 > \alpha > 0.4$; the soliton velocity cannot exceed about 1.6 times the ion sound velocity.

Can cold plasma electron oscillations produce solitons? The answer is no, as we are about to show. Since ions are motionless, Poisson's equation gives in the wave frame

$$\frac{1}{2}\left(\frac{V}{\omega_p}\right)^2 \psi'' = \frac{1}{\sqrt{1 - \psi}} - 1 \qquad (9\text{-}152)$$

where $\psi = 2e\phi/mV^2$. Multiply by ψ' and integrate

$$\left(\frac{V}{2\omega_p}\right)^2 \psi'^2 = 2\sqrt{1 + \psi} - \psi + c_1 = -(\sqrt{1 + \psi} - 1)^2 + c^2 \qquad (9\text{-}153)$$

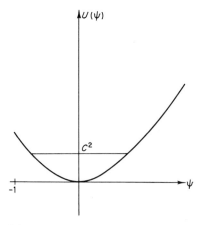

FIG. 9-16. Potential well from the integration of nonlinear electron plasma oscillations for cold electrons.

The plot of the "potential energy" function $U = (\sqrt{1 + \psi} - 1)^2$ in Fig. 9-16, shows only nonlinear oscillatory solutions for ψ and no soliton formation. As $C^2 \rightarrow 1$, the electron density becomes singular at the point $\psi = -1$, singnaling the beginning of wave breaking and turbulence. We may further integrate (9-153) by introducing the variable $z = \sqrt{1 + \psi} - 1$, to find

$$\frac{V}{2\omega_p} \int \frac{d\psi}{[C^2 - (\sqrt{1 + \psi} - 1)^2]^{1/2}} = \frac{V}{\omega_p} \int \frac{(1 + z)\,dz}{\sqrt{C^2 - z^2}} = \xi - \xi_0 \quad (9\text{-}154)$$

where ξ is the coordinate in the wave frame. This yields

$$\xi - \xi_0 = \frac{V}{\omega_p} \left[\sin^{-1} \frac{z}{C} - \sqrt{C^2 - z^2} \right] \quad (9\text{-}155)$$

This equation can be solved numerically, and the results for $C \ll 1$ and $C = 1$ are shown in Fig. 9-17. For the first case, the potential is nearly sinusoidal, as expected, while at the breaking threshold it is substantially deformed. It is an interesting consequence of (9-155) that the period of z (hence also the period of ψ) is independent of the amplitude. As z goes from $-C$ to $+C$, $\xi - \xi_0$ varies by the half wavelength $\lambda/2 = \pi V/\omega_p$. Expressing the wave velocity V in terms of the fundamental frequency ω and wave number k, $V = \omega/k$, one finds $k = 2\pi/\lambda = (\omega_p/\omega)k$, or $\omega = \omega_p$ independent of the amplitude. Such a simple result must have a simple physical reason. One-dimensional plasma oscillations can be pictured as charged electron sheets moving in the direction of the sheet normal, in a positive background. Consider the equation of motion of such a sheet $\ddot{X} = -(e/m)E$, where X is the

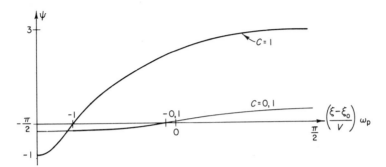

Fig. 9-17. Waveforms of cold electron oscillations in the wave frame. The curve $C = 0, 1$ corresponds to small amplitude; the curve $C = 1$ to the verge of breaking.

electron displacement from its equilibrium position. The electric field at the position of the sheet is simply Q/ε_0, where Q is the total charge in a cylinder of unit cross section from $x = -\infty$ to the momentary sheet position X. As the electron sheet is displaced by X and sheets do not cross, the total number of sheets to the left does not change. The only change of charge results from the neutralizing ion background left behind, namely, eNX. The equation of motion is $\ddot{X} = -\omega_p^2 X$. So in the absence of sheet crossing (wave breaking), the electrons follow simple harmonic motion with frequency ω_p. Of course, the wave form in nonharmonic (Fig. 9-17) and a stationary observer sees harmonics to the fundamental frequency. It should be noted that for other than plane (i.e., spherical or cylindrical) waves the fundamental frequency is amplitude dependent (Exercise 9-34).

One can pose now a far more general question: Under what conditions do the combined nonlinear Vlasov and Poisson's equations give traveling wave solutions? Since the electron and ion distributions can now be arbitrarily chosen, subject only to the condition (in one dimension) that

$$f(x,v) = f(\tfrac{1}{2}mv^2 + q\phi) \tag{9-156}$$

in the wave frame, one expects a large variety of possible solutions. This problem has been investigated by Bernstein, Green, and Kruskal, and the modes were named BGK waves. In addition to (9-156), Poisson's equation

$$\frac{d^2\phi}{dx^2} = \frac{e}{\varepsilon_0}\left[\int_{-\infty}^{+\infty} f_e\, dv - \int_{-\infty}^{+\infty} f_i\, dv\right] \tag{9-157}$$

must be satisfied. With regard to the boundary conditions, the problem may be posed in various ways. One possibility is to prescribe ϕ and ϕ' at a point (say $x = 0$), as well as the distribution functions f. In choosing the latter, one is confronted with the problem of trapped particles. Obviously, by prescribing a distribution function at A, the distribution function at B is not completely determined, since there may be particles at B without enough energy to "climb the potential hill" to reach A. Similarly, some of the particles at A will not arrive at C, and the phase space density of particles with energy E may be quite different to the right of point C than it is to the left (see Fig. 9-18a). Looking at the $E = $ const lines in phase space, one sees islands with closed curves representing trapped particles (Fig. 9-18b). The distribution function is stationary only if each $E = $ const line is populated with the same density of phase points. There is no reason, however, why two different unconnected lines should have the same population density, even if they correspond to the same E.

Consequently, in addition to prescribing the distribution functions at

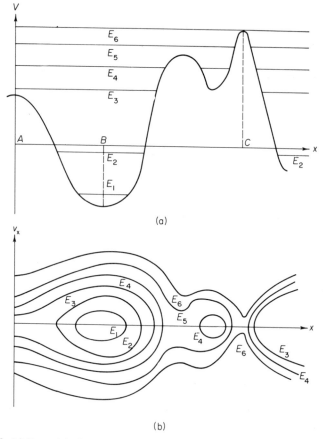

(a)

(b)

FIG. 9-18. (a) Potential of an arbitrarily prescribed traveling wave. (b) Phase plane with energy = const curves.

$x = 0$, one has the option of prescribing trapped distribution functions in every potential trough, for both species. A possible choice is to leave the troughs empty, and describe a distribution function for particles moving to the right

$$f_{\text{right}}(E) \qquad \text{for} \qquad E > q\phi_{\text{max}}$$

and another one to the left

$$f_{\text{left}}(E) \qquad \text{for} \qquad E > q\phi_{\text{max}}$$

where ϕ_{max} is the potential of the highest peak for ions, and of the deepest well for electrons.

By excluding trapped particles, the problem is well defined, and by specifying $\phi(0)$, $\phi'(0)$, and the two f functions (9-156) and (9-157) can be (numerically) integrated. One may choose, e.g., a Maxwellian distribution, properly shifted in velocity space, so that the wave frame moves (with the phase velocity) in the laboratory frame. It can also be shown that in the linear limit, and with high phase velocities, one recovers the corresponding linear dispersion relation with no Landau damping.

If one decides, on the other hand, to include trapped particles, this additional freedom can be used to produce almost any wave form. One can prescribe a desired potential function at will, for example the one plotted in Fig. 9-18a. In addition, one prescribes, again at will, four distribution functions: the untrapped electrons moving to the right, those moving to the left, and the two corresponding untrapped ion distribution functions. In general, of course, one does not get self-consistency that way. From the prescribed potential $\phi(x)$ one calculates the charge distribution $(1/\varepsilon_0)\phi''(x)$ $= -\rho(x)$ necessary to produce it. From the four untrapped distribution functions and $\phi(x)$, by integration over velocity space, one obtains ρ_{untr}. The charge imbalance $\rho - \rho_{\text{untr}}$ can be supplied now by "filling up" the potential wells with the proper distribution of trapped particles—ions around the potential minima, electrons around the maxima. Mathematically, this means the solving of integral equations

$$\sum_{+-} q \int_0^{v_{\text{max}}} f_{\text{tr}} \, dv = \rho - \rho_{\text{untr}} \qquad (9\text{-}158)$$

for f_{tr} in each trough for ions and on each hill for electrons. One concludes that with the help of trapped particles almost any stationary wave form, traveling with any prescribed velocity, can exist.

The stationary wave with trapped particles produced in the long time limit, by a large amplitude Landau damped wave (the asymptotic portion of Fig. 9-3) is such a BGK mode. Which ones of the infinite variety of possible BGK waves are favored by stability is not known.

It is interesting to investigate what happens in the limit of small amplitudes. In this case, the still-finite compensating charge density is to be accommodated in troughs of vanishing depth. As a result, the trapped distribution functions, with all particles in the velocity band of zero width, become δ functions. We have met these stationary solutions of the linear equation before: They are the "false" modes, where a "stream" is singled out to propagate the mode. The singular trapped-particle-distribution function, with all trapped particles moving exactly with the phase velocity, is just this stream.

Finally, a few words about turbulence are in order. When a single wave breaks, or waves of such magnitude interact that our expansion procedures for wave interaction lose validity, we do not know how to treat the process analytically, and one applies the words turbulence to describe the ensuing "messy" state of the plasma. In the absence of analytic description, computer simulation is about the only guide to understanding these processes to some degree. In computer simulation, particles with some initial positions and velocities are launched. The self-consistent fields at each particle position are calculated by the computer at this instant, and the particle positions and velocities are updated some time δt later, using the equations of motion. The new self-consistent fields are then computed to serve for the next updating of particle positions and velocities, and so on. Externally imposed fields (driving waves, etc.) can also be added. These computer "experiments" have the advantage over laboratory experiments of providing direct easy diagnostics. The computer can plot fields, phase space distribution, etc., on

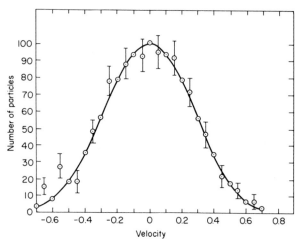

FIG. 9-19. Velocity distribution of an initially cold plasma, launched to carry out breaking oscillations, after many oscillation periods. [From J. M. Dawson, *Phys. Fluids* **5**, 445 (1962).]

command. On the other hand, particle numbers are limited by computer time and memory, they are rarely three dimensional (again computing time rears its head), and realistic boundary conditions are difficult to incorporate (interactions with walls, producing impurity influx, etc.).

Many analytic results described in this chapter have been confirmed, and some anticipated by computer experiments. When computer experiments are used to investigate strong turbulence, one often finds that, in a sense, turbulence acts like collisions; it tends to smooth out and Maxwellize distribution functions. For instance, breaking of a large amplitude cold plasma wave results in a seemingly Maxwellian velocity distribution as shown in Fig. 9-19.

One knows, of course, that a gas, where the velocity distribution is initially different from the Maxwellian, will thermalize as a result of collisions. The interesting fact is, however, that in the calculations of nonlinear plasma oscillations collisions are completely neglected. It seems, therefore, that collective electrostatic interactions have in a sense the same effect as collisions. In some cases (large amplitude waves or unstable distributions) they destroy ordered motion and establish a thermal velocity distribution in the plasma even in the absence of collisions.

This statement seems to contradict a previous theorem regarding the constancy of entropy in a collisionless plasma. We know that the Maxwellian distribution has a larger entropy than any other, and, in the case of collisions, thermalization is achieved by increasing the entropy of the system.

To reconcile these seemingly paradoxical statements, we have to investigate again what was meant by the constancy of entropy. Since the distribution function satisfies the Liouville equation $df/dt = 0$, it follows that any set of particles in a region of phase space will always occupy the same phase volume.

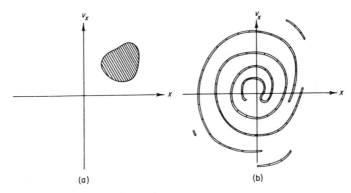

(a) (b)

FIG. 9-20. (a) Initial distribution function. (b) Distribution function spread out in phase space without an increase in entropy.

It is, however, entirely possible that (taking a one-dimensional example) a distribution like the one in Fig. 9-20a goes over into the one shown in Fig. 9-20b, as long as the shaded areas are equal.

Although the phase volumes are equal, the particles in Fig. 9-20b are much better "smeared out" in phase space; they are closer to a Maxwellian. Finally, phase space will be covered in a "coarse grain" manner, and one can find particles in the close vicinity of any phase point, with the proper Maxwellian density, without violating the constancy of occupied space volume or entropy. It is in this sense that thermalization (or quasi-thermalization) takes place in collective interactions.

9-6. Summary

The linearized wave equations are strictly valid only for infinitesimally small amplitude. For a single wave, Landau damping (or growth) is modified once the bounce frequency $\omega_B \sim \sqrt{E}$ becomes larger than the damping (growth) rate, or the inverse lifetime of the wave. While the real part of the dispersion relation stays virtually unchanged, Landau damping oscillates between damping and growth, due to resonant trapped particles, until the wave amplitude becomes stationary. The breakdown of linear theory affects primarily resonant particles.

In the presence of many simultaneously growing waves in a narrow range of phase velocities (like the instability of a bump on tail distribution) the zero-order distribution function gets slowly modified due to the rearrangement of resonant particles. This velocity space diffusion is described by the quasi-linear theory, and leads to the saturation of the instability, once the distribution function gradient in the critical region is wiped out.

In the Landau echo phenomenon, two waves launched at a time (or space) interval, such that the field of the first wave is damped out when the second wave is launched. While the field of both waves dies out, a perturbation of the distribution function remains. The combination of the remnants of the perturbed distribution functions produces a wave pulse at some later time.

Several (undamped) waves can beat to produce sum and difference frequency and wave-number waves, which need not be eigenmodes of the system. The coupling coefficients for this interaction are the higher-order dielectric functions. Particles can Landau damp on the beat waves, extracting energy from the higher frequency mode, and feeding some to the lower-frequency on (nonlinear Landau damping).

When the combination frequency is an eigenmode (or close to it), excitation is strong. If a strong pump wave interacts with two small eigenmodes, such that the ω,k matching conditions are satisfied, exponential growth of the small modes can result. Such "parametric processes" include the stimulated scattering of light waves on electron waves (Raman), ion waves (Brillouin), the oscillating two-stream instability, filamentation, and nonlinear Landau growth. An electron wave can also serve as pump wave. The nonlinear process is described by the ponderomotive force pointing away from regions of high field intensity. In the decay of an electromagnetic wave ($\omega_0 = 2\omega_p$) into two plasma waves, the nonlinearity of the equation of continuity is also important.

The ponderomotive force can produce self contained propagating electron (or e.m.) wave packets, by creating a well in the ion distribution, propagating along with the packets. These envelope solitons are described by solutions of the nonlinear Schrödinger equation. Ion acoustic waves on the linear portion of the ion wave curve interact strongly, since they automatically satisfy ω,k matching conditions. A nonlinear wave equation (the KDV equation) can be derived to describe the wave evolution. The solutions include nonlinear waves propagating slower than c_s, and stable self-contained solitary pulses (KDV solitons), propagating faster. The inclusion of higher nonlinearities shows that solitons can exist only for a limited region of Mach numbers. Both waves and solitons can break if the amplitudes are too large, leading to turbulence.

Electron oscillations with initially cold electrons do not form solitons. The nonlinear oscillations of plane waves produce a nonsinusoidal wave form, whose fundamental frequency is ω_p. If the amplitude becomes too large the wave breaks, and the ensuing turbulence leads to a Maxwellian-like particle distribution function even in the absence of collisions. It can be shown that the full set of the Vlasov and Poisson equations, when potentials and untrapped electron and ion distribution functions are freely chosen, leads to an infinite variety of traveling wave forms (BGK waves).

EXERCISES

9-1. Derive (9-18) and (9-19).

9-2. Use the result of Exercise 7-15 to show that $\gamma = \sqrt{3}kV$ in the linear theory. Integrate (9-20) to verify (9-21).

9-3. Show that (9-21) leads to the $\gamma \approx \omega_b$ condition for the breakdown of linear theory.

9-4. Show that the linear temporal growth (damping) rate γ and the spatial rate Λ are related by $\gamma = v_g\Lambda$, where v_g is the wave group velocity. Use energy conservation to prove.

9-5. Considerations for temporal damping of a large amplitude wave, leading to the behavior shown in Fig. 9-3 were valid as long as $\gamma/\omega_b \ll 1$. What is the corresponding inequality for spatial damping?

9-6. Prove that the quasi-linear theory conserves particle number, momentum, and particle plus field energy.

9-7. Estimate the saturation amplitude for a single growing wave using the quasi-linear equation from integrating $(\partial/\partial t)(\partial F/\partial v)(\omega_r/k)$ in time, assuming γ and F to be constant. Prove that saturation occurs when $\omega_b \approx \gamma$.

9-8. Evaluate the function $\psi(v)$ in (9-35).

9-9. Carry out the calculation for the spatial Landau echo and find the position where the echo appears.

9-10. Prove that if $\omega(\mathbf{k})$ satisfies the linear dispersion relation, $\omega^*(-\mathbf{k})$ also does, with $\mathbf{E}_{-\mathbf{k}} = \mathbf{E}_{\mathbf{k}}^*$ and $f_1(-\mathbf{k}) = f_1^*(\mathbf{k})$.

9-11. Derive (9-48) and (9-49).

9-12. Evaluate (9-57), using $\delta x(\omega_i' + \omega_j')$ from (9-54) and the appropriate first-order particle densities and displacements, to obtain (9-58).

9-13. Prove that $\varepsilon^{(2)}(\omega_2, \Delta\omega, k_2, \Delta k) = -\varepsilon^{(2)}(\omega_1, -\omega_2, k_1, -k_2)$.

9-14. Prove (9-70)–(9-72).

9-15. Evaluate the imaginary part of $\varepsilon^{(3)}$ and show that (9-73) holds.

9-16. Use E_{eff} from (9-82) and the energy exchange rate from (7-63) to derive the nonlinear Landau damping rate (9-75).

9-17. Estimate the nonlinear Landau damping rate of electron waves on electrons to show that $\varepsilon^{(2)}/\varepsilon^{(1)}$ nearly cancels $q\,\Delta k/m\omega^2$.

9-18. Calculate the nonlinear Landau damping rate of electron waves on ions.

9-19. Draw vector diagrams in ω,k space (like Fig. 9-9) to investigate the possibility of different parametric decay processes. Under what conditions can an electron wave decay into an ion wave and another electron wave propagating in the same direction (forward scattering)?

9-20. Show that the equation describing the propagation of an electron wave in slow density fluctuations is described by (9-102).

9-21. Derive the nonlinear equation for ion waves interacting with fast electron waves, and show that the result is (9-101).

9-22. Prove that the point ω_2, k_2 in Fig. 9.9 is characterized by $c_s k = k v_g - \delta$.

9-23. Use (9-102) to calculate the electric field produced by the interaction of a density perturbation, with wave number k and a space-independent pump field parallel to k, at frequency ω_0. Combine this result with (9-101) to derive a dispersion relation for the oscillating two-stream instability, and calculate the growth rate.

9-24. Incorporate damping in the decay modes (not the pump) to derive a modified form of (9-109). You can assume that the Γ's are small compared to the corresponding frequencies.
   (a)   Find the threshold for stimulated Brillouin scattering.
   (b)   Find the threshold for the filamentation instability.
   (c)   Incorporate damping into Exercise (9-23) to find the threshold for OTS.

9-25. Derive (9-117). Prove that in the fluid limit, and for $kc_s \ll \omega_{pi}$, this equation reduces to (9-105).

9-26. Use (9-130) to find k_x which maximizes the growth rate of the decay instability of an electromagnetic wave into two electrostatic waves, and the corresponding growth rate.

9-27. Derive the ponderomotive force acting on an electron in a uniform magnetic field subjected to a pump wave ($\omega_0 \ll \omega_{ce}$) and low frequency decay products.

9-28. Show that $\int_{-\infty}^{+\infty} |E|^2\, dx$ is conserved by (9-131).

9-29. Substitute the trial solution $A(x - Vt) \exp[i(k_0 x - \omega_0 t)]$ into (9-133) and show that (9-134) and (9-135) result.

9-30. Find the correct form of (9-102) and (9-131) for three-dimensional modes. Show that $\partial^2/\partial x^2 \to \nabla \nabla \cdot$.

9-31. Evaluate 9-145 for cold ions.

9-32. Expand $n_e = N \exp(e\phi/T)$ for $e\phi/T \ll 1$ and calculate the electronic susceptibility. Show that its approximate value is $\chi_e = 1/k^2 \lambda_D^2$.

9-33. Expand (9-149) for small ϕ and show that the ion wave dispersion relation results.

9-34. Derive the equations for nonlinear plasma oscillations for spherical and cylindrical modes in a cold electron plasma. (*Hint*: Use spherical and cylindrical electron sheets in an ion background to derive the equations of motion.)

9-35. Estimate the wave amplitude at which the linear theory of the two-stream instability breaks down.

X

Waves and Instabilities in Bounded Plasmas

10-1. Waves in a Plasma Slab

Plasmas found in nature, or produced in the laboratory, are necessarily bounded. The treatment of such systems is simple in principle—one solves the self-consistent equations inside the plasma, the equations for the electromagnetic field outside the plasma, and takes care of the boundary conditions —but the computational complexities are usually prohibitive. For this reason much of this important field is not yet systematically explored.

At first one may ask the following qualitative questions: To what extent can one trust the results obtained for an infinite uniform plasma to hold for the finite case, and what essentially new phenomena are expected to come up which were not present in the uniform configuration? One may be inclined to guess that as long as the wavelength is short compared to the dimensions of the plasma, waves and oscillations will obey a dispersion relation only slightly different from the one obtained in the infinite case. We shall learn soon that this guess is not always correct; in fact the finite size, however large, may bring about profound changes in the dispersion relation. As far as the second question is concerned, one easily finds effects, obviously absent in the uniform plasma, but suspected to appear in the bounded one. For instance, emission, absorption, and reflection of radiation by the plasma may be present in a finite system, even though they were meaningless for the plane-wave solutions used to treat the infinite plasma. Furthermore just as a hydromagnetic fluid may be stable or unstable, depending on the shape of the fluid and curvature of the confining magnetic field, similar results are expected to hold for a collisionless plasma as well. These instabilities are to be added to the ones found in a uniform plasma due to non-Maxwellian velocity distributions, for an estimate of the over-all stability of a plasma system.

First we are going to investigate waves and oscillations of a plasma column of infinite length, bounded, but of arbitrary cross section, embedded in a uniform magnetic field. The z axis of our coordinate system is chosen along the axis of the column, which is also the direction of the field $\mathbf{B}_0$. A cold plasma which carries no currents is chosen for simplicity. This choice also serves to exclude unstable solutions, since no energy is available to feed instabilities.

In the plasma interior the spatial part of the electric field obeys the same equation as in the uniform plasma, namely,

$$\nabla \times \nabla \times \mathbf{E} = \frac{\omega^2}{c^2} \overset{\leftrightarrow}{\kappa} \cdot \mathbf{E} \tag{8-14}$$

where the time dependences of $e^{i\omega t}$ was assumed and the dielectric tensor has the form

$$\overset{\leftrightarrow}{\kappa} = \begin{pmatrix} \kappa_T & i\kappa_H & | & 0 \\ & & | & \\ -i\kappa_H & \kappa_T & | & 0 \\ \hline 0 & 0 & | & \kappa_{\parallel} \end{pmatrix} \tag{8-10}$$

Owing to the translational symmetry in the z direction, the solutions of (8-14) can be Fourier-analyzed into waves with a z dependence e^{-ikz}. It is convenient to write

$$\nabla = \nabla_z + \nabla_{\perp} = -ik_{\parallel}\mathbf{e}_3 + \nabla_{\perp} \tag{10-1}$$

where $\mathbf{e}_3$ is the unit vector in the z direction and $\nabla_{\perp}$ is the del operator in the plane perpendicular to z. Using $\nabla \times \nabla \times = \nabla(\nabla \cdot) - \nabla^2$, (8-14) becomes

$$(\nabla_{\perp} - ik_{\parallel}\mathbf{e}_3)(\nabla_{\perp} \cdot \mathbf{E} - ik_{\parallel}E_z) - \nabla_{\perp}^2\mathbf{E} + k_{\parallel}^2\mathbf{E} = \frac{\omega^2}{c^2} \overset{\leftrightarrow}{\kappa} \cdot \mathbf{E} \tag{10-2}$$

Since $\nabla_{\perp} \cdot \mathbf{E} = \nabla_{\perp} \cdot \mathbf{E}_{\perp}$, (10-2) becomes after expansion

$$\nabla_{\perp}\nabla_{\perp} \cdot \mathbf{E}_{\perp} - ik_{\parallel}\nabla_{\perp}E_z - ik_{\parallel}\mathbf{e}_3\nabla_{\perp} \cdot \mathbf{E}_{\perp} - k_{\parallel}^2\mathbf{e}_3E_z - \nabla_{\perp}^2\mathbf{E} + k_{\parallel}^2\mathbf{E} = \frac{\omega^2}{c^2} \overset{\leftrightarrow}{\kappa} \cdot \mathbf{E} \tag{10-3}$$

This equation splits up into a z and $\perp$ part, to yield

$$-ik_{\parallel}\nabla_{\perp} \cdot \mathbf{E}_{\perp} - \nabla_{\perp}^2E_z = \frac{\omega^2}{c^2}\kappa_{\parallel}E_z \tag{10-4}$$

for the z part and

$$\nabla_{\perp}\nabla_{\perp} \cdot \mathbf{E}_{\perp} - ik_{\parallel}\nabla_{\perp}E_z - \nabla_{\perp}^2\mathbf{E}_{\perp} + k_{\parallel}^2\mathbf{E}_{\perp} = \frac{\omega^2}{c^2} \overset{\leftrightarrow}{\kappa}_{\perp} \cdot \mathbf{E}_{\perp} \tag{10-5}$$

for the $\perp$ part. The meaning of the 2×2 submatrix $\kappa_{\perp}$ is clear from (8-10).

At this point it is useful to specify the geometry. Consider the simple case of a slab of arbitrary thickness in the x direction and infinite in the y direction. Cartesian coordinates are best suited for describing the variables, and the solutions inside the plasma can be Fourier-analyzed into $\exp[-i(k_x x + k_y y)]$

modes. As a result (10-5) becomes

$$\left(\frac{\omega^2}{c^2}\overset{\leftrightarrow}{\kappa}_\perp + \mathbf{k}_\perp\mathbf{k}_\perp - k_\perp^2 - k_\parallel^2\right) \cdot \mathbf{E}_\perp = \overset{\leftrightarrow}{\rho}^{-1} \cdot \mathbf{E}_\perp = -k_\parallel \mathbf{k}_\perp E_z \qquad (10\text{-}6)$$

and multiplication by $\overset{\leftrightarrow}{\rho}$ yields $\mathbf{E}_\perp$ expressed as a function of E_z:

$$\mathbf{E}_\perp = -k_\parallel \overset{\leftrightarrow}{\rho} \cdot \mathbf{k}_\perp E_z \qquad (10\text{-}7)$$

Substituting this expression of $\mathbf{E}_\perp$ into (10-4) leads to the dispersion relation

$$k_\parallel^2 \mathbf{k}_\perp \cdot \overset{\leftrightarrow}{\rho} \cdot \mathbf{k}_\perp + k_\perp^2 = \frac{\omega^2}{c^2}\kappa_\parallel \qquad (10\text{-}8)$$

Clearly, since we have employed the same method of Fourier analysis as in the case of the infinite magnetoplasma in Sec. 8-1 this dispersion relation must be identical with (8-19). This is easily shown with some algebraic manipulations and is left as an exercise to the reader (Exercise 10-1). The difference between the infinite case and the present one lies in the application of the boundary conditions.

In the vacuum, all the field quantities obey the wave equation

$$\nabla^2 u + \frac{\omega^2}{c^2} u = 0 \qquad (10\text{-}9)$$

where the solutions can be Fourier-analyzed into $\exp[-i(k_x^v x + k_y^v y + k_z^v z)]$ modes, and the dispersion relation becomes simply

$$k^v = \pm \omega/c \qquad (10\text{-}10)$$

From $\nabla \cdot \mathbf{E} = \nabla \cdot \mathbf{B} = 0$ follows

$$\mathbf{k}^v \cdot \mathbf{E} = \mathbf{k}^v \cdot \mathbf{B} = 0 \qquad (10\text{-}11)$$

which expresses the transverse nature of electromagnetic waves.

The wave vector in the plasma, to be called $\mathbf{k}^p$, is of course not identical with the wave vector in vacuum, to be distinguished as $\mathbf{k}^v$. The boundary conditions require that E_z, E_y, and $\mathbf{B}$ be continuous (but not E_x). This yields four independent boundary conditions, since the continuity of B_x is a consequence of the continuity of E_y and E_z through $\nabla \times \mathbf{E} = -i\omega\mathbf{B}$. Continuity of any one of these quantities all over the boundary can only be assured for a Fourier mode if $k_y^v = k_y^p$ and $k_z^v = k_z^p$.

Consider, for example, the case of an infinite half-space filled with plasma, and an incoming plasma wave with wave vector $\mathbf{k}^{in}$ (Fig. 10-1a). There will be a reflected wave on the vacuum side with the same values of k_y and k_z, owing to the boundary conditions. Because of (10-10) k_x^{refl} must have the same absolute value as k_x^{in}, so clearly $k_x^{refl} = -k_x^{in}$. Inside the plasma k_y and k_z (and of course ω) are again the same as outside, and the value of k_x is determined

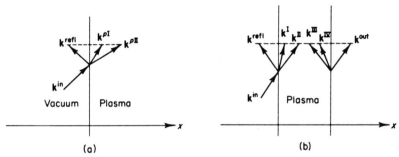

FIG. 10-1. (a) Electromagnetic wave impinging on a half-space filled with magnetoplasma (b) Transmission of electromagnetic wave through a magnetoplasma slab.

from the dispersion relation (10-8), from the equivalent (8-19), or from either one of the more convenient forms (8-20) and (8-24). Since the dispersion equation is of the fourth order in k, there are four solutions. In our case only the two with $k_x > 0$ are of interest. The resulting wave vectors are $\mathbf{k}^{pI}$ and $\mathbf{k}^{pII}$. After all the wave vectors are determined, we are left with the following four unknown quantities: E_z^I, E_z^{II}, E_z^{refl}, and E_y^{refl}. The four boundary conditions just suffice to determine these unknowns. All other quantities can be determined from these; inside the plasma (10-7) serves to determine $\mathbf{E}_\perp$ from E_z and $\mathbf{k}$, while in the vacuum we have (10-11) to find E_x. The magnetic field can be calculated once the electric field is given. $\mathbf{E}^{in}$ is, of course, known in advance.

In the case of transmission through a plasma slab (Fig. 10-1b), reflection from the second plasma-vacuum boundary is to be expected and all four values of $\mathbf{k}^v$ should be considered. This and the outgoing wave raise the number of unknowns to eight but also provide the eight continuity equations to find them.

As a specific example we work out the case of normal incidence $k_y = k_z = 0$ on a plasma-filled half-space. Inside the plasma this corresponds to the $\theta = \pi/2$ case in the notations of Sec. 8-1, with the two modes given by

$$\kappa_\parallel = n^2 \qquad (8\text{-}33)$$

and

$$\kappa_R \kappa_L / \kappa_T = n^2 \qquad (8\text{-}34)$$

with $n = (c/\omega)k$. We have also seen that the first of these modes involves an electric field vector in the z direction, while in the second $\mathbf{E} = \mathbf{E}_\perp$. [Using (10-7) leads of course to the same results; in the first case $k_\parallel = 0$ and $\overset{\leftrightarrow}{p} \cdot \mathbf{k}_\perp$ is finite, while in the second $\overset{\leftrightarrow}{p} \cdot \mathbf{k}_\perp$ diverges sufficiently so that $E_\perp/E_z \to \infty$. This is the subject of Exercise 10-2]. It can also be shown that in the second mode the electric field is elliptically polarized in the perpendicular plane with $E_x/E_y = -i\kappa_H/\kappa_T$ (Exercise 10-2).

The boundary conditions yield

$$E_z^{in} + E_z^{refl} = E_z^{P} \tag{10-12}$$

and since the continuity of **B** requires the continuity of $\partial E_z / \partial x$,

$$k_x^v E_z^{in} - k_x^v E_z^{refl} = k_x^P E_z^P = k_x^v \sqrt{\kappa_{\|}} E_z^P \tag{10-13}$$

Consequently,

$$\frac{E_z^P}{E_z^{in}} = \frac{2}{1 + \sqrt{\kappa_{\|}}} \tag{10-14}$$

and

$$\frac{E_z^{refl}}{E_z^{in}} = \frac{1 - \sqrt{\kappa_{\|}}}{1 + \sqrt{\kappa_{\|}}} \tag{10-15}$$

When the frequency of the incoming radiation is below the plasma frequency, $\kappa_{\|}^{1/2}$ is imaginary $|E_z^{refl}/E_z^{in}| = 1$, as expected. Equations similar to (10-12) and (10-13) hold for the y component except that now $k_x^P = k_x^v (\kappa_R \kappa_L / \kappa_T)^{1/2}$ and one finds

$$\frac{E_y^P}{E_y^{in}} = \frac{2\sqrt{\kappa_T}}{\sqrt{\kappa_T} + \sqrt{\kappa_R \kappa_L}} \tag{10-16}$$

and

$$\frac{E_y^{refl}}{E_y^{in}} = \frac{\sqrt{\kappa_T} - \sqrt{\kappa_R \kappa_L}}{\sqrt{\kappa_T} + \sqrt{\kappa_R \kappa_L}} \tag{10-17}$$

The computation of the transmission-frequency bands is the subject of Exercise 10-3. Since the y and z components behave quite differently, one expects that the reflected wave will have a polarization different from the incoming one. The slab case can be treated in a similar manner (see Exercise 10-4).

Another case of interest is that of a magnetoplasma in which oscillations or waves have been generated by some means, e.g., by using some of the micro-instabilities discussed in the previous chapter to convert kinetic energy into wave energy. The question can be raised whether such a plasma wave, when it impinges on a plasma-vacuum boundary, radiates electromagnetic energy into the vacuum region.

Whether the plasma wave radiates, or is totally reflected at the boundary, the vacuum fields satisfy the Helmholtz equation (10-9), with the corresponding solutions

$$u = u_0 \exp[i(\omega t - \mathbf{k}^v \cdot \mathbf{r})] \tag{10-18}$$

where the magnitude of the vacuum-wave vector $k^v = \omega/c$, according to (10-10). If the unit vector normal to the plasma-vacuum interface directed into the vacuum is **n**, one may write

$$\mathbf{k}^v = \mathbf{n} k^v \cdot \mathbf{n} + \mathbf{k}_t^v \tag{10-19}$$

where $\mathbf{k}_t^v$ is the component parallel to the boundary. The boundary conditions require that $\mathbf{k}_t^p = \mathbf{k}_t^v = \mathbf{k}_t$, and it follows from (10-10) and (10-19) that

$$\mathbf{k}^v \cdot \mathbf{n} = (\omega^2/c^2 - k_t^2)^{1/2} \tag{10-20}$$

If $|\mathbf{k}_t| > \omega/c$, $\mathbf{k}^v \cdot \mathbf{n}$ is imaginary, and the vacuum solution

$$u = u_0 \exp[i(\omega t - \mathbf{k}_t \cdot \mathbf{r})] \exp(-\alpha \mathbf{n} \cdot \mathbf{r}), \qquad \alpha^2 = k_t^2 - (\omega^2/c^2) \tag{10-21}$$

decays exponentially with the distance from the interface, the plasma wave is totally reflected. If on the other hand, $|\mathbf{k}_t| < \omega/c$, the solutions are periodic in the n direction and the plasma waves radiate.

Assume that a plasma wave is generated for which on the interface a-a $|\mathbf{k}_t| > \omega/c$; hence it suffers total reflection (Fig. 10-2). In order to get radiation

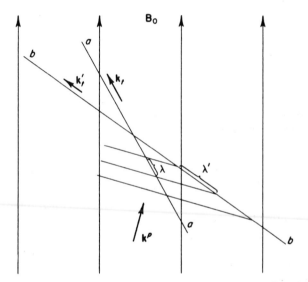

FIG. 10-2. Plasma wave with wave vector $\mathbf{k}^p$ impinging on a vacuum interface. If the interface is a-a the wave suffers total reflection; if it is b-b, part of the energy is radiated out as electromagnetic radiation.

from the same wave, it suffices to alter the interface to, say, b-b so that $|\mathbf{k}_t|_{b-b} < \omega/c$ and radiation is obtained. Note that the criterion for radiation requires that the phase velocity of the plasma waves parallel to the interface exceed the speed of light.

It is of special interest to investigate the radiation condition for the case where the interface is parallel to the magnetic field. The results can be read immediately from the phase-velocity surfaces of the Allis diagrams. As an illustration we have plotted in Fig. 10-3 the phase-velocity diagram for a cold plasma and a wave whose frequency is larger than the plasma frequency,

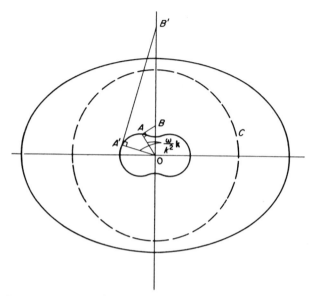

FIG. 10-3. Construction from an Allis diagram to obtain radiation conditions on the plasma-vacuum interface. The wave characterized by wave vector OA is totally reflected, while OA' radiates.

lower than the electron cyclotron frequency, and situated between the cyclotron cutoffs. The dotted circle corresponds to the free-space velocity diagram. It is easy to see that $OB = \omega/k_z$ and radiation occurs when B falls outside the light sphere provided $\mathbf{k}$ has no y component (the plasma-vacuum interface lies in the yz plane). If $k_y \neq 0$, a similar simple geometrical construction on the three-dimensional phase-velocity surface yields the radiation condition. In the above example, the left polarized wave is always partially transmitted while the right polarized one is totally reflected, if the phase-velocity vector of the plasma wave is OA, but radiates into the vacuum if it is OA'. It should be noted that the sharp boundary assumption has not been used. If the plasma density falls off gradually the same conditions obtain. The intensity of radiation is determined in every case by matching the solutions at the boundary.

10-2. Oscillations in a Plasma Column in a Strong Magnetic Field

We turn now to the problem of waves and oscillations along a plasma column with a longitudinal magnetic field. In an infinite plasma, waves propagating along the field are represented by a wave vector $\mathbf{k} = \mathbf{e}_3 k_z$ pointing in the direction of the field. It is easy to see that such waves cannot exist in a plasma column, since boundary conditions at the plasma-vacuum interface cannot be satisfied. Inside the plasma all fields are independent of x, so the matching of derivatives (e.g., $\partial E_y/\partial x$) leads to $k_x^v = 0$. Since $k_y^v = 0$ also, it follows that the matching vacuum fields would also propagate in the z direction. However, the difference of the dispersion equations on both sides of the boundary precludes this possibility.

The closest thing to plane waves traveling in the z direction is the super-position of two waves of equal amplitude and k_z, and equal but opposite $\mathbf{k}_\perp$ components. The resulting wave propagates in the z direction, but the field quantities depend on the xy coordinates as sine or cosine functions in the slab.

Since $k_z^p = k_z^v$ and $k_y^p = k_y^v$, k_x^v can be calculated from (10-10), if the plasma-wave vectors are given by

$$k_x^v = \pm[(\omega^2/c^2) - k_y^2 - k_z^2]^{1/2} \tag{10-22}$$

Two cases ensue, depending on the sign of $\omega^2/c^2 - (k_y^2 + k_z^2)$. If this expression is positive, k_x^v real, and the matching solutions on the vacuum side are outgoing electromagnetic waves (incoming ones must be excluded because of the nature of our physical assumptions), the plasma "radiates." If, on the other hand, $(k_y^2 + k_z^2) > \omega^2/c^2$, k_x^v is imaginary and the plasma oscillations or waves give rise only to exponentially decaying (in space) fringing fields in the vacuum.

The essential features of the solution of such a problem can be demonstrated, without getting involved in algebraic complexities, for the simplified case of a plasma immersed in a magnetic field strong enough to suppress particle motion in other than the z direction. One expects that in such an arrangement only longitudinal plasma oscillations along the magnetic field will survive. If ω_c^+, $\omega_c^- \gg \omega_p$, ω, it follows that $\kappa_R = \kappa_L = \kappa_T = 1$, $\kappa_H = 0$, and $\overset{\leftrightarrow}{\kappa}_\perp$ becomes the unit matrix. Consequently (10-5) reduces to

$$\nabla_\perp \times \nabla_\perp \times \mathbf{E}_\perp - ik_\parallel \nabla_\perp E_z = (\omega^2/c^2 - k_\parallel^2)\mathbf{E}_\perp \tag{10-23}$$

Taking the $\nabla_\perp \cdot$ of (10-23) and inserting it in (10-4), this becomes

$$\nabla_\perp^2 E_z + (\omega^2/c^2 - k_\parallel^2)\kappa_\parallel E_z = 0 \tag{10-24}$$

Focusing our interest again on the slab geometry, located symmetrically about $x = 0$, and choosing $k_y = 0$ for propagation along the field, the general solution of (10-24) assumes the form

$$E_z = E_z^a \sin k_x^p x \exp(-ik_z z) + E_z^s \cos k_x^p x \exp(-ik_z z) \tag{10-25}$$

The dispersion equation follows from inserting (10-25) in (10-24),

$$(k_x^p)^2 - (\omega^2/c^2 - k_z^2)\kappa_\parallel = 0 \tag{10-26}$$

It is easy to see that the same results obtain by superimposing two traveling waves, with equal k_z but opposite k_x, as indicated earlier (Exercise 10-5).

Consider separately the symmetric and antisymmetric solution in (10-25). Fitting the symmetric one to the vacuum solution $E_z^v \exp[-(k_x^v x + k_z z)]$ at the boundary $x = a$ one obtains

$$E_z^s \cos k_x^p a = E_z^v \exp(-ik_x^v a) \tag{10-27}$$

and from the continuity of $\partial E_z / \partial x$,

$$E_z^s k_x^p \sin k_x^p a = ik_x^v E_z^v \exp(-ik_x^v a) \tag{10-28}$$

Dividing (10-28) by (10-27),

$$k_x^p \tan k_x^p a = ik_x^v \tag{10-29}$$

and combining relations (10-22) and (10-26),

$$k_x^p = \kappa_{\parallel}^{1/2} k_x^v = (1 - \omega_p^2/\omega^2)^{1/2} k_x^v \tag{10-30}$$

where $\omega_p^2 = (\omega_p^+)^2 + (\omega_p^-)^2$, one finds

$$(\omega_p^2/\omega^2 - 1)^{1/2} \tan k_x^p a = 1 \tag{10-31a}$$

or the equivalent,

$$(\omega_p^2/\omega^2 - 1)^{1/2} \tan[(\omega^2/c^2 - k_z^2)(1 - \omega_p^2/\omega^2)]^{1/2} a = 1 \tag{10-31b}$$

This expression gives ω as a function of k_z, hence it is analogous to the corresponding dispersion relation for an infinite uniform plasma. Owing to the multivaluedness of the tan function, one expects an infinite number of ω's associated with a single k_z. Furthermore, there are two groups of solutions. One is obtained when the oscillations couple with an exponentially decaying fringing field in the vacuum; hence k_x^v is imaginary, while k_x^p is real. It is seen from (10-30) that this is the case whenever $\omega < \omega_p$. Whenever $\omega > \omega_p$, however, radiation arises in the vacuum.

So far we looked at the $x > 0$ region and only the symmetric part of the solution. Clearly the other cases are analogous to this one, and we shall restrict ourselves to the case we started to investigate.

To understand the dispersion relation better, some asymptotic solutions can be obtained without difficulty. Consider first the $\omega/\omega_p \ll 1$ case. Equation (10-31b) becomes in this limit

$$\frac{\omega_p}{\omega} \tan\left(k^2 - \frac{\omega^2}{c^2}\right)^{1/2} \frac{\omega_p}{\omega} a = 1 \tag{10-32}$$

where the z index has been dropped. This can also be written as

$$\tan^{-1} \frac{\omega}{\omega_p} = \left(k^2 - \frac{\omega^2}{c^2}\right)^{1/2} \frac{\omega_p}{\omega} a \tag{10-33}$$

Using the first term of the power-series expansion of $\tan^{-1}$ one finds

$$\left(n\pi + \frac{\omega}{\omega_p}\right) = \left(k^2 - \frac{\omega^2}{c^2}\right)^{1/2} \frac{\omega_p}{\omega} a, \qquad n = \ldots, -2, -1, 0, 1, 2, \ldots \quad (10\text{-}34)$$

which becomes, after squaring and neglecting higher-order terms in ω/ω_p,

$$n^2\pi^2 \frac{\omega^2}{\omega_p^2} = a^2\left(k^2 - \frac{\omega^2}{c^2}\right) \qquad (10\text{-}35)$$

to yield, finally,

$$\omega = \frac{ak}{[(n^2\pi^2/\omega_p^2) + (a^2/c^2)]^{1/2}} \qquad (10\text{-}36)$$

The meaning of the various n "modes" becomes clear on comparing (10-34) with (10-31a). One finds that the right side of (10-34) is $k_x{}^p a$ in the $\omega/\omega_p \ll 1$ limit, hence n is proportional to the number of waves across the plasma. We are mostly interested in the case which is closest to the plane wave, hence the principal $n = 0$ mode. Equation (10-36) becomes, for $n = 0$,

$$\omega = kc \qquad (10\text{-}37)$$

Low frequencies propagate with the speed of light along the plasma.

The other interesting limit is the one where the frequency is close to the plasma frequency. For the nonradiating case, $\omega < \omega_p$, and we introduce the small positive quantity

$$\varepsilon^2 = \omega_p{}^2 - \omega^2 \approx 2\omega_p(\omega_p - \omega) \qquad (10\text{-}38)$$

In this approximation the dispersion relation (10-31b) becomes

$$\frac{\varepsilon}{\omega_p} \tan\left(k^2 - \frac{\omega_p{}^2}{c^2}\right)^{1/2} \frac{\varepsilon}{\omega_p} a = 1 \qquad (10\text{-}39)$$

or

$$\cot^{-1} \frac{\varepsilon}{\omega_p} = \left(k^2 - \frac{\omega_p{}^2}{c^2}\right)^{1/2} \frac{\varepsilon}{\omega_p} a \qquad (10\text{-}40)$$

Power-series expansion of the left side leads to

$$\left(\frac{\pi}{2} + n\pi\right) - \frac{\varepsilon}{\omega_p} = \left(k^2 - \frac{\omega_p{}^2}{c^2}\right)^{1/2} \frac{\varepsilon}{\omega_p} a \qquad (10\text{-}41)$$

or

$$\frac{\varepsilon}{\omega_p} = \frac{\pi/2 + n\pi}{a[k^2 - (\omega_p{}^2/c^2)]^{1/2} + 1} \qquad (10\text{-}42)$$

The oscillation frequency is close to the plasma frequency only if both a and k are large enough. For very short wavelengths, $k \to \infty$, the frequency of the principal mode becomes identical with the frequency of an infinite plasma.

From these asymptotic solutions one can construct the shape of the dispersion curves (Fig. 10-4). At $k \to 0$, $\omega \to 0$, and $\omega/k \to c$ for the principal mode, while for $k \to \infty$, ω approaches ω_p from below.

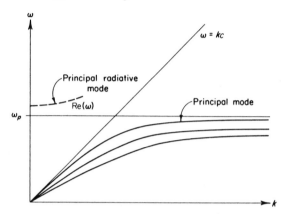

FIG. 10-4. Dispersion curves for a plasma slab in a strong magnetic field.

Furthermore, it is easy to see from (10-31b) that the curves do not cross the $\omega/k = c$ line. The $n \neq 0$ modes are located below the principal one. As expected, the frequency differs from ω_p for long wavelengths, but contrary to expectations the deviation becomes significant at about $k \approx (a^{-2} + \omega_p^2/c^2)^{1/2}$ and not at $k \approx a^{-1}$, where the wavelength is comparable to the slab thickness. We shall return to this point soon.

Turning now to the radiative oscillations, we look for frequencies near ω_p with $k^2 < \omega^2/c^2$. Equation (10-42) now becomes

$$\frac{\varepsilon}{\omega_p} = \frac{\pi/2 + n\pi}{ia[\omega_p^2/c^2 - k^2]^{1/2} + 1} \tag{10-43}$$

where the square root is real. Multiplying by the complex conjugate of the denominator

$$\frac{\varepsilon}{\omega_p} = \frac{1 - ia[(\omega_p^2/c^2) - k^2]^{1/2}}{1 + [(\omega_p^2/c^2) - k^2]a^2} \left(\frac{\pi}{2} + n\pi\right) \tag{10-44}$$

and

$$\frac{\omega_p - \omega}{\omega_p} = \frac{\varepsilon^2}{2\omega_p^2} = \frac{\pi^2(1 + 2n)^2[1 - a^2(\omega_p^2/c^2 - k^2) - 2ia(\omega_p^2/c^2 - k^2)^{1/2}]}{8[1 + a^2(\omega_p^2/c^2 - k^2)]^2} \tag{10-45}$$

The approximation is valid only when ε/ω_p is small, that is, if the slab is thick enough so that $a\omega_p/c$ is large, and k and n are both small. The real part of $\omega_p - \omega$ is of course negative, as expected of radiative oscillations. Since energy is radiated away, the oscillations decay, as expressed by the imaginary part of the frequency. In fact, since ω is complex, the normal-mode analysis is no longer justified, but it can be shown that the alternative treatment, based on the initial-value problem employing Laplace transform techniques, leads to the same result (see Exercise 10-6). It follows from (10-22) and (10-26) that if ω is complex so are $k_x^{\,p}$ and $k_x^{\,v}$. In particular it is interesting to note that $k_x^{\,v}$ represents exponentially *increasing* waves away from the plasma (Exercise 10-7). The radiation to be found far away from the slab was emitted some time ago, when the amplitude of oscillations, hence the amplitude of radiation, was larger than at present. Clearly the oscillations must have started some finite time ago. This is the physical reason why the initial-value treatment is the appropriate form of description for such a problem.

As k approaches ω_p/c the validity of (10-45) breaks down. Re ω as a function of k has been plotted in Fig. 10-4 for the principal radiative branch. The physically more interesting case of a plasma cylinder in a strong magnetic field can also be calculated without difficulty, and leads to qualitatively similar results. The reader is advised to carry out this calculation as an exercise (Exercise 10-8).

Turning now to the physical reasons behind the behavior of the dispersion curves calculated above, one considers the charge distribution along the x axis in a slab excited to a principal ($n = 0$) nonradiating oscillation mode (Fig. 10-5). The charge density ρ varies sinusoidally with z at any moment, while it moves along with the phase velocity $v_{\text{ph}} = \omega/k$. To be sure, no charged particle travels with the wave; each one just oscillates about its equilibrium position, and it is the phasing of these individual oscillators which amounts to the indicated macroscopic behavior. A mechanical model of such a traveling wave, consisting of a row of oscillators with the same characteristic frequency, is sketched in Fig. 10-5a. The density variation of these pendulums in space and time is the same as that of ρ in Fig. 10-5b.

The charge distribution (in space and time) determines the electromagnetic fields, which in turn provide the driving (or rather restoring) forces determining the particle motion and charge distribution in a self-consistent way. These electromagnetic fields can be found by use of the electromagnetic potentials, but it is more illuminating if one employs the "trick" of a suitable coordinate transformation. In a coordinate system moving with the phase velocity of the wave, the charge distribution is static and the electric fields can be more easily found. As long as $v_{\text{ph}} \ll c$, the results are not very striking. If the phase velocity approaches the speed of light, however, the relativistic effects are to be considered; notably, the wavelength in the co-moving system λ' exceeds λ, while the width of the slab appears the same for both observers. This situation is sketched in Fig. 10-5c and d. Even if the slab is thick compared to the wavelength in the rest frame, for sufficiently

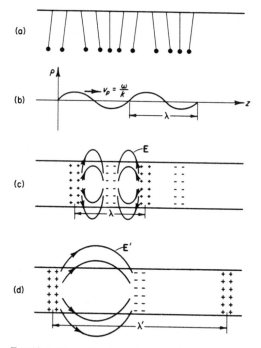

(a) Pendulums simulating plasma oscillations. (b) Space charge wave. (c) Charges and fields in rest frame. (d) Charges and static fields in co-moving frame.

FIG. 10-5. Illustration to explain the physics of plasma oscillations in a magnetoplasma slab.

large phase velocities this relationship can be reversed in the co-moving system. In this case the field in the plasma is reduced and considerable fringing fields arise in the primed system. (Note that charges are invariant under Lorentz transformation.)

To describe the oscillations one has to go back to the rest frame. The z component of the oscillating electric field, which sustains the oscillations, is an invariant of the transformation, $E'_z = E_z$. As a result, if $\omega/k \approx c$, the electrons experience only a weak driving field; consequently the oscillation frequency is reduced. As $\omega/k \to c$, all electric fields appear as fringing fields (hence the inverse penetration depth $k_v \to 0$), and the oscillation frequency drops to zero. This behavior occurs at any finite value of the slab thickness, in sharp contrast with the infinite plasma case.

On the basis of this picture one expects that the deviation of the actual oscillation frequency from ω_p becomes appreciable when $k' \approx 1/a$, where k' is the wave number in the co-moving system. Since

$$k' = k\left[1 - \left(\frac{v_{\text{ph}}}{c}\right)^2\right]^{1/2} = k\left(1 - \frac{\omega^2}{c^2 k^2}\right)^{1/2} \tag{10-46}$$

one finds that ω is close to ω_p only if

$$k \gg (a^{-2} + \omega^2/c^2)^{1/2} \tag{10-47}$$

in agreement with our conclusion from (10-42).

We have seen that if $\omega/k > c$ the plasma radiates. It is known that charges traveling faster than the speed of light in the surrounding medium radiate, a phenomenon known as *Cerenkov radiation*. If $\omega/k > c$, the macroscopic charge density actually "moves" faster than the vacuum light velocity, hence radiation is to be expected. This physically reasonable but qualitative statement can be confirmed simply by finding the electromagnetic field resulting from the charge distribution of Fig. 10-5b (Exercise 10-9). One finds that for $\omega/k > c$ radiation is emitted, while for $\omega/k < c$ only convective fields are present. The ever-growing fringing fields, as ω/k approaches c, "break loose" from the sources when ω/k exceeds the speed of light.

It should be mentioned that no co-moving Lorentz frame can be found for $\omega/k > c$; the system cannot be transformed to rest. One can find another interesting coordinate system, however, the one moving with the velocity $V = c^2/v_{\rm ph}$. In this frame the spatial variation along the z axis is transformed away, and the observer is confronted with a spatially (along z) uniform current oscillating in time. No macroscopic charges are to be found in this coordinate system. Space takes the place of time and current that of charge, compared to the $\omega/k < c$ case (constant in space $\leftrightarrow$ constant in time, charge transformed away $\leftrightarrow$ current transformed away, etc.). This exercise in the special theory of relativity is the subject of Exercise 10-10. The emission of radiation follows immediately from these considerations as well. The oscillating uniform current necessarily radiates, and a system radiating in one Lorentz frame radiates in all the others.

While these arguments reveal the nature of the radiation they do not explain why such oscillations take place in the plasma. In fact, these oscillations are quite different from electrostatic ones, and they could more appropriately be called *electromagnetic oscillations*. A physical picture for the $k = 0$ case can be presented as follows.

Think of the plasma slab as made up of thin sheets of thickness dx. If such a sheet were to oscillate with the frequency ω in the z direction it would give rise to a radiation field with the electric vector polarized in the z direction, the magnetic vector in the y direction, propagating along the $\pm x$ axis (Fig. 10-6). This radiation field can excite similar oscillations in other sheets, which in turn give rise to similar plane electromagnetic waves. These sheets can keep each other oscillating while radiating away energy through the slab boundaries.

The oscillating electric field at x resulting from the oscillating current sheet

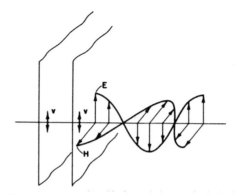

Fɪɢ. 10-6. Electromagnetic wave generated by oscillating electron sheets.

at ξ with thickness $d\xi$ is

$$dE(x,\xi,t) = -\left(\frac{\mu_0}{\varepsilon_0}\right)^{1/2} \frac{1}{2} J(\xi) \exp\left[i\omega\left(t - \frac{1}{c}|x - \xi|\right)\right] d\xi \qquad (10\text{-}48)$$

which adds up to the total electric field

$$E(x,t) = -\left(\frac{\mu_0}{\varepsilon_0}\right)^{1/2} \frac{1}{2} \int_{-\infty}^{+\infty} J(\xi) \exp\left[i\omega\left(t - \frac{1}{c}|x - \xi|\right)\right] d\xi \qquad (10\text{-}49)$$

This field drives a current of the density

$$J(x) = \frac{q^2 n(x)}{i\omega m} E(x) \qquad (10\text{-}50)$$

where $n(x)$ is the electron density. Self-consistency requires that the integral equation

$$J(x) = \frac{i}{2} \frac{q^2 n}{m\varepsilon_0 \omega c} \int_{-\infty}^{+\infty} J(\xi) \exp\left(-i\frac{\omega}{c}|x - \xi|\right) d\xi \qquad (10\text{-}51)$$

be satisfied. For a slab with uniform density,

$$n(x) = \begin{cases} n_0 & \text{for } |x| < a \\ 0 & \text{for } |x| > a \end{cases} \qquad (10\text{-}52)$$

equation (10-51) can be solved, looking for solutions which vary as $\sin k_x x$ and $\cos k_x x$, to lead to an equation for ω (see Exercise 10-11),

$$\left(\frac{\omega_p^2}{\omega^2} - 1\right)^{1/2} \tan \frac{1}{c}(\omega^2 - \omega_p^2)^{1/2} a = 1 \qquad (10\text{-}53)$$

which is identical to (10-31b) with $k_z = 0$.

10-3. Zero and Finite Background Magnetic Field

In the absence of a background magnetic field the dielectric tensor reduces to a scalar:

$$\kappa_{ik} = (1 - \omega_p^2/\omega^2)\delta_{ik} \qquad (10\text{-}54)$$

and the wave equation to

$$\nabla \times \nabla \times \mathbf{E} = -\nabla^2\mathbf{E} = (\omega^2/c^2)(1 - \omega_p^2/\omega^2)\mathbf{E} \qquad (10\text{-}55)$$

The z dependence can again be assumed to have the $\exp(-ik_z z)$ form, resulting in the equation

$$\nabla_{\perp}^2\mathbf{E} + \left(\frac{\omega^2 - \omega_p^2}{c^2} - k_z^2\right)\mathbf{E} = 0 \qquad (10\text{-}56)$$

To demonstrate the general behavior of the solutions, we choose a slab geometry where the solutions of (10-56) again assume the simple form

$$\mathbf{E} = \mathbf{E}^a \sin k_x^p x \, \exp(-ik_z z) + \mathbf{E}^s \cos k_x^p x \, \exp(-ik_z z) \qquad (10\text{-}57)$$

where

$$k_x^p = \pm\left(\frac{\omega^2 - \omega_p^2}{c^2} - k_z^2\right)^{1/2} \qquad (10\text{-}58)$$

The vacuum dispersion relation (10-22) can be written by using (10-58) as

$$k_x^v = \pm\left(\frac{\omega^2}{c^2} - k_z^2\right)^{1/2} = \pm\left[(k_x^p)^2 + \frac{\omega_p^2}{c^2}\right]^{1/2} \qquad (10\text{-}59)$$

Nonradiative modes exist only if

$$(k_x^p)^2 + (\omega_p^2/c^2) < 0 \qquad (10\text{-}60)$$

which implies that k_x^p is pure imaginary. Consequently for these modes $\mathbf{E}$ varies as a hyperbolic sine or cosine function across the slab. These are surface modes penetrating the slab to the depth of the order

$$\delta = 1/|k_x^p| < c/\omega_p \qquad (10\text{-}61)$$

To obtain a dispersion relation we need the components of $\mathbf{E}$. If E_z is a symmetric function of x,

$$E_z^s = E_z^p \cos k_x^p x \, \exp(-ik_z z) \qquad (10\text{-}62)$$

and E_x becomes, from $\nabla \cdot \mathbf{E} = 0$,

$$E_x = i\frac{k_z}{k_x^p} E_z^p \sin k_x^p x \, \exp(-ik_z z) \qquad (10\text{-}63)$$

The continuity of E_z at the boundary yields

$$E_z^p \cos k_x^p a = E_z^v \exp(-ik_x^v a) \qquad (10\text{-}64)$$

Another boundary condition can be obtained from the continuity of a magnetic field component, or simpler yet, by using the continuity of κE_x, which leads to

$$i\left(1 - \frac{\omega_p^2}{\omega^2}\right)\frac{k_z}{k_x^p} E_z^p \sin k_x^p a = -\frac{k_z^v}{k_x^v} E_z^v \exp(-ik_x^v a) \qquad (10\text{-}65)$$

where use has been made of $k_x^v E_x^v + k_z^v E_z^v = 0$. Dividing (10-65) by (10-64) yields

$$\left(1 - \frac{\omega_p^2}{\omega^2}\right)k_x^v \tan k_x^p a = ik_x^p \qquad (10\text{-}66)$$

For nonradiative modes k_x^p and k_x^v are imaginary, and for $a > 0$ k_x^v is negative, for $a < 0$ positive-imaginary. Consequently (10-66) can be written with the help of (10-58) and (10-59) in the form

$$\left(\frac{\omega_p^2}{\omega^2} - 1\right)(c^2 k_z^2 - \omega^2)^{1/2} \, \text{th}\left((\omega_p^2 + c^2 k_z^2 - \omega^2)^{1/2}\frac{a}{c}\right)$$
$$= (\omega_p^2 + c^2 k_z^2 - \omega^2)^{1/2} \qquad (10\text{-}67)$$

Evaluating this dispersion relation one finds that for $k_z \to 0$, $\omega \to 0$, but contrary to the strong magnetic field case $\omega/k_z \neq c$. Since the th function is positive for $a > 0$, $\omega < \omega_p$. A better upper bound for ω can be found by noting that th $x \leq 1$. It is easy to show from (10-67) that

$$\omega \leq \omega_p/\sqrt{2} \qquad (10\text{-}68)$$

The proof is left as an exercise (Exercise 10-12). It was noted before that electromagnetic waves cannot propagate in a plasma in the absence of an external magnetic field unless $\omega > \omega_p$. Hence it is not surprising that one finds that low-frequency modes are surface modes where the response of surface electrons to the field is such as to shield the plasma interior.

The modes with $\omega/k > c$ are again radiating ones, not very different in character from the ones obtained for the case of large B_0. In fact one can prove that the dispersion relations are identical for $k_z = 0$ (Exercise 10-14). This is exactly what one expects on the basis of the physical nature of these oscillations as described in the previous section.

By plotting the principal branches of these dispersion curves (Fig. 10-7) for large values of a, one may be puzzled to note that the $\omega = \omega_p$ mode is not regained in the $a \to \infty$ limit. It can be seen, however, that in addition to the modes described by (10-66) there exists another set of modes not included in this dispersion relation.

Consider any disturbance of the electrons inside the slab (not extending to the plasma-vacuum interface). Since electrostatic oscillations at zero temperature in the absence of a background $\mathbf{B}_0$ do not propagate, any such disturbance is expected to oscillate with the frequency $\omega = \omega_p$, since the perturbed particles "do not know" that there is a boundary further away. For these trapped modes $\omega = \omega_p$ and $\kappa = 0$. The electric field in the plasma satisfies the Laplace equation, from (10-55),

$$\nabla^2 \mathbf{E} = 0 \qquad (10\text{-}69)$$

while at the boundary $E_z = E_y = 0$. The vacuum fields, as well as the magnetic field both inside and outside the plasma are zero. Typical solutions of (10-69) satisfying these boundary conditions are easy to find (see Exercise 10-15).

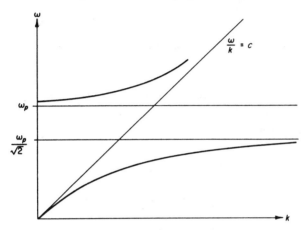

FIG. 10-7. Principal dispersion curves of a plasma slab in the absence of a magnetic field.

If $a \to \infty$ the surface modes disappear (recede to infinity), but the trapped modes survive, yielding the familiar electrostatic plasma oscillations. The radiative modes will also survive but the imaginary part of ω decreases to zero as $a \to \infty$. The remaining modes are the transverse waves discussed in Sec. 7-5.

At this point one may mention what happens if $a \to \infty$ in the case of a large background field. The result is shown in Fig. 10-8. As $a \to \infty$ the lower (static) curve approaches the light line between $\omega = k = 0$ and $\omega = \omega_p$ and the horizontal straight line $\omega = \omega_p$ for $k > \omega_p/c$. The radiative branch undergoes a similar transition, approaching $\omega = \omega_p$ for $k < \omega_p/c$ and the light line for $k > \omega_p/c$, while Im $\omega \to 0$. Hence a section of the radiative branch links up with a part of the static branch to form the longitudinal electrostatic oscillations, while the remaining parts join to form the branch which corresponds to a light wave traveling along the magnetic field.

We turn finally to the general case of a cylindrical plasma in a uniform magnetic field of arbitrary magnitude. The cylinder may be circular, rectangular,

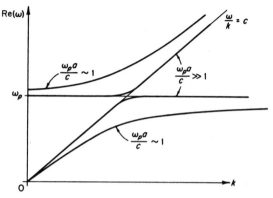

FIG. 10-8. Transition of the dispersion curves when the slab thickness $a \to \infty$ in a strong magnetic field.

a slab, or any other geometry describable in two-dimensional orthogonal curvilinear coordinates x_1, x_2 with the length of the line element in the form

$$(ds)^2 = (h_1\, dx_1)^2 + (h_2\, dx_2)^2 \qquad (10\text{-}70)$$

For the electric field we obtained previously,

$$-ik_\parallel \nabla_\perp \cdot \mathbf{E}_\perp - \nabla_\perp^2 E_z = \frac{\omega^2}{c^2} \kappa_\parallel E_z \qquad (10\text{-}4)$$

In order to eliminate $\mathbf{E}_\perp$ let us compute with the help of Appendix II:

$$\nabla_\perp \cdot (\overset{\leftrightarrow}{\kappa}_\perp \cdot \mathbf{E}_\perp) = \frac{1}{h_1 h_2} \left[\frac{\partial}{\partial x_1} h_2(\kappa_T E_1 + i\kappa_H E_2) + \frac{\partial}{\partial x_2} h_1(-i\kappa_H E_1 + \kappa_T E_2) \right]$$

$$= \kappa_T \frac{1}{h_1 h_2} \left[\frac{\partial}{\partial x_1}(h_2 E_1) + \frac{\partial}{\partial x_2}(h_1 E_2) \right]$$

$$+ i\kappa_H \frac{1}{h_1 h_2} \left[\frac{\partial}{\partial x_1}(h_2 E_2) - \frac{\partial}{\partial x_2}(h_1 E_1) \right]$$

$$= \kappa_T \nabla_\perp \cdot \mathbf{E}_\perp + i\kappa_H (\nabla \times \mathbf{E})_z \qquad (10\text{-}71)$$

With the help of

$$\nabla \cdot \overset{\leftrightarrow}{\kappa} \cdot \mathbf{E} = \nabla_\perp \cdot \overset{\leftrightarrow}{\kappa}_\perp \cdot \mathbf{E}_\perp - ik_\parallel \kappa_\parallel E_z = 0 \qquad (10\text{-}72)$$

and

$$(\nabla \times \mathbf{E})_z = -i\omega\mu_0 H_z \qquad (10\text{-}73)$$

one obtains

$$\nabla_\perp \cdot \mathbf{E}_\perp = i\frac{\kappa_\parallel}{\kappa_T} k_\parallel E_z - \frac{\kappa_H}{\kappa_T} \omega\mu_0 H_z \qquad (10\text{-}74)$$

Inserting (10-74) into (10-4) yields finally

$$\nabla_\perp^2 E_z + \left(\frac{\omega^2}{c^2}\kappa_\| - k_\|^2\frac{\kappa_\|}{\kappa_T}\right)E_z = ik_\|\frac{\kappa_H}{\kappa_T}\omega\mu_0 H_z \qquad (10\text{-}75)$$

This equation contains both E_z and H_z. Another expression similar to (10-4) can be obtained if one takes the z component of the curl of Maxwell's $\nabla \times \mathbf{H} = i\omega\overset{\leftrightarrow}{\kappa} \cdot \mathbf{E}$ (Exercise 10-16):

$$\nabla_\perp^2 H_z - k_\|^2 H_z = -i\omega\varepsilon_0(\nabla \times \overset{\leftrightarrow}{\kappa}_\perp \cdot \mathbf{E}_\perp)_z \qquad (10\text{-}76)$$

The right side can again be evaluated with the help of Appendix II to yield, after some manipulation (see Exercise 10-17),

$$\nabla_\perp^2 H_z + \left(\frac{\omega^2}{c^2}\frac{\kappa_T - \kappa_H}{\kappa_T} - k_\|^2\right)H_z = -ik_\|\varepsilon_0\omega\frac{\kappa_\|\kappa_H}{\kappa_T}E_z \qquad (10\text{-}77)$$

The two second-order differential equations (10-75) and (10-77) can be solved for E_z and H_z. The fact that these quantities are coupled implies that in general the axial components of both the electric and magnetic fields appear simultaneously. One may separate the variables by expressing H_z from (10-75) to insert in (10-77) and E_z from (10-77) in (10-75) to arrive at two fourth-order differential equations.

The appearance of the fourth-order equations is not unexpected. We know that in Cartesian coordinates four independent plane waves are associated with a given pair of ω and k_z. Indeed, setting $\nabla_\perp = -i\mathbf{k}_\perp$ the resulting fourth-order algebraic equation should be identical with (8-19), the dispersion relation for an infinite magnetoplasma (see Exercise 10-18).

The recognition of the nature of the expected solutions facilitates finding them. We seek solutions that satisfy the two-dimensional wave equation, to find in Cartesian coordinates, for example,

$$E_z \sim H_z \sim \begin{Bmatrix}\sin\\\cos\end{Bmatrix}k_x x \cdot \begin{Bmatrix}\sin\\\cos\end{Bmatrix}k_y y \qquad (10\text{-}78)$$

or, in cylindrical coordinates,

$$E_z \sim H_z \sim e^{\pm im\varphi}\begin{Bmatrix}J_m\\N_m\end{Bmatrix}k_r r \qquad (10\text{-}79)$$

It is immediately clear that such solutions indeed satisfy (10-75) and (10-77), resulting in fourth-order algebraic equations for k_x, k_y, or m and k_r.

To fit these solutions to the solutions of the vacuum-wave equation at the boundary, $\mathbf{E}_\perp$ and $\mathbf{H}_\perp$ are needed. These can be expressed with the help of E_z and H_z from Maxwell's equations,

$$(\nabla_\perp - ik_\|\mathbf{e}_3) \times (\mathbf{E}_\perp + \mathbf{e}_3 E_z) = -i\omega\mu_0\mathbf{H} \qquad (10\text{-}80)$$

and

$$(\nabla_\perp - ik_\parallel \mathbf{e}_3) \times (\mathbf{H}_\perp + \mathbf{e}_3 H_z) = i\omega\varepsilon_0 \overset{\leftrightarrow}{\kappa} \cdot \mathbf{E} \tag{10-81}$$

Taking the transverse parts one obtains

$$\nabla_\perp \times \mathbf{e}_3 E_z - ik_\parallel \mathbf{e}_3 \times \mathbf{E}_\perp = -i\omega\mu_0 \mathbf{H}_\perp \tag{10-82}$$

and

$$\nabla_\perp \times \mathbf{e}_3 H_z - ik_\parallel \mathbf{e}_3 \times \mathbf{H}_\perp = i\omega\varepsilon_0 \overset{\leftrightarrow}{\kappa}_\perp \cdot \mathbf{E}_\perp \tag{10-83}$$

Applying the $\mathbf{e}_3 \times$ operator to (10-82) one finds

$$\nabla_\perp E_z + ik_\parallel \mathbf{E}_\perp = -i\omega\mu_0 \mathbf{e}_3 \times \mathbf{H}_\perp \tag{10-84}$$

Inserting $\mathbf{e}_3 \times \mathbf{H}_\perp$ from (10-84) into (10-83) yields

$$\omega\mu_0 \nabla_\perp \times \mathbf{e}_3 H_z + k_\parallel \nabla_\perp E_z = i\left(\frac{\omega^2}{c^2}\overset{\leftrightarrow}{\kappa}_\perp - k_\parallel^2\right) \cdot \mathbf{E}_\perp = \overset{\leftrightarrow}{\Lambda} \cdot \mathbf{E}_\perp \tag{10-85}$$

Consequently $\mathbf{E}_\perp$ can be determined from

$$\mathbf{E}_\perp = \overset{\leftrightarrow}{\Lambda}^{-1}[\omega\mu_0 \nabla_\perp \times \mathbf{e}_3 H_z + k_\parallel \nabla_\perp E_z] \tag{10-86}$$

and $\mathbf{H}_\perp$ with the help of (10-82).

The computations of the dispersion curves $\omega(k_\parallel)$ for a plasma column in an arbitrary background magnetic field is now straightforward but rather tedious. The resulting curves are plotted in Fig. 10-9a and b. It was again assumed that $k_y = 0$ for the slab or $m = 0$ for the circular cylindrical column, and only the principal lines corresponding to the smallest k_x^p or k_r^p have been drawn. Two situations arise, depending on whether ω_p or ω_c is bigger. On the lower (nonradiative) side of the light line there are two branches: a slow or static one starting from $\omega = k_z = 0$ and approaching the smaller one of ω_p or ω_c for $k_z \to \infty$, and a fast or dynamic mode which runs into the light line somewhere above ω_p. On the radiating side there are three branches: one originates above ω_p at $k_z = 0$ and behaves very much like the radiating branch found in the infinite and zero field case for small k_z but joins the dynamic mode on the light line; the other two originate above ω_1 and ω_2 where

$$\omega_{1,2} = [\omega_p^2 + (\omega_c/2)^2]^{1/2} \pm (\omega_c/2) \tag{10-87}$$

and approach the light line for $k_z \to \infty$. It is easy to see that $\omega_{1,2}$ are the cutoffs for waves propagating along the field in an infinite plasma (see Exercise 10-19). While these modes do not change much as $a \to \infty$, it is again interesting to see how the $\omega = \omega_p$ branch is formed. For $\omega_p > \omega_c$ the dynamic branch with its continuation in the radiative region approaches ω_p for all values of k_z. If $\omega_c > \omega_p$, a phenomenon similar to that found for $\mathbf{B}_0 \to \infty$ occurs. Both the static and dynamic branch split at $\omega = \omega_p$ to link up with a part

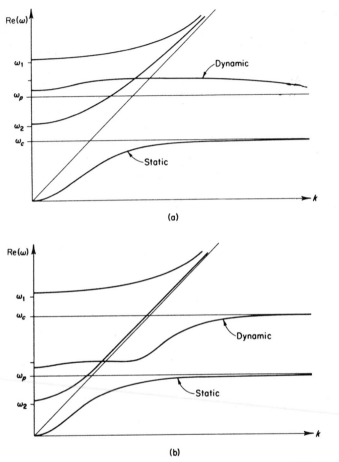

FIG. 10-9. Dispersion curves of a plasma column in a uniform magnetic field: (a) $\omega_p > \omega_c$, (b) $\omega_p < \omega_c$.

of the other branch, as indicated in Fig. 10-9b, to form the branches associated with electrostatic plasma oscillations and electron cyclotron waves in an infinite plasma.

10-4. Flute-Type Instabilities

We turn our attention now to instabilities in finite magnetically confined plasma systems. Instabilities in confined magnetofluid systems have been studied in Chapter V. On the basis of the discussed similarities between the behavior of hydromagnetic fluids and collisionless magnetoplasmas (guiding-center plasmas), one expects some similarities in stability conditions as well.

This will be demonstrated for the plasma analogue of the Kruskal-Schwartz-schild instability of a fluid suspended by a magnetic field against a gravitational field. In fact, since the geometry is two-dimensional, with the parallel velocity and pressure components playing no role in the process, we will find (as expected on the basis of Sec. 3-6) that the instability is described by the same differential equation as the one occurring in the fluid system with the same geometry.

The instability of a plasma suspended against a gravitational field serves as a prototype for a large range of important instabilities, including the mirror and pinch types, as will be demonstrated in this section. Therefore most of our discussion will be devoted to the study of this particular arrangement.

While in hydromagnetic fluids the sharp boundary case corresponds to a realistic physical situation, this is no longer true for a magnetoplasma. Therefore, we will analyze boundaries with gradually decreasing plasma densities and will find increased stability compared to the sharp boundary. It is also worthwhile to go somewhat beyond the guiding-center approximation in describing particle motion. The result is improved stability, in agreement with experimental results.

The geometry of the sharp boundary case is sketched in Fig. 10-10. The plasma is situated in region I $(x > 0)$ and the gravitational acceleration $\mathbf{g}$ points in the negative x direction. The magnetic field is directed into the paper and is uniform throughout. This is equivalent to assuming that the plasma density and temperature are low enough so that β can be neglected. This assumption will be waived later.

Owing to the gravitational field, the guiding centers of ions drift in the y direction, with the velocity

$$\mathbf{w}_+{}^g = \frac{M}{q} \frac{\mathbf{g} \times \mathbf{B}}{B^2} \tag{10-88}$$

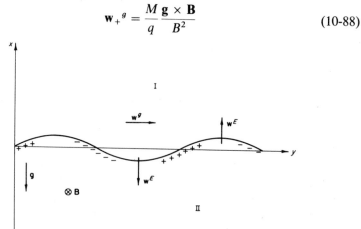

FIG. 10-10. Sketch illustrating flute instability in a magnetoplasma with a sharp boundary.

The electron guiding-center drift is in the $-y$ direction but m/M times smaller and will therefore be neglected.

The mechanism of the instability becomes clear if one considers a small sinusoidal perturbation of the boundary, as indicated in Fig. 10-10. The gravitational drift gives rise to a charge separation, and a corresponding electric field, which in turn produces an electric drift which tends to increase the original perturbation. If the equation of the perturbed interface is

$$X = a(t) \sin ky \qquad (10\text{-}89)$$

and $\alpha(y)$ the (small) angle between interface and the y axis, the rate of accumulation of the surface charge density is described by

$$\dot{\sigma} = \frac{\partial \sigma}{\partial t} = J_y \sin \alpha \approx J_y \frac{\partial X}{\partial y} = J_y ak \cos ky \qquad (10\text{-}90)$$

where J_y is the current density

$$J_y = qw^g n = \frac{Mng}{B} \qquad (10\text{-}91)$$

Here n is the ion (or guiding-center) density. From (10-90) and (10-91)

$$\sigma = \sigma_0(t) \cos ky \qquad (10\text{-}92)$$

where

$$\dot{\sigma}_0 = \frac{gMn}{B} ka \qquad (10\text{-}93)$$

Since the perturbation of the boundary is infinitesimal, one may compute the electric fields in regions I and II as if they were due to the surface charge density $\sigma(y,t)$ located on the unperturbed boundary (the y axis).

As no space charges arise outside the boundary, the electrostatic potential in region I satisfies Laplace's equation. The solution which remains finite for $x \to \infty$ and fits (10-92) is

$$\varphi = \varphi^I \cos ky e^{-kx} \qquad (x > 0) \qquad (10\text{-}94)$$

with the electric field components

$$E_x = \varphi^I k \cos ky e^{-kx} \qquad (10\text{-}95a)$$

and

$$E_y = \varphi^I k \sin ky e^{-kx} \qquad (10\text{-}95b)$$

In region II

$$\varphi = \varphi^{II} \cos ky e^{kx} \qquad (x < 0) \qquad (10\text{-}96)$$

and

$$E_x = -\varphi^{II} k \cos ky e^{kx} \qquad (10\text{-}97a)$$

$$E_y = \varphi^{II} k \sin ky e^{kx} \qquad (10\text{-}97b)$$

The constants φ^{I} and φ^{II} are to be determined from the boundary conditions. Since E_y is continuous at $x = 0$,

$$\varphi^{\mathrm{I}} = \varphi^{\mathrm{II}} = \varphi_0 \tag{10-98}$$

For the normal field components,

$$\kappa \varepsilon_0 E_x^{(\mathrm{I})}(x = 0) - \varepsilon_0 E_x^{(\mathrm{II})}(x = 0) = \sigma \tag{10-99}$$

where the dielectric permeability of the plasma

$$\kappa = 1 + \frac{Mn}{\varepsilon_0 B^2} \approx \frac{Mn}{\varepsilon_0 B^2} \gg 1 \tag{10-100}$$

Using (10-92), (10-95a), and (10-97a) in (10-99) yields

$$\kappa \varepsilon_0 \varphi_0 k + \varepsilon_0 \varphi_0 k = \sigma_0 \tag{10-101}$$

resulting in

$$\varphi_0 = \frac{\sigma_0}{\varepsilon_0 k(\kappa + 1)} \approx \frac{\sigma_0}{\varepsilon_0 k \kappa} \tag{10-102}$$

The electric drift $\mathbf{w}^E$ produces a mass motion of the plasma, with the velocity

$$w_x = \frac{E_y}{B} = \frac{\sigma_0}{\varepsilon_0 \kappa B} \sin ky\, e^{-kx} \tag{10-103a}$$

and

$$w_y = -\frac{E_x}{B} = -\frac{\sigma_0}{\varepsilon_0 \kappa B} \cos ky\, e^{-kx} \tag{10-103b}$$

It is easy to show with the help of (10-103a) and (10-103b) that the guiding-center plasma moves like an incompressible fluid (Exercise 10-20).

The velocity of the boundary as calculated from (10-103a) can be compared with (10-89),

$$\dot{X} = w_x(x = 0) = \frac{\sigma_0}{\varepsilon_0 \kappa B} \sin ky = \dot{a} \sin ky \tag{10-104}$$

to yield with the help of (10-93) and (10-100) the differential equation for the amplitude

$$\ddot{a} = \frac{\dot{\sigma}_0}{\varepsilon_0 \kappa B} = gka \tag{10-105}$$

with the solution

$$a(t) = a_0 \exp[\pm(gk)^{1/2} t] \tag{10-106}$$

The development of the perturbation is identical with what we found for a magnetofluid in Sec. 5-3. The fact that in the hydromagnetic case a high β

configuration was considered is not significant, since the result was independent of β and the magnetic field; consequently it also applies in the $\beta \to 0$ limit.

The driving mechanism of this instability is the gravitational drift. Clearly other particle drifts parallel to the boundary-producing surface charges and corresponding electric fields "in the right direction" would give rise to similar surface instabilities. Consider, for instance, a sharp boundary plasma confined by curved magnetic field lines, as in the case of the mirror geometry. The guiding centers are subject to the centrifugal and ∇B drift,

$$\mathbf{w} = \frac{m}{qB^4}\left(w_\parallel{}^2 + \frac{v_\perp{}^2}{2}\right)\left[\mathbf{B} \times \nabla \frac{B^2}{2}\right] \qquad (2\text{-}67)$$

in the low β limit. This drift is parallel to the boundary and produces charge separation on the ripples or "flutes," which makes the perturbations grow. For a rigorous treatment the three-dimensional configuration ought to be considered, but for wavelengths much shorter than the characteristic dimensions (k large grows fastest anyhow), the previous two-dimensional analysis applies. This amounts to replacing g in (10-106) by

$$g \to \frac{w_\parallel{}^2 + (v_\perp{}^2/2)}{R} \qquad (10\text{-}107)$$

where R is the radius of curvature of the field lines. It is easy to see (Exercise 10-21) that in configurations where the magnetic field lines curve *away* from the plasma (cusp geometry) the electric field produced by the surface charges tends to restore equilibrium and stable oscillations ensue. We arrive therefore at a stability condition similar to the one in the magnetohydrodynamic case. Magnetically confined plasmas with sharp boundaries are stable if the field lines curve everywhere away from the plasma, unstable if they curve toward the plasma. Unstabilized pinches are therefore again unstable. Mirror geometries require more care, since R varies along a field line and may even change sign. Since a given particle oscillates between the turning points, exploring regions of varying R, the growth rate is determined by some proper average R. All these instabilities result in growing flutes on the surface, and are known as *flute instabilities*.

However, it was found experimentally that mirror machines can be quite stable. Also the outer boundary of the Van Allen radiation belt was found to be much more stable than was expected on the basis of this theory. This (lucky) discrepancy between theory and experiment is quite unusual—theories usually err on the optimistic side in plasma physics. Its resolution is provided by considering more sophisticated models.

Consider first a continuous boundary with a small gradient of particle

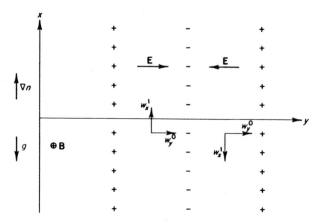

FIG. 10-11. Particle drifts associated with flute instabilities in a finite boundary layer.

density (Fig. 10-11). The gravitational drift which drives the instability is, again,

$$w_y{}^0 = \frac{Mg}{qB} = \frac{g}{\Omega} \qquad (10\text{-}108)$$

where $\Omega = \omega_c{}^+$ is on the ion cyclotron frequency. The electron drift will be neglected as before. Consider now a perturbation electric field $E_y \sim \exp[i(\omega t + ky)]$, $E_x = E_z = 0$.

[The assumption that the x dependence of the electrostatic potential may be neglected is a reasonable one, provided that the wavelength of the perturbation is much shorter than a characteristic distance in the x direction (say the boundary thickness).]

The resulting particle drifts are the electric drift

$$w_x{}^1 = E/B \qquad (10\text{-}109)$$

and the inertial drift from (2-73),

$$w_y{}^1 = \frac{m}{qB^2}\left(\frac{\partial E_y}{\partial t} + w_y{}^0 \frac{\partial E_y}{\partial y}\right) = \frac{i}{\Omega B}(\omega + w_y{}^0 k)E \qquad (10\text{-}110)$$

In the following analysis terms containing E will be considered small perturbation quantities. The application of the guiding-center approximation requires that $\omega/\Omega \ll 1$, with $(\omega/\Omega)^2$ terms neglected.

If the equilibrium guiding-center density for a species is n_0, the perturbed value $n = n_0 + \delta n$, one may write

$$\frac{\partial}{\partial t}\delta n + n_0 \frac{\partial w_x}{\partial x} + w_x \frac{\partial n_0}{\partial x} + n_0 \frac{\partial w_y}{\partial y} + w_y \frac{\partial}{\partial y}\delta n = 0 \qquad (10\text{-}111)$$

One substitutes $\partial/\partial t \rightarrow i\omega$, $\partial/\partial y \rightarrow ik$, $\partial n_0/\partial x = n'$, recognizes that $\partial w_x/\partial x = 0$, and obtains after linearization,

$$i(\omega + w_y{}^0 k)\,\delta n + w_x{}^1 n' + ikn_0 w_y{}^1 = 0 \tag{10-112}$$

The first-order guiding-center density is, therefore,

$$\delta n = \frac{i}{\omega + kw_y{}^0}\left(\frac{E}{B}n' + ikw_y{}^1 n_0\right) \tag{10-113}$$

The resulting charge density can be obtained as

$$\rho = q(\delta n^+ - \delta n^-) = \frac{iq}{\omega + kw_y{}^0}\left(\frac{E}{B}n' + ikw_y{}^1 n_0\right) - iq\frac{E}{B}\frac{n'}{\omega} \tag{10-114}$$

since w_y for electrons is neglected. This charge density is responsible for the electric field, hence

$$\rho = \varepsilon_0 ikE = \frac{iq}{\omega + kw_y{}^0}\left[\frac{E}{B}n' - \frac{Ekn_0}{\Omega B}(\omega + kw_y{}^0)\right] - iq\frac{E}{B}\frac{n'}{\omega} \tag{10-115}$$

iE drops out, and after multiplying by $\varepsilon_0{}^{-1}\omega(\omega + kw_y{}^0)$ and rearranging terms, one arrives at once at the dispersion relation

$$\omega^2\left(1 + \frac{qn_0}{\Omega B\varepsilon_0}\right) + \omega k\frac{g}{\Omega}\left(1 + \frac{qn_0}{\Omega B\varepsilon_0}\right) + \frac{qn'}{B\varepsilon_0}\frac{g}{\Omega} = 0 \tag{10-116}$$

The expression in parentheses is κ. The frequency

$$\omega = \tfrac{1}{2}\left[-k\frac{g}{\Omega} \pm \left(k^2\frac{g^2}{\Omega^2} - 4\frac{g}{\Omega}\frac{qn'}{B\varepsilon_0}\kappa^{-1}\right)^{1/2}\right] \tag{10-117}$$

is, in general, complex. Stable traveling waves ensue if

$$k^2 \geq 4\frac{qn'}{B\varepsilon_0}\frac{\Omega}{g}\kappa^{-1} \tag{10-118}$$

If $\kappa \gg 1$, the stability condition becomes

$$k^2 \geq 4\frac{\Omega^2}{g}\frac{n'}{n_0} \tag{10-119}$$

Short wavelengths, which gave the fastest-growing waves in the previous calculation, are entirely stable according to the present analysis. The two calculations differ not only in the shape of the boundary but more significantly also in the ordering of terms. If one neglects (ω/Ω), the linear term in (10-116) vanishes and all k values lead to exponential growth.

The mechanism of the instability is illustrated in Fig. 10-12. The $\mathbf{w}^E$ drift moves up lower-density plasma in region a to (say) the y axis, while in region b

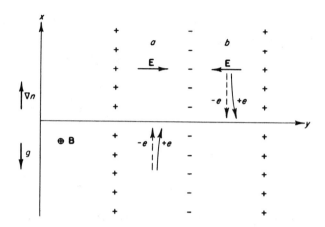

FIG. 10-12. Mechanism of flute instability in a finite layer.

it moves down higher-density plasma. The ion drift in region a tends to neutralize the space charge which causes the electric field, but it enhances the space charge in region b. The latter effect wins out because of the higher-density plasma in region b due to vertical plasma motion.

Superimposed on this effect is the motion of ions to the right due to the zero-order gravitational drift. This is responsible both for the traveling-wave character of the perturbation, as reflected in the real part of ω, and the damping, as seen from (10-117). The traveling wave arises because the density perturbation is carried bodily by the ions (Exercise 10-22). The damping can be understood as follows.

While the plasma density decreases in region a and increases in b, the space charge wave travels to the right, and after a while the direction of $\mathbf{E}$ changes sign in both regions. The result is a neutralization of space charges. Note that this happens in every case, even if the oscillations are unstable; there is no exponential runaway as in the hydromagnetic case, but an oscillatory over-stability. For short wavelengths (large k) $\mathbf{E}$ oscillates rapidly at a given point in space and stable oscillations result, while for small k much space charge is accumulated in a half-cycle and the oscillation overshoots.

In the absence of the linear damping term in (10-116), the characteristic time of growth would be

$$\tau_{\text{MHD}} \approx \left(\kappa \frac{B\varepsilon_0\Omega}{qn'g} \right)^{1/2} \tag{10-120}$$

The time required for an ion to travel a wavelength is of the order of

$$T \approx \Omega/gk \tag{10-121}$$

One expects stability roughly when $\tau_{\text{MHD}} > T$. This is the meaning of the stability condition (10-118).

It is clear from the previous discussion that any effect which adds to the speed of the space charge wave also has a stabilizing effect. So far it was assumed that the ionic charge is located at the guiding center itself or in other words that the gyroradius is zero. This is a rather crude assumption for most experimentally interesting configurations, and we will now investigate the effect of waiving this restriction.

First we show qualitatively how the finite gyroradius affects the speed of the space charge wave. There are two effects to note:

1. The electric drift now defines the velocity of the guiding center in the x direction, not the velocity of the ion itself. Figure 10-13 shows the drift of an

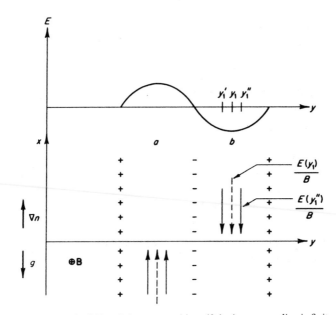

FIG. 10-13. Electric drifts of electrons and ions if the ion gyroradius is finite.

electron situated at y_1 compared to ions at the same point. The guiding centers of these ions, however, are situated at points other than y_1, for instance, at y_1' and y_1''. Since the electric field at these points is smaller than at y_1, the average ion drift velocity in the x direction is smaller than the electron drift at this point. It is easy to see that this is true for any other point as well (see Exercise 10-23). Hence there will be more ions arriving at the y axis in region a (they come from a region of higher density) and more electrons in b. This results in a shift of space charges to the right, producing a traveling wave.

2. The ion motion in the x direction is slower than the electric drift that is associated with the ion guiding center. While it gyrates, the ion samples the electric field on its path. Its value at a point $\mathbf{R}$ (gyroradius) away from the guiding center can be written by expansion:

$$E = E_0 + (\mathbf{R} \cdot \nabla)E + \tfrac{1}{2}R_i R_k \frac{\partial^2 E}{\partial x_i \, \partial x_k} + \cdots \tag{10-122}$$

Where E_0 is the field at the guiding center. The actual drift velocity will be determined by the average value of E over a full gyration. The linear as well as the mixed quadratic terms vanish due to their symmetry properties, and one is left with two quadratic terms, such as

$$\frac{1}{2}\frac{1}{2\pi} \int_0^{2\pi} R^2 \cos^2\varphi \, \frac{\partial_0^2 E}{\partial y^2} \, d\varphi = \tfrac{1}{4}R^2 \frac{\partial_0^2 E}{\partial y^2} \tag{10-123}$$

Neglecting higher-order terms in the expansion, one finds that the relevant electric field is

$$\langle E \rangle = E_0 + \tfrac{1}{4}R^2 \nabla_0^2 E = E_0 - \tfrac{1}{4}k^2 R^2 E_0 \tag{10-124}$$

instead of E_0, with the resulting drift smaller than E_0/B.

For a quantitative treatment of the first effect, one compares the particle distribution function $f(x, y, v_x, v_y)$, with the guiding-center distribution function $F(x, y, u_\perp)$, which is defined in the following manner. Let $\mathbf{u}_\perp$ be the perpendicular particle velocity vector in the guiding-center frame gyrating momentarily about the x,y point. Then $F(x, y, u_\perp)$ is the density of guiding centers around the point x,y, associated with particles gyrating about this point with the speed $u_\perp$.

The two functions are related by

$$f(x, y, v_x, v_y) = F[x - R_x, y - R_y, u_\perp] \tag{10-125}$$

where $\mathbf{R}$ is the gyroradius and $\mathbf{u} = \mathbf{v} - \mathbf{w}$. Since $R_x = -u_y/\Omega$ and $R_y = u_x/\Omega$ one may expand about x,y to find

$$f(x, y, v_x, v_y) = F\left(x + \frac{u_y}{\Omega}, y - \frac{u_x}{\Omega}, u_\perp\right)$$

$$= F(x, y, u_\perp) + \frac{\partial F}{\partial x}\frac{u_y}{\Omega} - \frac{\partial F}{\partial y}\frac{u_x}{\Omega} + \frac{1}{2}\frac{\partial^2 F}{\partial x^2}\left(\frac{u_y}{\Omega}\right)^2$$

$$+ \frac{1}{2}\frac{\partial^2 F}{\partial y^2}\left(\frac{u_x}{\Omega}\right)^2 - \frac{\partial^2 F}{\partial x \, \partial y}\frac{u_x u_y}{\Omega^2} + \cdots \tag{10-126}$$

One may drop higher-order terms and integrate over velocity space. The linear as well as the $u_x u_y$ term vanishes after integration and one finds

$$N = n + \frac{1}{2\Omega^2}\frac{\partial^2}{\partial x^2} n\langle u_y^2\rangle + \frac{1}{2\Omega^2}\frac{\partial^2}{\partial y^2} n\langle u_x^2\rangle \tag{10-127}$$

where N is the particle and n the guiding density.

If the perturbed ion density is $N^+ = N_0^+ + \delta N^+$, the electron density $N^- = N_0^- + \delta N^-$, with $N_0^+ = N_0^- = n_0$, Poisson's equation requires

$$q(\delta N^+ - \delta N^-) = \varepsilon_0 ikE \tag{10-128}$$

δN^+ can be computed from (10-127) to yield

$$\delta N^+ = \delta n^+ - \frac{\delta n^+}{2\Omega^2} k^2 \langle u_x^2 \rangle = \delta n^+ \left(1 - \frac{k^2 u_T^2}{4\Omega^2}\right) \tag{10-129}$$

where $\langle u_x^2 \rangle = \langle u_y^2 \rangle = u_T^2/2$ and it was assumed as previously that the perturbed quantities are independent of x and $n'' = 0$. u_T is the "thermal speed" in the generalized sense.

Equation (10-129) expresses the ion density in terms of the ion guiding-center density, whose motion is described by equation (10-113) if one replaces $E \rightarrow E - \frac{1}{4}k^2(u_T^2/\Omega^2)E$ according to (10-124). Since $\frac{1}{4}k^2 u_T^2/\Omega^2 \ll 1$, and the second term in (10-113) is much smaller than the first (w_y^1 is ω/Ω times smaller than $w_x = E/B$), one may write

$$\delta N^+ = \frac{i}{\omega + kw_y^0}\left(\frac{E}{B}n' - \frac{1}{4}\frac{k^2 u_T^2}{\Omega^2}\frac{E}{B}n' + ikw_y^1 n_0\right)\left(1 - \frac{k^2 u_T^2}{4\Omega^2}\right)$$

$$\approx \frac{i}{\omega + kw_y^0}\left[\frac{E}{B}n' - \frac{E}{B}\frac{k^2 u_T^2}{2\Omega^2}n' - \frac{E}{\Omega B}(\omega + kw_y^0)kn_0\right] \tag{10-130}$$

where (10-110) has also been used. The electronic gyroradius is small compared to the ionic one and one takes $\delta N^- = \delta n^-$. Equation (10-128) now becomes

$$\varepsilon_0 k = \frac{q}{\omega + kw_y^0}\left[\frac{n'}{B} - \frac{k^2 u_T^2 n'}{2B\Omega^2} - \frac{1}{B\Omega}(\omega + w_y^0 k)kn_0\right] - \frac{q}{\omega}\frac{n'}{B} \tag{10-131}$$

or after multiplication by $(\varepsilon_0 k)^{-1}\omega(\omega + kw_y^0)$ and rearrangement of terms,

$$\omega^2\left(1 + \frac{qn_0}{\varepsilon_0\Omega B}\right) + \omega k\frac{g}{\Omega}\left(1 + \frac{qn_0}{\varepsilon_0\Omega B}\right) + \omega k\frac{qu_T^2 n'}{2B\Omega^2\varepsilon_0} + \frac{qn'g}{\varepsilon_0 B\Omega} = 0 \tag{10-132}$$

This dispersion relation differs from (10-116) only in the appearance of an additional "damping" term due to the ion gyroradius u/Ω. The frequency now becomes

$$\omega = \frac{1}{2}\left\{-k\left(\frac{g}{\Omega} + \frac{qu_T^2 n'\kappa^{-1}}{2B\Omega^2\varepsilon_0}\right) \pm \left[k^2\left(\frac{g}{\Omega} + \frac{qu_T^2 n'\kappa^{-1}}{2B\Omega^2\varepsilon_0}\right)^2 - 4\frac{qn'g\kappa^{-1}}{\varepsilon_0 B\Omega}\right]^{1/2}\right\}$$

$$\tag{10-133}$$

Stable traveling waves arise if the expression under the radical sign is positive. For $\kappa \gg 1$ the dispersion relation becomes

$$\omega^2 + \frac{\omega k}{\Omega}\left(g + \frac{u_T^2 n'}{2n_0}\right) + g\frac{n'}{n_0} = 0 \tag{10-134}$$

and the stability condition

$$\frac{k^2}{\Omega^2}\left(g + \frac{u_T^2 n'}{2n_0}\right)^2 \geq 4g\frac{n'}{n_0} \tag{10-135}$$

It is instructive to compare the orders of magnitude of the stabilizing terms for a mirror geometry. Again using $g \sim (w_{\parallel}^2 + v_{\perp}^2/2)R^{-1}$, assuming a nearly thermal plasma $(w_{\parallel}^2 = u_T^2/2)$, and writing $n'/n_0 = 1/L$ (R is the radius of curvature, and L is the thickness of boundary layer), (10-135) becomes

$$\frac{k^2}{\Omega^2}\left(\frac{3}{2}\frac{u_T^2}{R} + \frac{u_T^2}{2L}\right)^2 \geq 4\frac{3}{2}\frac{u_T^2}{RL} \tag{10-136}$$

where the finite gyroradius term is roughly R/L times greater than the old stabilizing term. Stability is ensured for short wavelengths and large ion cyclotron radii, namely, when

$$k^2 R_c^2\left(\frac{3}{2}L + \frac{R}{2}\right)^2 \geq 6RL \tag{10-137}$$

where $R_c = u_T/\Omega$ is the average ion cyclotron radius. According to (10-136) and (10-137), stability increases if the density gradient is steeper. Our approximations limit the density gradient, however, to $L \gg R_c$.

10-5. Collisionless Drift Waves

A plasma confined by a magnetic field exhibits instabilities even in the absence of gravity or field line curvature. They are drift wave instabilities and, because they affect all magnetically confined plasmas, the name *universal instability* is often applied.

An important example is the collisionless drift wave, an electrostatic mode that propagates almost perpendicular to the magnetic field, and perpendicular to the density and temperature gradient. The frequency in the mode considered is low compared to the cyclotron frequencies, so the guiding center approximation can be invoked for motion perpendicular to **B**.

For such low-frequency waves it is convenient to introduce a guiding-center phase space, with a guiding-center distribution function $f(\mathbf{r}, \mu_m, v_{\parallel}, t)$ It is clear that for given electric and magnetic fields, $\mathbf{r}$, μ_m, $v_{\parallel}$, and t uniquely determine the guiding-center distribution. The conservation of guiding-centers leads to the *drift kinetic equation*

$$\frac{\partial f}{\partial t} + \nabla_\perp \cdot (\mathbf{v}_\perp f) + v_\parallel \nabla_\parallel f + \frac{\partial}{\partial v_\parallel} \left(\frac{F_\parallel}{m} f \right) = 0 \qquad (10\text{-}138)$$

Since $d\mu_m/dt = 0$, the term containing μ_m does not appear. Note that $v_\parallel$ is a coordinate of the drift phase space, while $\mathbf{v}_\perp$ is not.

For an electrostatic wave propagating in a uniform magnetic field, the linearized version of the drift kinetic equation yields

$$\frac{\partial f_1}{\partial t} + \nabla_\perp \cdot (\mathbf{v}_\perp f_0) + v_\parallel \nabla_\parallel f_1 + \frac{q}{m} E_\parallel \frac{\partial f_0}{\partial v_\parallel} = 0 \qquad (10\text{-}139)$$

where

$$\mathbf{v}_\perp = \frac{\mathbf{E} \times \mathbf{B}}{B^2} + \frac{m}{qB^2} (\dot{\mathbf{E}}_\perp + v_\parallel \nabla_\parallel \mathbf{E}_\perp) \qquad (10\text{-}140)$$

is linear in the small wave electric field. If $\mathbf{B}$ points in the z direction and ∇f_0 points in the x direction, one may Fourier-decompose all quantities in terms of $\Phi(x) \exp[i(k_\parallel z + k_\perp y - \omega t)]$. If f_0 varies gently, such that the scale length of its variation in the x direction L is large compared to the wavelength in the y direction ($k_\perp L \gg 1$) one expects $\Phi(x)$ to vary little on the wavelength scale. One may then adopt the *local approximation*, neglecting $\partial \Phi/\partial x$ compared to $ik_\perp \Phi$, and the linearized drift kinetic equation yields

$$i(k_\parallel v_\parallel - \omega) f_1 + \frac{E_\perp}{B} \frac{\partial f_0}{\partial x} + \frac{m}{qB^2} k_\perp (\omega - k_\parallel v_\parallel) E_\perp f_0 + \frac{q}{m} E_\parallel \frac{\partial f_0}{\partial v_\parallel} = 0 \qquad (10\text{-}141)$$

Here we have used the fact that $\nabla \cdot (\mathbf{E} \times \mathbf{B}) = 0$ (since $\nabla \times \mathbf{E} = 0$), and that $\mathbf{E} = -\nabla \varphi$ is a vector in the yz plane, in the local approximation. For electrons $q = -e$ and one obtains for the electron guiding-center density perturbation

$$n_e = \int f_{1e} \, dv_\parallel \, d\mu_m$$

$$= i \left[\frac{E_\perp}{B} \frac{\partial}{\partial x} \int \frac{f_{0e} \, dv_\parallel}{k_\parallel v_\parallel - \omega} + \frac{m}{eB^2} k_\perp E_\perp N - \frac{e}{m} E_\parallel \int \frac{\partial f_{0e}/\partial v_\parallel}{k_\parallel v_\parallel - \omega} \, dv_\parallel \right] \qquad (10\text{-}142)$$

where $N(x)$ is the equilibrium density. The $f_{0e}(v_\parallel)$ are reduced distribution functions, where the μ_m integration has already been performed.

For a Maxwellian distribution $f_0 = \alpha \exp(-\beta v_\parallel^2)$ with

$$\alpha(x) = N(x)\sqrt{\frac{\beta(x)}{\pi}}$$

and $\beta(x) = m/2T_e(x)$, the integrals in (10-142) reduce to types encountered before. For numerical evaluation it is useful to introduce the Z *function*, or *plasma dispersion function*

$$Z(\xi) = \frac{1}{\sqrt{\pi}} \int_{-\infty}^{+\infty} \frac{\exp(-x^2)\,dx}{x - \xi} \tag{10-143}$$

whose values are tabulated; computer programs to generate the function are also available.

For analytic work one can expand the integrals directly. The most interesting regime is the one where $v_e \gg \omega/k_\parallel \gg v_i$ (v_e and v_i are the electronic and ionic thermal speeds). When evaluating the real parts of the two integrals we simply ignore ω compared to $k_\parallel v_\parallel$ and find no contribution from the real part of the first integral, whereas the second is trivial. The pole contributions from both integrals give the imaginary parts. This works out to give

$$n_e \approx i\left[\frac{m}{eB^2} k_\perp E_\perp N + \frac{eE_\parallel N}{T_e k_\parallel} + \frac{i\pi}{k_\parallel}\left(\frac{E_\perp}{B}\frac{\partial}{\partial x} - \frac{e}{m} E_\parallel \frac{\partial}{\partial v_\parallel}\right) f_{0e}\left(\frac{\omega}{k_\parallel}\right)\right] \tag{10-144}$$

The second term is larger than the first one by a large factor $(k_\perp R_e)^{-2}$, where $R_e = \sqrt{T_e m}/eB$ is the average electron gyroradius. Since $E_\perp = Ek_\perp/k$ and $E_\parallel = Ek_\parallel/k$, the electron density becomes approximately

$$n_e \approx i\left[\frac{eN}{T_e k} + \frac{i\pi}{k_\parallel kB}\left(k_\perp \frac{\partial}{\partial x} - \Omega_e k_\parallel \frac{\partial}{\partial v_\parallel}\right) f_{0e}\left(\frac{\omega}{k_\parallel}\right)\right] E \tag{10-145}$$

where Ω_e is again the electron gyrofrequency.

The perturbed ion density

$$n_i = i\left[\frac{E_\perp}{B}\frac{\partial}{\partial x}\int \frac{f_{0i}\,dv_\parallel}{k_\parallel v_\parallel - \omega} - \frac{M}{eB^2}k_\perp E_\perp N + \frac{e}{M}E_\parallel \int \frac{\partial f_{0i}/\partial v_\parallel}{k_\parallel v_\parallel - \omega}\,dv_\parallel\right] \tag{10-146}$$

can be simplified by ignoring ion temperature entirely, to find

$$n_i \approx i\left[-\frac{k_\perp N'}{B\omega} - \frac{k_\perp^2 N}{B\Omega_i} + \frac{e}{M}\frac{k_\parallel^2}{\omega^2} N\right]\frac{E}{k} \tag{10-147}$$

where $N' = dN/dx$. From Poisson's equation $i\mathbf{k} \cdot \mathbf{E} = (e/\varepsilon_0)(n_i - n_e)$ the dispersion relation follows immediately,

$$\varepsilon = 1 + \frac{1}{k^2 \lambda_D^2} + \frac{\omega_{pi}^2}{\Omega_i^2} \frac{k_\perp^2}{k^2} - \frac{\omega_{pi}^2}{\omega^2} \frac{k_\parallel^2}{k^2} + \frac{\omega_{pe}^2}{\Omega_e \omega} \frac{k_\perp}{k^2} \frac{d}{dx} \ln N$$

$$+ i \frac{\pi e}{k_\parallel k^2 B \varepsilon_0} \left(k_\perp \frac{\partial}{\partial x} - \Omega_e k_\parallel \frac{\partial}{\partial v_\parallel} \right) f_{0e}\left(\frac{\omega}{k_\parallel}\right) = 0 \qquad (10\text{-}148)$$

We evaluate first the real part of the dispersion relation. Since $(k\lambda_D)^2 \ll 1$ one finds

$$\omega^2 \left(1 + \frac{c_s^2 k_\perp^2}{\Omega_i^2} \right) - \omega k_\perp v_D - c_s^2 k_\parallel^2 = 0 \qquad (10\text{-}149)$$

where $c_s = \lambda_D \omega_{pi}$ is the ion acoustic velocity and

$$v_D = - \frac{\omega_{pe}^2 \lambda_D^2}{\Omega_e} \frac{d}{dx} \ln N = -v_e R_e \frac{d}{dx} \ln N \qquad (10\text{-}150)$$

is the so-called diamagnetic drift velocity of the electrons. This is the average electron velocity due to the density gradient, and should not be confused with a guiding-center drift.

Introducing

$$\sigma = 1 + \frac{c_s^2 k_\perp^2}{\Omega_i^2} \qquad (10\text{-}151)$$

the solution is

$$\omega = \frac{1}{2\sigma} \left[k_\perp v_D \pm \sqrt{(k_\perp v_D)^2 + 4\sigma c_s^2 k_\parallel^2} \right] \qquad (10\text{-}152)$$

The two branches of $\omega(k_\parallel)$ for fixed $k_\perp$ are plotted in Fig. 10-14.

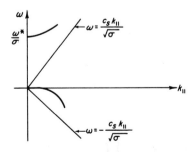

FIG. 10-14. Drift wave dispersion curve.

The upper branch is the drift wave branch, the lower a modified ion acoustic branch. The quantity $k_\perp v_D = \omega^*$ is the drift frequency.

When the imaginary part of ε is small, the growth rate can be evaluated in the usual way from

$$\gamma = -\left.\frac{\text{Im }\varepsilon}{\partial\varepsilon/\partial\omega}\right)_{\omega_0} \tag{10-153}$$

where Re $\varepsilon(\omega_0,k) = 0$. For simplicity, we evaluate the growth rate for the case when no temperature gradient is present and leave the more general case as an exercise.

We may now evaluate Im ε:

$$\text{Im }\varepsilon = \frac{\pi e}{k_\parallel k^2 B\varepsilon_0}\,\alpha\,\exp\left[-\beta\left(\frac{\omega}{k_\parallel}\right)^2\right]\left(k_\perp\frac{d}{dx}\ln N + \frac{m\omega_0\Omega_e}{T_e}\right) \tag{10-154}$$

The two terms in the brackets have opposite signs. If $(d/dx)\ln N < 0$, v_D and ω_0 are positive, while if $(d/dx)\ln N > 0$, $\omega_0 < 0$ for the drift branch. We evaluate ω_0 from (10-152) for $(k_\perp v_D)^2 \gg 4\sigma c_s^2 k_\parallel^2$ to find

$$k_\perp\frac{d}{dx}\ln N + \frac{m\omega_0\Omega_e}{T_e} \approx k_\perp\frac{d}{dx}\ln N\left(1 - \frac{1}{\sigma}\right) \tag{10-155}$$

and

$$\text{Im }\varepsilon \approx \frac{\pi e}{k_\parallel k^2 B\varepsilon_0}\,\alpha k_\perp\frac{d}{dx}\ln N\left(1 - \frac{1}{\sigma}\right)\exp\left[-\beta\left(\frac{\omega}{k_\parallel}\right)^2\right] \tag{10-156}$$

For $v_D > 0$, $(d/dx)\ln N < 0$ and so is Im ε.

One computes now the real part of $\partial\varepsilon/\partial\omega$. Form (10-148) and (10-150)

$$\begin{aligned}
\left.\frac{\partial\varepsilon}{\partial\omega}\right)_{\omega_0} &= \frac{2\omega_{pi}^2\,k_\parallel^2}{\omega_0^3\,k^2} - \frac{\omega_{pe}^2}{\Omega_e\omega_0^2}\frac{k_\perp}{k^2}\frac{d}{dx}\ln N \\
&= \frac{1}{\omega_0^2 k^2}\left(\frac{2\omega_{pi}^2 k_\parallel^2}{\omega_0} + \frac{k_\perp v_D}{\lambda_D^2}\right)
\end{aligned} \tag{10-157}$$

For $(kv_D)^2 \gg \sigma c_s^2 k_\parallel^2$ the second term dominates with $\omega_0 \approx k_\perp v_D/\sigma \approx kv_D/\sigma$ so approximately

$$\left.\frac{\partial\varepsilon}{\partial\omega}\right)_{\omega_0} \approx \left(\frac{\sigma}{k\lambda_D}\right)^2\frac{1}{k_\perp v_D} \tag{10-158}$$

and

$$\gamma \approx \sqrt{\beta\pi}\,\frac{1}{k_\parallel}\left(\frac{k_\perp v_D}{\sigma}\right)^2\left(1 - \frac{1}{\sigma}\right)\exp\left[-\beta\left(\frac{\omega}{k_\parallel}\right)^2\right] \tag{10-159}$$

To estimate the growth rate we first maximize this expression with respect to $k_\perp$. This is left for Exercise 10-26, and one finds that $(k_\perp c_s/\Omega_i)^2 = 2$ or $\sigma = 3$ gives the maximum growth rate. The fastest growth occurs when the perpendicular wavelength is comparable to the gyroradius that ions would have if electron and ion temperatures were comparable. If the temperatures of the two species are comparable, finite ion gyroradius effects must be taken into account, and the maximum growth rate is somewhat reduced.

The growth rate is still a function of $k_\parallel$. In deriving the dispersion relation we have assumed that $\beta(\omega/k_\parallel)^2 \ll 1$. Nevertheless, one may estimate the order of magnitude of the largest growth rate as $k_\parallel$ is reduced. For $k_\parallel$ small, from (10-149) one may estimate $\omega \approx k_\perp V_D/\sigma$, and write (with $\sigma = 3$)

$$\frac{\gamma}{\omega} \approx \frac{2}{3} \sqrt{\pi} x \exp(-x^2) \tag{10-160}$$

where

$$x = \sqrt{\beta}\, \omega/k_\parallel \tag{10-161}$$

The maximum of $x e^{-x^2}$ is where $x = 2^{-1/2}$, and its value is 0.43. So for maximum growth γ and ω are comparable. This estimate is borne out by numerically evaluating the exact dispersion relation in terms of Z functions.

It is easy to see that the other branch of the dispersion relation is stable. The calculation is left for Exercise 10-27.

In order to understand the physical processes involved, note that the ions perform $\mathbf{E} \times \mathbf{B}$ drift motion in the x direction, leading to a displacement $\delta x_i = E_\perp/B\omega$ along the density gradient. This results in an ion density perturbation as seen in the first term of (10-147). Since the electrons move fast along the magnetic field, their displacement is to be calculated from the Doppler-shifted frequency, and gives $\delta x_e = E_\perp/B(\omega - k_\parallel v_\parallel)$ for electrons with velocity $v_\parallel$. Since $\omega \ll k_\parallel v_\parallel$ for most electrons, the average displacement nearly vanishes, since there are the same number of electrons moving with $v_\parallel$ as with $-v_\parallel$. The corresponding electron density perturbation is therefore small, as seen from the near vanishing of the first integral in (10-142). The net ion density produces a space charge that is responsible for the wave. The net space charge is, of course, considerably less, since electrons, moving freely along $\mathbf{B}$, tend to neutralize it, resulting in the shielding term $(k\lambda_D)^{-2}$ in ε.

The resonant electrons moving with the parallel velocity $v_\parallel \approx \omega/k_\parallel$ suffer large $\mathbf{E} \times \mathbf{B}$ displacements, and are responsible for the instability. Their displacement at time t after the wave has been switched on is $\delta x_{res} = (E_\perp/B)t$. The number of such particles per unit volume is $f_0\,(v_\parallel = \omega/k_\parallel, x)\,\Delta v_\parallel$, where $\Delta v_\parallel$ is the velocity spread for resonant electrons. An electron stays at resonance at time t if it has moved only a fraction of a wavelength along $\mathbf{B}$ during that time; so for the resonant velocity width at time t, one may estimate

$t\,\Delta v_\| \approx \frac{1}{2}\lambda_\| = \pi/k_\|$. Since these electrons move along the density gradient they give rise to an excess density. This is shown in Fig. 10-15, where constant phase surfaces are sketched in the wave frame. In region a electrons move up from a region of higher electron density, producing an excess of electrons, while in region b electrons came from a lower density region resulting in a local depletion of electrons. In the laboratory frame the wave moves to the right along $\mathbf{B}$, so resonant electrons in a do work against the field while those in b are accelerated by it. Since an excess of resonant electrons is being pumped in region a into the yz plane by the $\mathbf{E} \times \mathbf{B}$ drift, more electrons give energy to the field than take energy from it (in region b). The net effect is wave growth.

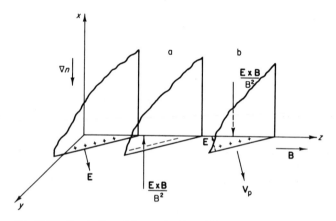

Fig 10-15. Illustration of the mechanism of the drift wave instability.

Quantitatively the excess resonant electron density is

$$\delta n_{res} = -\delta x_{res}\,\frac{d}{dx}f_0\!\left(\frac{\omega}{k_\|},x\right)\Delta v_\| = -\frac{E_\perp}{B}\frac{d}{dx}f_0\!\left(\frac{\omega}{k_\|},x\right)\frac{\pi}{k_\|} \quad (10\text{-}162)$$

The power delivered by the particles to the wave is

$$P_D = eE_\|\,\delta n_{res}\,\frac{\omega}{k_\|} = -\frac{e\pi E_\perp E\omega}{k_\| kB}\frac{d}{dx}f_0 \quad (10\text{-}163)$$

Energy is also taken out from the wave by the usual Landau damping process at the rate

$$P_L = -\pi\,\frac{e^2 E_\|^{\,2}}{m}\frac{\omega}{k_\|^{\,2}}\frac{\partial f_0}{\partial v_\|} = -\pi\,\frac{e^2 E^2}{m}\frac{\omega}{k^2}\frac{\partial f_0}{\partial v_\|} \quad (10\text{-}164)$$

The net power delivered to the wave $P = P_d - P_L$ results in wave amplitude growth, described by the energy equation

$$P_d - P_L = 2\gamma \frac{\varepsilon_0 E_2}{2} \omega \frac{\partial \varepsilon}{\partial \omega} \tag{10-165}$$

The growth rate γ, calculated from the last three equations, yields the same answer as obtained before, verifying the validity of the physical picture (see Exercise 10-28).

In order to investigate the mode structure in the x direction, one has to go beyond the local approximation to find a differential equation for the scalar potential $\varphi(x)$

$$\left[\frac{1}{k^2 \lambda_D^2} + \frac{\omega_{pi}^2}{\Omega_i^2} \frac{k_y^2}{k^2} - \frac{\omega_{pi}^2}{\Omega_i^2 k^2} \frac{\partial^2}{\partial x^2} + \frac{k_y}{k^2} \frac{\omega_{pe}^2}{\Omega_e \omega} \frac{\partial}{\partial x} \ln N - \frac{\omega_{pi}^2}{\omega^2} \frac{k_\parallel^2}{k^2} \right] \varphi$$

$$- \frac{\omega_{pi}^2}{\Omega_i^2} \frac{1}{k^2} \frac{\partial}{\partial x} \ln N \frac{\partial \varphi}{\partial x} + \frac{i\pi e \varphi}{k^2 k_\parallel \varepsilon_0} \left(\frac{k_y}{B} \frac{\partial}{\partial x} - \frac{e}{m} k_\parallel \frac{\partial}{\partial v_\parallel} \right) f_0 \left(\frac{\omega}{k_\parallel} \right) = 0 \tag{10-166}$$

where the potential is $\phi = \varphi(x) \exp[i(k_\parallel z + k_y y - \omega t)]$. The derivation of this equation is left for Exercise 10-29. When $\varphi = \text{const}$ one recovers the local approximation (10-148). With the substitution $\varphi = N^{-1/2} \psi$ this equation can be transformed into

$$\psi'' + F(x)\psi = 0 \tag{10-167}$$

where F is also a function of the parameters $k_\parallel$, k_y, and ω. Since the solutions must be bounded in x, this becomes an eigenvalue problem for ω. The frequency corresponding to given $k_\parallel$ and k_y will take on discrete values, thus leading to dispersion relations. The modes are standing waves in x, propagating in the yz direction.

A far more important nonlocal problem involves magnetic shear. In toroidal devices (e.g. tokamaks), as a result of current flowing along the magnetic field lines the pitch angle of the magnetic field lines change across the plasma cross section. Instead of the complicated toroidal geometry, consider one with straight field lines, where the field direction varies along the x axis, $\mathbf{B} = B_0(\hat{e}_z + \hat{e}_y x/L_s)$, where L_s is the characteristic shear length. One can now look for electrostatic waves where the potential is given by $\phi = \varphi(x) \exp[i(k_y y - \omega t)]$. For such a wave $k_\parallel \approx \mathbf{k} \cdot \mathbf{B}/B_0 = k_y x/L_s$; hence $k_\parallel$ changes with x.

From local theory we know that growth occurs for a certain range of $k_\parallel$, peaking at about $k_\parallel = \omega/v_e$, while for $k_\parallel \gg \omega/v_e$ the growth is zero. Hence one expects the wave growth to be restricted to the corresponding range in x. Outside this range the waves propagate away, carrying energy outward. In the region where $|k_{\parallel} v_i/\omega| = |k_y(x/L_s)(v_i/\omega)| \approx v_i$ they are absorbed by ion Landau damping. Hence the drift modes are stabilized when the energy generated in the inner x region by the instability is balanced by the energy carried away in the outer region in a steady state. Indeed, it can be shown that magnetic shear stabilizes drift modes, so the universal instability is not quite universal. The sheared field, however, is produced by an electron current flowing along $\mathbf{B}$; hence the electron distribution function is a shifted Maxwellian. This may result in destabilizing the drift waves again. More important, however, while shear may stabilize drift waves, it gives rise to the electromagnetic tearing mode instability, as we have seen in Sec. 5-6. This illustrates again the uncanny ability of plasma to escape confinement.

10-6. Minimum-B Geometries

Consider a plasma confined in a magnetic mirror geometry, where the conditions of the guiding-center approximation are satisfied. A particle confined between the end mirrors in equilibrium is unable to escape through these mirrors, whatever adiabatic (μ_m conserving) perturbation is applied. This is the reason why instabilities of mirror configurations carry the plasma *across* the field, where the constants of motion do not prevent the plasma from escaping. On the contrary, the magnetic field intensity decreases radially away from the axis; hence from the constancy of μ_m, the perpendicular kinetic energy of an outward moving particle decreases as well, and the excess energy becomes available as drift and field energy to feed an instability.

This consideration suggests trying field configurations, where the field intensity has a minimum somewhere inside the machine and increases outward in every direction. Here it is assured that any adiabatic perturbation which moves particles outward increases their gyration energy. If the second adiabatic invariant J is also conserved, determining $v_{\parallel}$ on any given field line, one may hope to find that a perturbation moving the plasma outward exhausts its energy by increasing $v_{\parallel}$ and $v_{\perp}$, resulting in stable oscillation, not an instability.

Such "minimum-B" geometries can be constructed by combinations of mirror and cusp fields (Fig. 5-12) or by using the radially diverging field lines of cusp-type fields (Fig. 10-16). If the walls of the coils facing each other were parallel, the magnetic field would decrease as $1/r$ in the gap. If the gap narrows for larger values of r the field can be made to increase again, forming a minimum at some r_0.

It is important to note that cusp geometries, although they possess a minimum of B, do not fall into this category. If this minimum is $B = 0$ the particle motion is nonadiabatic and our arguments do not hold. Hence the expression minimum B is reserved for configurations where this minimum value of $B \neq 0$.

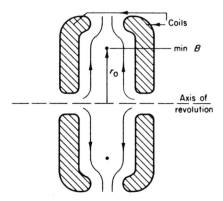

FIG. 10-16. A minimum-B geometry produced by two coaxial, opposite current-carrying coils.

For the formal treatment of a general minimum-B configuration we introduce a magnetic coordinate system, with the coordinates α and β constant along field lines, so that specifying a pair (α,β) specifies a field line and can thus be regarded as the coordinates of the field line. $\nabla\alpha$ and $\nabla\beta$ are clearly perpendicular to the field at any point, and the scale of the coordinates α,β can be chosen so that

$$\mathbf{B} = \nabla\alpha \times \nabla\beta \qquad (10\text{-}168)$$

The vector potential may be written

$$\mathbf{A} = \alpha\,\nabla\beta \qquad (10\text{-}169)$$

since the $\nabla \times$ of (10-169) results in (10-168). To locate a point P on a field line whose coordinates (α,β) have been specified, one introduces

$$\chi = \int_{S}^{P} \mathbf{B} \cdot d\mathbf{s} \qquad (10\text{-}170)$$

where S is a reference surface perpendicular to the field lines.

Consider now a guiding-center plasma with the particles replaced by magnetic dipoles of moment μ_m placed at the guiding centers of respective particles. We wish to express the Lagrangian and the Hamiltonian of such a dipole in the (α,β,χ) coordinate system. The Lagrangian can be written (see Exercise 10-30) as

$$L = \tfrac{1}{2}mv^2 + q(\mathbf{A} \cdot \mathbf{v}) - qV - \mu_m B \qquad (10\text{-}171)$$

where $\mu_m B$ plays the role of potential energy as expected. Taking the time derivative of (10-170),

$$\dot{\chi} = \left[\frac{d}{d\mathbf{s}} \int_{S}^{P} \mathbf{B} \cdot d\mathbf{s} \right] \cdot \frac{d\mathbf{s}}{dt} = Bv_{\parallel} \qquad (10\text{-}172)$$

Hence the Lagrangian in the new coordinates becomes

$$L = \frac{m\dot{\chi}^2}{2B^2} + q\alpha\dot{\beta} - qV(\alpha,\beta,\chi) - \mu_m B(\alpha,\beta,\chi) + \varepsilon_\perp \qquad (10\text{-}173)$$

where the drift kinetic energy $\varepsilon_\perp$ will be neglected.

To calculate the Hamiltonian one needs the canonical momenta

$$P_\alpha = \partial L/\partial\dot{\alpha} = 0 \qquad (10\text{-}174)$$

$$P_\beta = \partial L/\partial\dot{\beta} = q\alpha \qquad (10\text{-}175)$$

and

$$P_\chi = \partial L/\partial\dot{\chi} = m\dot{\chi}/B^2 \qquad (10\text{-}176)$$

The Hamiltonian in the new coordinates becomes

$$\mathscr{H} = P_i\dot{Q}_i - L = \frac{B^2}{2m}P_\chi^{\;2} + qV(\alpha,\beta,\chi) + \mu_m B(\alpha,\beta,\chi) \qquad (10\text{-}177)$$

We wish to introduce the second adiabatic invariant,

$$J = \oint mv_\parallel \, ds = \oint P_\chi \, d\chi = \oint [2m(\mathscr{H} - qV - \mu_m B)]^{1/2} \, ds \quad (10\text{-}178)$$

where the integral is to be performed along a field line characterized by the pair of values (α,β), between the two turning points at the mirrors. J is a function of $\mathscr{H}$, α, β, and μ_m, or the Hamiltonian may be thought of as a function of the new variables, α, β, J, and μ_m. In other words, the energy of a particle is uniquely defined if its field line (α,β) and constants μ_m and J are specified. We denote this Hamiltonian function by $K(\alpha, \beta, \mu_m, J)$. The old equations of motion (10-174) and (10-175) are unaffected by this transformation.

From Hamilton's canonical equations

$$\dot{\alpha} = q^{-1}\dot{P}_\beta = -q^{-1} \, \partial K(\alpha, \beta, \mu_m, J)/\partial\beta \qquad (10\text{-}179)$$

and

$$\dot{\beta} = q^{-1} \, \partial K(\alpha, \beta, \mu_m, J)/\partial\alpha \qquad (10\text{-}180)$$

Now consider the motion in the $(\alpha, \beta, \mu_m, J)$ "phase space." Since J and μ_m are both constants of motion, this problem is immediately reduced to motion in (α,β) planes with μ_m and J fixed. This motion is that of an incompressible fluid, since from (10-179) and (10-180),

$$\nabla \cdot \mathbf{v} = \frac{\partial\dot{\alpha}}{\partial\alpha} + \frac{\partial\dot{\beta}}{\partial\beta} = 0 \qquad (10\text{-}181)$$

It is evident from (10-177) that the Hamiltonian function expresses the energy of a particle, also a constant of the motion for a stationary electric field-free plasma. With the three constants of motion known, the most general stationary distribution function can be described by

$$f_{st} = f[K_{\mu,J}(\alpha, \beta), \mu_m, J] \qquad (10\text{-}182)$$

where the number of particles in the element $(d\alpha, d\beta, d\mu_m, dJ)$ is defined by

$$f[K_{\mu,J}(\alpha,\beta), \mu_m, J] \, d\alpha \, d\beta \, d\mu_m \, dJ \qquad (10\text{-}183)$$

It remains to select the stable distribution functions from all those described in (10-183). This can easily be done as follows:

Consider $K_{\mu, J} = $ const loops on the (α, β) planes. These are, of course, also $f_{st} = $ const curves. If each plane is populated in such a way that f_{st} is largest where the energy $(K_{\mu, J})$ is smallest, and decreases everywhere with increasing K, i.e.,

$$(\partial f/\partial K)_{\mu, J} < 0 \qquad (10\text{-}184)$$

the configuration exhibits absolute stability within the (assumed) restriction that μ_m and J are constants of motion. The proof is entirely analogous to that given in Sec. 3-5 for the absolute stability of a uniform plasma with the distribution function decreasing with v^2 and need not be repeated here. It is also easy to calculate the maximum energy available for instability if (10-184) is not satisfied. This is left for Exercise 10-31.

Since in a confined plasma one expects that f decreases toward the periphery, one may see from (10-184) that in stable configurations K must increase toward the periphery. *Hence a stable minimum-B configuration is a minimum-K configuration as well.* For a given magnetic field configuration the $K = $ const contours can be easily mapped without the knowledge of plasma parameters. If K has a minimum on a field line located inside the machine for any pair of values μ_m and J, then every stationary distribution that satisfies (10-184) is a stable one. This seems to be the case for a large class of minimum-B fields.

Note two important restrictions in the theory:

1. It does not predict stability against perturbations with a fast enough time scale such that μ_m and J are not conserved.

2. It was assumed that $\beta \approx 0$.

10-7. Summary

Plasma waves propagating along a cold magnetoplasma column exhibit properties quite different from those found in an infinite uniform plasma, unless the wavelength *measured in the Lorentz frame moving with the phase velocity* is much smaller than the transverse dimension of the column. For longer waves $\omega \to 0$ as $k \to 0$ in such a way that $\omega/k \leq c$. In addition to these pure oscillatory solutions, radiating modes appear. These are characterized by $\omega/k \geq c$. If the background magnetic field is large, the dispersion relation of a slab is given by (10-31b). The general behavior of the dispersion curves is sketched in Fig. 10-4; the principal branches for an arbitrary magnetic field are shown in Fig. 10-9. The radiative modes can be viewed (from a proper frame) as electromagnetic plasma oscillations, while the mechanism of radiation is similar to that of Cerenkov radiation. The zero-background magnetic field is treated separately, and the resulting dispersion relation for surface modes is given by (10-67). Nonpropagating pure oscillatory volume modes with $\omega = \omega_p$ still persist for this special case.

The treatment of reflection, refraction, and transmission follows the same lines as in optics. The dielectric constant is, however, a tensor (as in crystal optics), which results in some complication of the computation and the effect of double refraction. Waves generated inside the plasma suffer total reflection on the boundary if $\omega/k_t < c$; some of their energy is converted into pure radiation on the vacuum side if $\omega/k_t > c$, where k_t is the component of the wave vector tangential to the interface.

Surface instabilities of bounded magnetoplasmas, when studied in the guiding-center approximation, exhibit many similarities to those encountered in magnetofluids. A plasma suspended by a magnetic field in a gravitational field, the pinch and mirror geometries are basically unstable, while the cusp geometry is again found to be stable. There are, however, some notable differences. The short-wavelength perturbations are now stable, while for long wavelengths one finds overstability instead of the exponential runaway type of instability which afflicts magnetofluids of similar geometry. The growth rates of the overstable oscillations are reduced compared to those of the fluid model. Stability for shorter wavelengths may assure complete stability for certain mirror geometries, where the circumference sets a limit to the wavelength of the perturbation.

In studying the stability conditions of various geometries, the behavior of a plasma suspended against a gravitational field serves as a convenient model. Here the instability is driven by the gravitational drift, not dissimilar to the centrifugal and ∇B drifts, which play a similar role in mirror and pinch geometries. The sharp boundary models well suited for conductive fluids are replaced by the more realistic continuous boundary models for plasmas.

One may go a step beyond the guiding-center plasma by recognizing that the ion gyroradius is not zero. While the motion of the guiding center is computed as before (only the guiding-center field is to be replaced by the average field on the particle orbit), in calculating the ionic space charge account is taken of the fact that the ion is located a finite distance away from its guiding center. The result of the inclusion of this effect is improved stability for all wavelengths. The dispersion relation for the $\beta \simeq 0$ case is shown in (10-132).

A plasma confined by a uniform magnetic field is subject to drift-type instabilities even in the absence of equivalent gravity. Low-frequency electrostatic waves with a **k** vector perpendicular to the density gradient, with a small component along **B**, acquires energy from resonant electrons to drive this mode. It is stabilized by magnetic shear.

One may find a large class of stable low-β configurations in the guiding-center approximation. If the adiabatic invariants μ_m and J are conserved, any minimum-B geometry (B has a nonzero minimum value inside) with a minimum value of particle energy for any fixed pair of μ_m, J values inside the field is absolutely stable, for certain classes of distribution functions. This simple stability condition is expressed in (10-184).

EXERCISES

10-1. Prove that in (10-6)
(a)
$$\overset{\leftrightarrow}{\rho}{}^{-1} = \begin{pmatrix} (\omega^2/c^2)\kappa_T - k_y{}^2 - k_{\parallel}{}^2 & i(\omega^2/c^2)\kappa_H + k_x k_y \\ -i(\omega^2/c^2)\kappa_H + k_x k_y & (\omega^2/c^2)\kappa_T - k_x{}^2 - k_{\parallel}{}^2 \end{pmatrix}$$

and the inverse matrix

$$\overset{\leftrightarrow}{\rho} = \frac{1}{\det \overset{\leftrightarrow}{\rho}{}^{-1}} \begin{pmatrix} (\omega^2/c^2)\kappa_T - k_x{}^2 - k_{\parallel}{}^2 & -i(\omega^2/c^2)\kappa_H - k_x k_y \\ i(\omega^2/c^2)\kappa_H - k_x k_y & (\omega^2/c^2)\kappa_T - k_y{}^2 - k_{\parallel}{}^2 \end{pmatrix}$$

(b) The dispersion relations (10-8) and (8-19) are identical.

10-2. Show that for $k_y = k_z = 0$, if
(a) $\kappa_{\parallel} = n^2 : \mathbf{E}_{\perp} = 0$

(b) $\dfrac{\kappa_R \kappa_L}{\kappa_T} = n^2 : \dfrac{E_z}{E_{\perp}} = 0$ and $\dfrac{E_x}{E_y} = -\dfrac{i\kappa_H}{\kappa_T}$

10-3. Evaluate (10-16) and (10-17) as functions of frequency. Calculate the transmission and reflection bands.

10-4. An electromagnetic wave impinges at right angles on a magneto-plasma slab of thickness a. Calculate the reflected, transmitted, and plasma fields.

10-5. Prove that (10-25) and (10-26) can also be obtained from the superposition of two traveling waves with identical k_z and equal but opposite k_x values.

10-6. Treat the problem of plasma oscillations in a slab along a strong magnetic field as an initial-value problem, employing the techniques of Laplace transformation. Show that the same dispersion relations obtain as in Sec. 10-2.

10-7. Prove that radiating plasma oscillations emit waves exponentially increasing away from the plasma.

10-8. Carry out the calculation to obtain the dispersion relation for an oscillating plasma cylinder in a strong magnetic field, in analogy with the procedure in Sec. 10-2 for the slab case.

10-9. Find the electromagnetic fields produced by the charge distribution $\rho = \rho_0 e^{i(\omega t - kz)} \delta(r)$ where r,z are cylindrical coordinates. Prove that radiation obtains for $\omega/k > c$ and calculate the fields in the radiation zone.

10-10. Investigate a space-charge wave propagating with the phase velocity $\omega/k > c$ from a coordinate frame traveling with the velocity $V = c^2 k/\omega$ in the direction of the wave propagation. Interpret your results on a Minkowsky diagram set up in the rest frame, plotting the Lorentz-invariant phase = const lines.

10-11. Prove that (10-51) can be satisfied with solutions varying as $\sin k_x x$ or $\cos k_x x$, and show that this leads to (10-53) as a condition for ω.

10-12. Prove that the frequency of the surface modes in a nonmagnetized plasma slab never exceeds $\omega_p/\sqrt{2}$.

10-13. Derive the dispersion relation for a circular plasma cylinder in the absence of a background magnetic field.

10-14. Show that the zero-field dispersion relation (10-67) yields the same radiating oscillations for $k_z = 0$ as (10-31), which describe oscillations of a plasma slab with a large background magnetic field.

10-15. Solve (10-69) subject to the boundary conditions that $\mathbf{E} = 0$ at
(a) the boundaries of a slab of thickness t.
(b) at the surface of a cylinder with radius R.

10-16. Derive (10-76).

10-17. Derive (10-77).

10-18. Find plane-wave solutions to (10-75) and (10-77) by setting $\nabla_\perp = i\mathbf{k}_\perp$ and prove that the resulting dispersion relation is identical with (8-19).

10-19. Prove that the cyclotron cutoffs (8-48) and (8-49) yield (10-87) if one neglects the ion motion.

10-20. Prove from (10-103a) and (10-103b) that $\nabla \cdot \mathbf{w} = 0$, hence that the guiding-center fluid under investigation behaves like an incompressible fluid.

10-21. Investigate the behavior of a sharp-boundary, low-β guiding-center plasma with respect to flutes when the gravitational field is directed into the plasma. What can you say about flute stability of a cusp-type geometry?

10-22. Demonstrate that in the absence of a density gradient, the dispersion relation (10-116) describes dispersionless waves traveling with the ion drift velocity.

10-23. Demonstrate on the basis of Fig. 10-13 that the average guiding-center drift ($\mathbf{w}^E$) for a particle with finite gyroradius never exceeds the drift velocity associated with the guiding-center field.

10-24. Prove that $f = F(x + (u_y/\Omega),\ y - (u_x/\Omega),\ u_\perp)$ in (10-126) is an exact solution of the collisionless Boltzmann equation, for a plasma in a uniform magnetic field.

10-25. Evaluate the growth rate of a drift wave with a density and temperature gradient in the x direction.

10-26. Show that by maximizing (10-159) with respect to $k_\perp^2$ one finds that $(k_\perp c_s/\Omega_i)^2 = 2$ gives the $k_\perp$ that produces the fastest growth.

10-27. Show that for the negative frequency drift mode $\mathrm{Im}\ \varepsilon < 0$ and $\partial\varepsilon/\partial\omega < 0$, so the wave is damped.

10-28. Show that the drift wave growth rate, when calculated from the energy conservation equation (10-165) with the power exchange given by (10-163) and (10-164), agrees with the one obtained from the dispersion relation.

10-29. Derive (10-166) and find $F(x)$ in (10-167).

10-30. Demonstrate that the Lagrangian (10-171) leads to the correct equation of motion for a particle of mass m, charge q, and magnetic dipole moment μ_m.

10-31. Prove that in a minimum-B geometry the maximum energy available for instability is

$$\int K_{\mu,J}(f - f_0)\, d\alpha\, d\beta\, d\mu_m\, dJ$$

where $f = f[K_{\mu,J}(\alpha,\beta)\mu_m,J]$ is the distribution function and f_0 is obtained from f by area-preserving transformations on the α,β planes, and $\partial f_0/\partial K \leq 0$.

XI

Collisions in Plasmas

11-1. Collisions

So far we have neglected the effects of interparticle collisions in our studies of plasma properties. While this is a sensible approximation in many important cases, we have yet to show that this is true and to investigate the role of collision phenomena in plasma physics. Collisions can be expected to bring about effects such as thermalization, resistivity, diffusion, etc., not investigated in our previous studies.

In fact we have yet to say what we mean by collisions among plasma particles. The collision between two billiard balls is fairly well defined. While not in touch, the small (gravitational) interaction can be neglected, but as they are pushed against each other a strong repulsive force comes into action, which ceases to exist as soon as the balls are detached from each other. Charged particles, however, interact through a Coulomb force with an infinite range. The collision between two electrons consists of two point charges moving on hyperbolic paths in each other's electric fields.

One may argue at first that our self-consistent treatment, in which each particle moves in the field created by all the others, has already taken care of this effect. This would really be the case if we were in the position of locating each particle all the time and calculating the resulting fields *exactly*. Instead of carrying out this impossible task, we were satisfied with computing *macroscopic* charge and current densities, by simply *counting* particles and adding velocity vectors in relatively big volume elements containing a large number of particles.

This is essentially the same procedure one adopts in classical electromagnetic theory. In calculating the electrostatic field inside a parallel-plate capacitor, for example, one does not know where individual electrons and ions are located in the plates, and in describing the motion of a charged particle between these plates (Fig. 11-1a) no one would think of carrying out the cumbersome procedure of calculating the "collisions" with each individual charged particle on the electrodes. On the other hand, if our charged particle passes through one of the electrodes, its motion will be affected, in addition to the "smooth" fields produced by macroscopic charge densities, by "collisions" with individual electrons and ions as well (Fig. 11-1b).

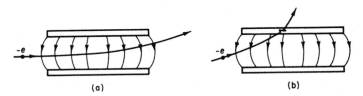

FIG. 11-1. Deflection of an electron in the (a) macroscopic, (b) microscopic, field of charged particles.

It seems that if our particle, to be called the *test particle*, is far away from those that produce the field, the *field particles* as we shall call them, the macroscopic description is sufficient, while for close encounters the individual particle nature of the field particle comes into play. What is close and what is far in this context is a question which cannot be easily answered in general. In the special case of a plasma, in or "close to" thermal equilibrium, an intuitive answer can be given. We have seen that the Coulomb field produced by a particle is shielded out by others at distances much larger than the Debye length, while it is unaffected at a distance much smaller than that length. Hence one may say roughly that a test particle passing through such a plasma "sees" individual particles that are closer than the Debye radius and notices all the others only through their smooth macroscopic fields.

Another interesting question may be asked (and answered) as follows: Under what conditions would the collisionless theory exactly hold? Think for a moment that electrons and ions are divisible. Take the plasma under consideration, divide each particle into equal parts, and distribute the parts uniformly about the original particle position in such a way that the resulting charge density varies smoothly between the original particle positions. It is easy to see that this new plasma is identical with the original one in all its macroscopic properties. Charge, mass, and current densities are obviously the same, and so are the macroscopic electromagnetic fields. Furthermore, all results derived on the basis of the collisionless theory still hold, since Maxwell's equation as well as the Vlasov equation are unaffected by the process. q/m remains unchanged, and the new f function differs from the old one only by an unimportant normalization factor. Since qn also remains unchanged, basic plasma characteristics like the Debye length, plasma frequency, dielectric tensor, β, and the particle cyclotron radii remain invariant

(Exercise 11-1). At the same time the microscopic properties of the plasma lose much of their significance. The smoother the charge distribution, the better the description of the actual fields given by the macroscopic plasma characteristics. The acceleration of a test particle due to macroscopic fields $F_{macr}/m = q/m(\mathbf{E} + \mathbf{v} \times \mathbf{B})$ remains unchanged, while the acceleration due to a single field particle (collision) decreases as q, ($F_{coll}/m \approx q^2/m \sim q$). "Liquefying" the plasma, by going to the limit $q \to 0$, $m \to 0$ (q/m, and qn unchanged) one arrives at a collisionless plasma that retains all the macroscopic properties of the real one. One usually describes this process by the dimensionless parameter $g = (n\lambda_D{}^3)^{-1}$, roughly the reciprocal value of the number of particles in the Debye sphere. If $g \to 0$, one expects to arrive at the collisionless theory, while small but finite values of g ought to yield collisional correction terms.

The rigorous but rather lengthy proof of these statements can be obtained with the help of the methods of statistical mechanics. We shall only sketch the main lines of reasoning.

The correct way of treating a many-body system is to solve simultaneously the equations of motion of all particles as they move in the fields of all the others. For a plasma with a large number of particles, this is obviously a hopeless task, even with the fastest computers. In addition, the solution requires the knowledge of the initial conditions for all particles, in fact if radiation phenomena are included, their past history as well. Fortunately, however, just as we do not know (and do not care) what the initial positions and momenta of all particles are, we are not interested in their present status either. It would be entirely sufficient if one could find out the average number of particles one can expect to find at a certain time in small space-velocity space intervals.

To this end one constructs (mentally) a so-called *ensemble of plasmas*. This consists of a large number of macroscopically identical plasmas at $t = 0$, which are, however, quite different in the values of coordinates and momenta of individual particles. Assume, for example, that one knows the particle, momentum, and energy density of a plasma. Clearly there is an infinite number of ways one can arrange individual particles and their momenta to give the same macroscopic data. All these arrangements are members of the ensemble. The construction of an ensemble, corresponding to a given macrosystem, is not a simple matter. Care must be taken that all microarrangements be represented with the proper weight. As a trivial example, one may mention the interchange of electron i with electron j. For every configuration in the ensemble, there must be one (and not more) system which differs from the first one only in the interchange of electrons i and j. In general, statistical mechanics provides methods for constructing ensembles for macrosystems with the "least bias."

The initially macroscopically identical microsystems develop in their own way in time, leading to macroscopically different states at a time t. Since we have no way of knowing which member of the ensemble corresponded to our real laboratory plasma at $t = 0$, we cannot tell with certainty which macroscopic state will describe it at t either. It is reasonable to assume, on the other hand, that if one finds many members of the ensemble in the neighborhood of a macrostate, then this state is more *probable* than one compatible only with few members of the ensemble. Thus one is led to a probabilistic interpretation of the ensemble.

For the mathematical treatment of the problem, it is convenient to introduce a $6N$-dimensional coordinate system, called Γ space, where N is the number of particles. In this phase space each particle is allotted six coordinates for the three positions and three velocity components. Each member of the ensemble is represented by one point in Γ space, whose motion describes the *exact* development of the many-body system. The ensemble itself is represented by a distribution of phase points characterized by the density function $D(\mathbf{r}_1, \mathbf{v}_1, ..., \mathbf{r}_N, \mathbf{v}_N, t)$, whose value at $t = 0$ is to be chosen so as to describe the macroscopic system most properly. It is easy to see that D satisfies Liouville's equation (Exercise 11-2)

$$\frac{\partial}{\partial t} D + \sum_1^N \mathbf{v}_i \frac{\partial}{\partial \mathbf{r}_i} D + \sum_1^N \frac{q}{m} (\mathbf{E}_i + \mathbf{v}_i \times \mathbf{B}_i) \cdot \frac{\partial D}{\partial \mathbf{v}_i} = 0 \qquad (11\text{-}1)$$

where $\mathbf{E}_i$ and $\mathbf{B}_i$ are the fields produced by all particles except the ith, and are very complicated functions of the positions and momenta of all particles. Although one notices the formal similarity between (11-1) and the collisionless Boltzmann equation, the two are clearly very different in content. The solution of Liouville's equation would be equivalent to the exact solution of an infinite number of many-body problems. (If D is chosen a δ function at $t = 0$, its time development describes the exact motion of a plasma.)

In accordance with our previous arguments, one ascribes a statistical interpretation to D. $D\, d\mathbf{r}_1\, d\mathbf{v}_1 \cdots d\mathbf{r}_N\, d\mathbf{v}_N$ is the probability that at time t particle 1 will be found in the volume element $d\mathbf{r}_1$ velocity range $d\mathbf{v}_1$, particle 2 in volume element $d\mathbf{r}_2$ velocity range $d\mathbf{v}_2$, etc. We cannot solve (11-1), and one finds again that its solution would provide us with too much unnecessary information in any case. It would be entirely satisfactory if we knew the probability of finding any one of the particles in $d\mathbf{r}\, d\mathbf{v}$ (taking for simplicity only one particle species, say electrons). Since $\int D\, d\mathbf{r}_2\, d\mathbf{v}_2 \cdots d\mathbf{r}_N\, d\mathbf{v}_N$ represents the probability of finding the first electron around a given point in phase space regardless of all the others, and due to the aforementioned symmetry of D with respect to the interchange of particles,

$$Nf^{(1)}(\mathbf{r}, \mathbf{v}, t) = N \int D\, d\mathbf{r}_2\, d\mathbf{v}_2 \cdots d\mathbf{r}_N\, d\mathbf{v}_N \qquad (11\text{-}2)$$

expresses the probability of finding any one of the electrons in a unit phase-space volume about $\mathbf{r}, \mathbf{v}$ at time t. $f^{(1)}$ is called the one-particle distribution function and the index 1 has been dropped from $\mathbf{r}_1, \mathbf{v}_1$.

This suggests seeking an equation which describes f_1 instead of D. One may hope to find this by integrating (11-1) over all particles except one. In the last Σ, however, $\mathbf{E}_i$ and $\mathbf{B}_i$ depend on all particle coordinates, hence one winds up with

$$\frac{\partial}{\partial t} f^{(1)} + \mathbf{v} \cdot \frac{\partial}{\partial \mathbf{r}} f^{(1)} + \frac{q(N-1)}{m} \int (\mathbf{E}_{12} + \mathbf{v} \times \mathbf{B}_{12})$$

$$\cdot \frac{\partial}{\partial \mathbf{v}} f^{(2)}(\mathbf{r}, \mathbf{v}, \mathbf{r}_2, \mathbf{v}_2) \, d\mathbf{r}_2 \, d\mathbf{v}_2 = 0 \qquad (11\text{-}3)$$

where $\mathbf{E}_{12}$ and $\mathbf{B}_{12}$ are the fields produced by particle 2 at the position of particle 1.

The vanishing of D for large values of $\mathbf{r}_i, \mathbf{v}_i$ has been used in the integration as well as the symmetry of D in particle coordinates. The two-particle distribution function $f^{(2)} = \int D \, d\mathbf{r}_1 \cdots d\mathbf{v}_N$ expresses the joint probability of finding one particle about $\mathbf{r}, \mathbf{v}$, while another one is about $\mathbf{r}_2, \mathbf{v}_2$. Equation (11-3) can only be solved for $f^{(1)}$ if $f^{(2)}$ is known. If one tries to find an equation for $f^{(2)}$ by integrating (11-1) over all particles except two, $f^{(3)}$ enters, and so on. One arrives at a hierarchy of equations where $f^{(s)}$ depends on $f^{(s+1)}$. This chain can be broken by going to the limit $g \to 0$, where $f^{(1)}$ may be shown to obey the Vlasov equation while $f^{(2)} = f^{(1)}(\mathbf{r}, \mathbf{v}) \cdot f^{(1)}(\mathbf{r}_2, \mathbf{v}_2)$.

Even more significantly, a series expansion in g can be carried out and the correction to $f^{(1)}$ from the linear term in g obtained. The resulting equation yields the Vlasov equation on the left side, while the new terms arising on the right side may be identified with $\partial f/\partial t)_{\text{coll}}$ in the small g approximation. They are known as the *Fokker-Planck terms* and the equation the Fokker-Planck equation. Instead of following this cumbersome procedure, we simply state that it can be done and derive the Fokker-Planck equation via a more heuristic argument.

Before closing this section a remark is in order regarding the interpretation of the distribution functions f and $f^{(1)}$. We defined f as the actual number density of phase points, and $f^{(1)}$ was the probability of finding a phase point in the unit volume element in phase space. For a large number of particles $(n \to \infty)$, the number of particles found in a phase-volume element becomes exactly proportional to the probability of finding a particle there. Hence not unexpectedly, f and $f^{(1)}$ satisfy the same equation. In this chapter the two interpretations of f will be used alternately, and the index (1) will be dropped.

11-2. The Fokker-Planck Equation

The Coulomb collision between two charged particles a and b can be conveniently described in the center-of-mass system where the particles move with the relative velocity u. The path is hyperbolic, and introducing the reduced mass

$$M = \frac{m_a m_b}{m_a + m_b} \tag{11-4}$$

one may show that

$$\cot \frac{\theta}{2} = \tan \psi = 4\pi\varepsilon_0 \frac{M p u^2}{q_a q_b} \tag{11-5}$$

The angles ψ and θ, and the impact parameter p are defined in Fig. 11-2. Since

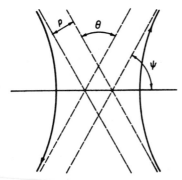

Fig. 11-2. Coulomb scattering.

our knowledge of the impact parameters in a plasma is only statistical, one introduces the concept of differential collision cross section:

$$d\sigma = |2p\pi\, dp| \tag{11-6}$$

From (11-5),

$$dp = -\frac{q_a q_b}{4\pi\varepsilon_0 M u^2} \frac{d\theta}{2 \sin^2(\theta/2)} \tag{11-7}$$

and

$$d\sigma = 2\pi \left(\frac{q_a q_b}{4\pi\varepsilon_0 M u^2}\right)^2 \frac{d\theta}{2 \sin^2(\theta/2) \tan(\theta/2)}$$

$$= \left(\frac{q_a^2 q_b^2}{4\pi\varepsilon_0 M u^2}\right)^2 \frac{2\pi \sin\theta\, d\theta}{4 \sin(\theta/2) \cos(\theta/2) \sin^2(\theta/2) \tan(\theta/2)} \tag{11-8}$$

If one introduces the differential solid angle

$$d\Omega = 2\pi \sin\theta\, d\theta \tag{11-9}$$

(11-8) can be written

$$d\sigma = \left(\frac{q_a q_b}{8\pi\varepsilon_0 Mu^2}\right)^2 \frac{d\Omega}{\sin^4(\theta/2)} = \sigma \, d\Omega \qquad (11\text{-}10)$$

If the initial conditions of the colliding particles (impact parameter) are unknown, then the probability that the deflection angle will fall between θ and $(\theta + d\theta)$ is proportional to (11-8) and scattering into the differential solid angle $d\Omega$ proportional to (11-10). One can see immediately that small-angle scattering is a far more probable process than large-angle deflection.

It is useful to define the impact parameter p_0 where the deflection is $\theta = \pi/2$. From (11-5),

$$p_0 = \frac{1}{4\pi\varepsilon_0} \frac{|q_a q_b|}{Mu^2} \qquad (11\text{-}11)$$

One may say roughly that for $p < p_0$ one obtains large-angle deflections, while $p_0 < p < \lambda_D$ yields small-angle deflections. If $p > \lambda_D$, the two particles "do not see" each other's Coulomb field any longer except for the smoothed-out many-particle fields described by the Vlasov equation. The ratio of large- to small-angle collisions goes as

$$\frac{p_0^2}{\lambda_D^2 - p_0^2} \approx \left(\frac{p_0}{\lambda_D}\right)^2 \ll 1 \qquad (11\text{-}12)$$

For typical particle energies, 10 eV or higher, p_0 is of the order of 10^{-10} m or less, much smaller than λ_D under usual laboratory conditions. One concludes that the overwhelming majority of particle encounters in a plasma lead to small deflections. A computational advantage follows immediately. If a particle suffers two collisions simultaneously, the deflection is $\theta \approx \theta_1 + \theta_2$. Since each particle "collides" simultaneously with all other particles within a Debye distance, the additive nature of scattering angles proves very useful.

Let $P(\mathbf{v}; \Delta\mathbf{v})$ be the probability that a particle changes its velocity coordinate from $\mathbf{v}$ to $\mathbf{v} + \Delta\mathbf{v}$, in the time interval Δt, due to multiple small-angle collisions. If only collisions are effective in changing the (spatially uniform) distribution function, its value at time t is given by

$$f(\mathbf{v},t) = \int f(\mathbf{v} - \Delta\mathbf{v}; t - \Delta t) P(\mathbf{v} - \Delta\mathbf{v}; \Delta\mathbf{v}) \, d^3 \Delta v \qquad (11\text{-}13)$$

Since for small-angle Coulomb collisions $\Delta\mathbf{v}$ is small if Δt is, one may expand (9-13) in the form

$$f(\mathbf{v},t) = \int \left\{ f(\mathbf{v},t)P(\mathbf{v}; \Delta\mathbf{v}) - \Delta t \frac{\partial f}{\partial t} P(\mathbf{v}; \Delta\mathbf{v}) - \Delta\mathbf{v} \frac{\partial}{\partial \mathbf{v}} [f(\mathbf{v},t)P(\mathbf{v}; \Delta\mathbf{v})] \right.$$
$$\left. + \tfrac{1}{2} \Delta v_i \, \Delta v_k \frac{\partial^2}{\partial v_i \, \partial v_k} [f(\mathbf{v},t)P(\mathbf{v}; \Delta\mathbf{v})] + \cdots \right\} d^3 \Delta v \qquad (11\text{-}14)$$

Noting that the probability that some transition takes place is one,

$$\int P(\mathbf{v}; \Delta \mathbf{v}) \, d^3 \, \Delta v = 1 \tag{11-15}$$

and defining the average velocity change per unit time as

$$\langle \Delta \mathbf{v} \rangle = \frac{1}{\Delta t} \int P(\mathbf{v}; \Delta \mathbf{v}) \, \Delta \mathbf{v} \, d^3 \, \Delta v \tag{11-16}$$

and

$$\langle \Delta v_i \, \Delta v_k \rangle = \frac{1}{\Delta t} \int P(\mathbf{v}; \Delta \mathbf{v}) \, \Delta v_i \, \Delta v_k \, d^3 \Delta v \tag{11-17}$$

then by neglecting higher-order terms, (11-14) yields

$$\left. \frac{\partial f}{\partial t} \right)_{\text{coll}} = - \frac{\partial}{\partial v_i} [\langle \Delta v_i \rangle f(\mathbf{v}, t)] + \frac{1}{2} \frac{\partial^2}{\partial v_i \, \partial v_k} [\langle \Delta v_i \, \Delta v_k \rangle f(\mathbf{v}, t)] \tag{11-18}$$

This is known as the *Fokker-Planck equation*. It is noteworthy that (11-18) may be cast in the form of an equation of continuity,

$$\left. \frac{\partial f}{\partial t} \right)_{\text{coll}} + \nabla_v \cdot \mathbf{g} = 0 \tag{11-19}$$

where ∇_v is the del operator in velocity space and

$$\mathbf{g} = \langle \Delta \mathbf{v} \rangle f - \tfrac{1}{2} \nabla_v \cdot [\langle \Delta \mathbf{v} \, \Delta \mathbf{v} \rangle f] \tag{11-20}$$

The effect of multiple small-angle collisions may be thought of as causing a continuous flow of phase points in velocity space, described by the flow vector $\mathbf{g}$.

To understand the physical significance of $\langle \Delta \mathbf{v} \rangle$ and $\langle \Delta \mathbf{v} \, \Delta \mathbf{v} \rangle$ consider the behavior of a stream of test particles with velocity $\mathbf{v}$ injected into a plasma. A unit time later they have slowed down as a result of collisions (in the average) to approach the average field particle velocity. This average change in velocity vector is $\langle \Delta \mathbf{v} \rangle$, and the quantity $m\langle \Delta \mathbf{v} \rangle$, which has the dimension of force, is named *dynamical friction*. In Fig. 11-3 we have sketched snapshots taken of

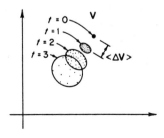

FIG. 11-3. Positions in velocity space of a stream of test particles at $t = 0, 1, 2, 3$ seconds.

the phase points in unit time intervals of test particles injected into a plasma in thermal equilibrium. The spreading out of the points is characterized by the diffusion term $\langle \Delta v \, \Delta v \rangle$.

The Fokker-Planck equation in the form given in (11-18) holds for all multiple collision processes which introduce only small changes in the velocity vectors of participating particles. To apply it to a plasma we have to evaluate $\langle \Delta v \rangle$ and $\langle \Delta v \, \Delta v \rangle$ for the special case of Coulomb collisions.

The probability that a particle moving with relative velocity u with respect to field particles of density $f_f(v) \, d^3v$ suffers a collision into the solid angle $d\Omega$ per unit time is

$$u\sigma(\Omega, u) f_f(v) \, d^3v \, d\Omega \tag{11-21}$$

Averages can be computed by adding up all collision processes (per unit time) with all field particles, leading to deflection. Hence

$$\langle \Delta v_i \rangle = \iint \Delta v_i \, u\sigma(\Omega, u) f_f(v) \, d^3v \, d\Omega \tag{11-22}$$

and

$$\langle \Delta v_i \, \Delta v_k \rangle = \iint \Delta v_i \, \Delta v_k \, u\sigma(\Omega, u) f_f(v) \, d^3v \, d\Omega \tag{11-23}$$

The collision of two particles can be viewed conveniently in velocity space (Fig. 11-4a). The center-of-mass velocity V remains unchanged, and the new

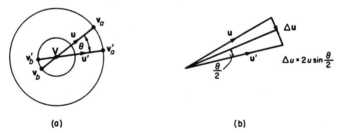

(a)   (b)

FIG. 11-4. Collision of two particles viewed in velocity space.

relative velocity vector u' subtends the angle θ with the old one. Particle velocities are v_a, v_b and v_a', v_b' before and after collision, respectively. Other possible loci of v_a', v_b' (depending on exact initial conditions) are located on the surfaces of concentric spheres centered around the center-of-mass velocity. The magnitude of relative velocities remains unchanged, $|u| = |u'|$.

One may easily compute the change in relative velocity vectors in a Cartesian coordinate system where one of the axes is directed along u. In this frame (see Fig. 11-4b)

$$\Delta u_1 = -2u \sin^2(\theta/2) \tag{11-24}$$

$$\Delta u_2 = 2u \sin(\theta/2) \cos(\theta/2) \cos \phi \tag{11-25}$$

and

$$\Delta u_3 = 2u \sin(\theta/2) \cos(\theta/2) \sin \phi \tag{11-26}$$

where ϕ is the polar angle. With the help of σ, as given in (11-10), the integration over the solid angle in (11-22) can be carried out:

$$\{\Delta u_1\} = \int_\Omega \Delta u_1 \, u\sigma \, d\Omega = -\left(\frac{q_a q_b}{8\pi\varepsilon_0 M u^2}\right)^2 4u^2\pi \int \frac{\sin^2(\theta/2) \sin \theta}{\sin^4(\theta/2)} \, d\theta$$

$$= -\left(\frac{q_a q_b}{2\varepsilon_0 M u}\right)^2 \frac{1}{\pi} \int_{\theta_{min}}^{\theta_{max}} \frac{\cos(\theta/2) \, d(\theta/2)}{\sin(\theta/2)} \tag{11-27}$$

while

$$\{\Delta u_2\} = \{\Delta u_3\} = 0 \tag{11-28}$$

by symmetry. The integral in (11-27) yields

$$\ln \sin \frac{\theta}{2}\Big]_{\theta_{min}}^{\theta_{max}} \approx \ln \frac{\theta}{2}\Big]_{\theta_{min}}^{\theta_{max}} \approx -\ln \frac{\theta_{min}}{2} \tag{11-29}$$

since $1 \gg \theta_{max} \gg \theta_{min}$. Assuming that θ_{min} corresponds to a collision with the impact parameter $p = \lambda_D$, with the help of (11-5) one finds

$$\frac{2}{\theta_{min}} \approx \frac{1}{\tan(\theta_{min}/2)} = 4\pi\varepsilon_0 \frac{M\lambda_D u^2}{q_a q_b} = \frac{\lambda_D}{p_0} = \Lambda \tag{11-30}$$

and

$$\{\Delta u_1\} = -\frac{1}{\pi}\left(\frac{q_a q_b}{2\varepsilon_0 M u}\right)^2 \ln \Lambda \tag{11-31}$$

As mentioned before, the ratio Λ is usually a large number, and $\ln \Lambda$ is roughly about 10 for typical plasmas. Since a great deal of arbitrariness is involved in choosing λ_D as the cutoff impact parameter (why not, e.g., $2\lambda_D$?), it is reassuring to find that the result is logarithmically insensitive to this choice.

The vector $\{\Delta \mathbf{u}\}$ [see (11-31) and (11-28)] can be expressed in the laboratory frame as

$$\{\Delta \mathbf{u}\} = -\frac{1}{\pi}\left(\frac{q_a q_b}{2\varepsilon_0 M}\right)^2 \ln \Lambda \frac{\mathbf{u}}{u^3} \tag{11-32}$$

Since we need the change of test-particle velocity $\Delta \mathbf{v}_a$ (not $\Delta \mathbf{u}$) we note that

$$\mathbf{v}_a = \mathbf{V} + \frac{m_b}{m_a + m_b} \mathbf{u} \tag{11-33}$$

and

$$\Delta \mathbf{v}_a = \frac{m_b}{m_a + m_b} \Delta \mathbf{u} \tag{11-34}$$

where $\mathbf{V}$ is the center-of-mass velocity. Consequently

$$\{\Delta \mathbf{v}_a\} = \int \Delta \mathbf{v}_a \, u\sigma \, d\Omega = -\frac{q_a^2 q_b^2 \ln \Lambda}{4\pi\varepsilon_0^2 M m_a} \frac{\mathbf{u}}{u^3} \tag{11-35}$$

is the average rate of change of velocity of test particles a due to collisions with field particles of velocity $\mathbf{v}_b$. We have yet to integrate over all field particles in (11-22) to find

$$\langle \Delta \mathbf{v}_a \rangle = -\Gamma_a \frac{m_a}{M} \int f_f(\mathbf{v}_b) \frac{\mathbf{v}_a - \mathbf{v}_b}{|\mathbf{v}_a - \mathbf{v}_b|^3} \, d^3 v_b \tag{11-36}$$

where

$$\Gamma_a = \frac{q_a^2 q_b^2 \ln \Lambda}{4\pi\varepsilon_0^2 m_a^2} \tag{11-37}$$

There is a striking resemblance between (11-36) and the equation expressing the electrostatic field at the point $\mathbf{v}_a$ resulting from a space-charge distribution proportional to $f(\mathbf{v}_b)$. One may introduce the "potential function"

$$h(\mathbf{v}_a) = \frac{m_a}{M} \int \frac{f(\mathbf{v}_b)}{|\mathbf{v}_a - \mathbf{v}_b|} \, d^3 v_b \tag{11-38}$$

and write

$$\langle \Delta \mathbf{v}_a \rangle = \Gamma_a \, \partial h / \partial \mathbf{v}_a \tag{11-39}$$

If (11-36) and (11-39) hold, so must "Poisson's equation,"

$$\nabla_v^2 h = -4\pi \frac{m_a}{M} f_f(\mathbf{v}) \tag{11-40}$$

Since the methods of electrostatics are highly developed, this formal analogy proves very useful in calculating the dynamical friction due to various distributions of field particles.

The calculation of the diffusion tensor $\langle \Delta v_i \, \Delta v_k \rangle$ follows lines similar to those leading to $\langle \Delta \mathbf{v} \rangle$, with one additional assumption: All terms not containing $\int \theta^{-1} \, d\theta \approx \ln \Lambda$ will be considered small and neglected. This leads to considerable simplification and is justified insofar as the neglected terms are roughly an order of magnitude smaller than the others. Such is the case with $\{\Delta u_1 \, \Delta u_1\}$, while mixed terms vanish when integrating over ϕ. The only remaining terms are (see Exercise 11-4):

$$\{\Delta u_2 \, \Delta u_2\} = \{\Delta u_3 \, \Delta u_3\} = \frac{1}{\pi} \left(\frac{q_a q_b}{2\varepsilon_0 M}\right)^2 \ln \Lambda \frac{1}{u} \tag{11-41}$$

In the laboratory frame this tensor has the form

$$\{\Delta u_i \, \Delta u_k\} = \frac{1}{\pi} \left(\frac{q_a q_b}{2\varepsilon_0 M}\right)^2 \ln \Lambda \frac{1}{u} \left[\delta_{ik} - \frac{u_i u_k}{u^2}\right] \tag{11-42}$$

obviously a tensor quantity, whose only two nonvanishing components in the $e_1 \parallel u$ frame reduce to (11-41). Furthermore (dropping subscript a),

$$\{\Delta v_i \, \Delta v_k\} = \frac{q_a{}^2 q_b{}^2 \ln \Lambda}{4\pi\varepsilon_0{}^2 m_a{}^2 u} \left[\delta_{ik} - \frac{u_i u_k}{u^2} \right] \tag{11-43}$$

which can be cast with the help of

$$\frac{\partial^2}{\partial v_i \, \partial v_k} u = \frac{\partial}{\partial v_i} \frac{u_k}{u} = \frac{u\delta_{ik} - (u_i u_k / u)}{u^2} \tag{11-44}$$

in the form

$$\{\Delta v_i \, \Delta v_k\} = \frac{q_a{}^2 q_b{}^2 \ln \Lambda}{4\pi\varepsilon_0{}^2 m_a{}^2} \frac{\partial^2 u}{\partial v_i \, \partial v_k} \tag{11-45}$$

The integration over the field particles may now be carried out. One defines the "potential"

$$g(\mathbf{v}_a) = \int f(\mathbf{v}_b) |\mathbf{v}_a - \mathbf{v}_b| \, d^3 v_b \tag{11-46}$$

and finds

$$\langle \Delta v_i \, \Delta v_k \rangle = \Gamma_a \frac{\partial^2 g}{\partial v_i \, \partial v_k} \tag{11-47}$$

where again $\mathbf{v}$ means $\mathbf{v}_a$.

One may easily prove (Exercise 11-5) that h and g are related through

$$\frac{m_a}{M} \nabla_v{}^2 g = 2h \tag{11-48}$$

Finally one arrives at the Fokker-Planck equation for a plasma,

$$\frac{1}{\Gamma_a} \left(\frac{\partial f}{\partial t} \right)_{\text{coll}} = -\frac{\partial}{\partial v_i} \left[f \frac{\partial h}{\partial v_i} \right] + \frac{1}{2} \frac{\partial^2}{\partial v_i \, \partial v_k} \left(f \frac{\partial^2 g}{\partial v_i \, \partial v_k} \right) \tag{11-49}$$

Here $f(\mathbf{v})$ is the test-particle distribution, while h and g are integrals over the field-particle distribution.

If one wishes to follow the time development of a plasma due to collisions, one needs to solve four coupled Fokker-Planck equations simultaneously. One describes the effect of electrons colliding with electrons (both f_t and f_f are the same electron distribution function), another the change of the electron distribution due to collisions with ions, the third the ionic distribution function charging due to collisions with electrons, and the last the effect of ionic self-collisions. The solution of four coupled, highly nonlinear integro-differential equations is a formidable task, to be attacked only by computers.

The development of a spatially uniform plasma, with an isotropic, but monoenergetic initial velocity distribution, was followed by a computer. The system was found to relax to a Maxwell-Boltzmann distribution.

Since (11-49) was derived via various approximations, it is not clear *a priori* that it satisfies conservation of particles, momentum, and energy. That the first one is true follows immediately from the equation of continuity (11-19). Momentum conservation may be proved by multiplying (11-49) by $m\mathbf{v}$ and integrating by parts over velocity space (see Exercise 11-6). A similar calculation may be carried out to prove conservation of energy (Exercise 11-7).

11-3. Relaxation Times

With the help of the potentials h and g one can compute the quantities $\langle\Delta\mathbf{v}\rangle$ and $\langle\Delta\mathbf{v}\,\Delta\mathbf{v}\rangle$ for given field-particle distribution functions. We shall carry out this calculation for the important case where the field particles have a Maxwellian distribution

$$f_f(v_b) = \frac{n_b l_b^3}{\pi^{3/2}} \exp(-l_b^2 v_b^2) \tag{11-50}$$

where

$$l_b^2 = m_b/2kT \tag{11-51}$$

The spherical symmetry of f_f in velocity space (space-charge distribution in configuration space) leads to substantial simplifications. Clearly the potentials exhibit the same spherical symmetry, and may be written as $h(v_b)$ and $g(v_b)$; $\langle\Delta\mathbf{v}_a\rangle$ (the "field strength") is proportional to $\Delta_v h$, a vector pointing in the $\mathbf{v}_a$ direction.

It is convenient to express all quantities in terms of components parallel to the test-particle velocity vector $\mathbf{v}_a$ as parallel components, and components perpendicular to $\mathbf{v}_a$ as perpendicular components. In this coordinate system the dynamical friction term has only one component $\langle\Delta v_\parallel\rangle$. Not surprisingly, friction tends to slow down the test particles. The diffusion tensor becomes diagonal in this coordinate system. The gradient of g is a parallel vector, its derivative in the parallel direction parallel; the derivatives in the perpendicular direction yield perpendicular vectors (Fig. 11-5), parallel to the direction of

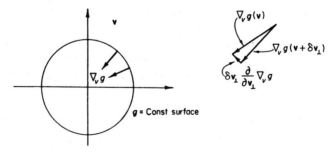

FIG. 11-5. Sketch to aid evaluation of $\langle\Delta v_i\,\Delta v_k\rangle$.

differentiation. The diffusion tensor therefore reduces to

$$\langle \Delta v_i \, \Delta v_k \rangle = \begin{pmatrix} \langle \Delta v_\parallel^2 \rangle & 0 & 0 \\ 0 & \tfrac{1}{2}\langle \Delta v_\perp^2 \rangle & 0 \\ 0 & 0 & \tfrac{1}{2}\langle \Delta v_\perp^2 \rangle \end{pmatrix} \tag{11-52}$$

where the sum of the two equal perpendicular terms has been denoted as $\langle \Delta v_\perp^2 \rangle$. Hence only three quantities need to be evaluated. Making use of the fact that the field strength produced by a spherically symmetric charge distribution may be calculated by concentrating the "inner" charges at the center and ignoring the rest, one finds, with the help of (11-36),

$$\langle \Delta v_\parallel \rangle = -\frac{m_a}{M}\,\Gamma_a \frac{1}{v^2} \int_0^v \frac{4\pi v_b^2 n_b l_b^3}{\pi^{3/2}} \exp(-l_b^2 v_b^2)\, dv_b, \tag{11-53}$$

The integral can be reduced, integrating by parts,

$$l_b \int_0^v v_b 2l_b^2 v_b \exp(-l_b^2 v_b^2)\, dv_b$$

$$= [-l_b v_b \exp(-l_b^2 v_b^2)]_0^v + \int_0^v l_b \exp(-l_b^2 v_b^2)\, dv_b \tag{11-54}$$

and introducing $y = l_b v_b$ and $x = l_b v$, to

$$\langle \Delta v_\parallel \rangle = -\frac{m_a}{M}\,\Gamma_a n_b \frac{2}{\pi^{1/2}} l_b^2 \frac{\displaystyle\int_0^x \exp(-y^2)\, dy - x(d/dx)\int_0^x \exp(-y^2)\, dy}{x^2} \tag{11-55}$$

Adopting the notation for the error function

$$\phi(x) = \frac{2}{\pi^{1/2}} \int_0^x \exp(-y^2)\, dy \tag{11-56}$$

defining

$$G(x) = \frac{\phi(x) - x\phi'(x)}{2x^2} \tag{11-57}$$

and introducing the "diffusion constant"

$$A_D = 2\Gamma_a n_b = \frac{q_a^2 q_b^2 n_b \ln \Lambda}{2\pi\varepsilon_0^2 m_a^2} \tag{11-58}$$

the dynamical friction becomes

$$\langle \Delta v_\parallel \rangle = -A_D l_b^2 \left(1 + \frac{m_a}{m_b}\right) G(l_b v) \tag{11-59}$$

To compute the components of the diffusion tensor we note first that

$$h = n_b l_b \left(1 + \frac{m_a}{m_b}\right) \frac{\phi(l_b v)}{l_b v} \tag{11-60}$$

since $\Gamma_a \, \partial h / \partial v$ yields (11-59). From (11-48), applying Gauss' theorem,

$$\frac{\partial g}{\partial v} = 2 \frac{M}{m_a} \frac{1}{v^2} \int_0^v h v_b^2 \, dv_b \tag{11-61}$$

One may now easily arrive at the coefficients (see Exercise 11-8)

$$\langle \Delta v_{\|}^2 \rangle = \frac{A_D}{v} G(l_b v) \tag{11-62}$$

and

$$\langle \Delta v_{\perp}^2 \rangle = \frac{A_D}{v} [\phi(l_b v) - G(l_b v)] \tag{11-63}$$

We investigate now the asymptotic limits of these expressions:

As $x \to 0$,

$$G(x) \to \frac{\phi'(0) - x\phi''(0) - \phi'(0)}{4x} = -\frac{\phi''(0)}{4} = 0 \tag{11-64}$$

$$\frac{G(x)}{x} \to -\frac{x\phi''(0)}{6x^2} = \frac{2}{3\sqrt{\pi}} \tag{11-65}$$

$$\frac{\phi(x)}{x} \to \phi'(0) = \frac{2}{\sqrt{\pi}} \tag{11-66}$$

and

$$\frac{\phi(x) - G(x)}{x} \to \frac{4}{3\sqrt{\pi}} \tag{11-67}$$

As $x \to \infty$:

$$\phi(x) \to 1 \tag{11-68}$$

and

$$G(x) \to \frac{1}{2x^2} \tag{11-69}$$

It follows from (11-64) that test particles of zero velocity suffer no dynamical friction, as expected. The diffusion coefficients do not vanish for such particles. From (11-65) and (11-67) $\langle \Delta v_{\perp}^2 \rangle = 2 \langle \Delta v_{\|}^2 \rangle$, which is reasonable, since for the zero-energy test particles all directions in velocity space are equivalent.

For very fast test particles (compared to the characteristic field-particle velocity), the dynamical friction decreases as $1/v^2$, the parallel diffusion coefficient even faster, $\langle \Delta v_\parallel^2 \rangle \sim 1/v^3$, while $\langle \Delta v_\perp^2 \rangle \sim 1/v$. Consequently very fast particles diffuse mostly sideways. If one increases the test particle mass m_a, keeping everything else the same, the dynamical friction tends to dominate, although all coefficients diminish, since $A_D \sim 1/m_a^2$. For example, for an ion moving in an electron gas, the dynamical friction is the dominant effect.

Although deflections are small and at random, one may introduce, following Spitzer, several relaxation times. The slowing-down time is defined as

$$t_s = -\frac{v}{\langle \Delta v_\parallel \rangle} = \frac{v}{[1 + (m_a/m_b)]A_D l_b^2 G(l_b v)} \tag{11-70}$$

For small velocities, t_s is nearly a constant, while for large velocities $t_s \sim v^3$ [see (11-65) and (11-69)].

The deflection time is defined as

$$t_D = \frac{v^2}{\langle \Delta v_\perp^2 \rangle} = \frac{v^3}{A_D[\phi(l_b v) - G(l_b v)]} \tag{11-71}$$

while for the energy-exchange time one sets

$$t_E = \frac{E^2}{\langle \Delta E^2 \rangle} \tag{11-72}$$

where in a single encounter

$$\Delta E = \frac{m}{2}[(v + \Delta v_\parallel)^2 + \Delta v_\perp^2] - \tfrac{1}{2}mv^2$$

$$= \frac{m}{2}[2v\,\Delta v_\parallel + \Delta v_\parallel^2 + \Delta v_\perp^2] \approx mv\,\Delta v_\parallel \tag{11-73}$$

Consequently,

$$t_E = \frac{m^2 v^4}{4m^2 v^2 \langle \Delta v_\parallel^2 \rangle} = \frac{v^3}{4A_D G(l_b v)} \tag{11-74}$$

For large velocities $t_D/t_E \sim 1/v^2$; consequently deflections dominate over energy exchange.

Now consider the fate of an average particle in a thermal distribution as it interacts with other particles of its own species. If $l_b v \approx 1.3$, i.e., the chosen particle travels with roughly the mean thermal speed, $\phi - G \approx 4G$ (see Fig. 11-6) and $t_D \approx t_E$. Thus such a particle exchanges energy with its fellow particles and is deflected by them at about the same rate. This characteristic time, the "self-collision time" $t_c \approx t_D \approx t_E$, is an important plasma characteristic. Expressing v in terms of temperature and evaluating the constants, its value is roughly

$$t_c \approx \frac{v^3}{0.7A_D} \approx \frac{T^{3/2}}{Z^4 n}\left(\frac{m}{m_p}\right)^{1/2} \times 10^6 \tag{11-75}$$

where Z is the charge number, m_p the proton mass, and T is given in degrees Kelvin. If a plasma deviates from thermal equilibrium, t_c marks the time scale required to regain it by collision. If the plasma experiences a process

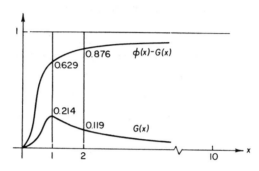

FIG. 11-6. $G(x)$ and $\phi(x) - G(x)$.

(instability, compression, etc.) with a time scale slow compared to t_c, it may be considered as collision-dominated, and it can be hoped that a hydromagnetic fluid model gives a good description of the process. However, if the time scale is much shorter than t_c, one should rather use the collisionless particle description.

It can be seen that the self-collision time increases strongly with temperature, and it is larger for ions than for electrons. For a typical fusion experiment, where the plasma is heated by rapid magnetic compression, the electrons are collision-dominated and behave like a hydromagnetic fluid, while a typical ion suffers only a few (or none) full-scale deflections during the process. This is one reason for the difficulties in finding an appropriate model for the description of such processes. The reader is advised to calculate the value of t_c for several plasma parameters, as given in Exercise 11-9.

After the thermalization of the electronic and ionic components of a plasma, the two species may wind up with a different temperature. Energy exchange by the two species through collisions is relatively slow, because of the large mass ratio. Assume that the temperatures of the two species are T_a and T_b. A particle of species a changes its energy per unit time, as a result of collisions with the other species according to (11-73) by

$$\langle \Delta E_a \rangle = \frac{m_a}{2} [2v_a \langle \Delta v_\parallel \rangle + \langle \Delta v_\parallel^2 \rangle + \langle \Delta v_\perp^2 \rangle]$$

$$= \frac{m_a}{2} \left[-2 A_D l_b^2 \left(1 + \frac{m_a}{m_b} \right) v_a G(l_b v_a) + \frac{A_D}{v_a} \phi(l_b v_a) \right] \quad (11\text{-}76)$$

where (11-59), (11-62), and (11-63) have been used. The rate of change of energy of species a is clearly

$$\int \langle \Delta E_a \rangle \frac{n_a l_a^3}{\pi^{3/2}} \exp(-l_a^2 v_a^2) 4\pi v^2 \, dv = \tfrac{3}{2} k n_a \frac{dT_a}{dt} \tag{11-77}$$

The integrals on the left side are straightforward but somewhat laborious and are left for Exercise 11-10. The result may be put in the form

$$\frac{dT_a}{dt} = \frac{T_b - T_a}{t_{eq}} \tag{11-78}$$

where

$$t_{eq} = \frac{2^{1/2} m_a m_b \pi^{3/2} k^{3/2} \varepsilon_0^2}{q_a^2 q_b^2 n_b \ln \Lambda} \left(\frac{T_a}{m_a} + \frac{T_b}{m_b} \right)^{3/2} \tag{11-79}$$

or evaluating the constants,

$$t_{eq} = 5.87 \frac{\lambda_a \lambda_b}{n_b Z_a^2 Z_b^2 \ln \Lambda} \left(\frac{T_a}{\lambda_a} + \frac{T_b}{\lambda_b} \right)^{3/2} \times 10^6 \tag{11-80}$$

where $\lambda_a = m_a/m_{\text{proton}}$; $\lambda_b = m_b/m_{\text{proton}}$. Comparison with (11-75) indicates that equipartition time is roughly $(m_+/m_-)^{1/2}$ times larger than the ion self-collision time and m_+/m_- times that of the electron self-collision time.

A word of caution is in order concerning all estimates based on collisional processes. As we have seen previously, a plasma not in thermal equilibrium is very often susceptible to (collisionless) instabilities and seeks to establish thermal equilibrium through a collisionless collective process. The time scale of an electrostatic (e.g., two-stream) instability is $1/\omega_p$, which is usually much shorter than the time needed for collisional interactions to be effective. This is the reason why many experiments show the plasma running away from a desired configuration (e.g., confinement) in the direction of thermal equilibrium, much faster than expected on the basis of collisional estimates.

Finally we establish the order of characteristic lengths in a plasma. In Exercise 11-11 we consider the mean interparticle distance d, the Debye length λ_D, the impact parameter for large deflections p_0, the mean free path due to multiple small-angle scatterings l, and the mean free path due to a single large-angle scattering l_σ. For most plasma parameters of interest one finds:

1. $\lambda_D \gg d$, an assumption we used all along.
2. $p_0 \ll d$, a sign that large-angle deflections are rare.
3. l is comparable to or bigger than the size of a laboratory plasma, and usually largely exceeds the length of the gyration orbit in a typical magnetic field.
4. $l_\sigma \gg l$, which justifies our neglect of large-angle collisions.

11-4. Transport Phenomena

Consider first the conductivity of a magnetic field-free plasma near thermal equilibrium. A small static electric field accelerates electrons and ions in opposite directions until an average transport velocity is reached, limited by the dynamical friction between the two species. This velocity can be easily calculated if one assumes that the electron and ion distributions remain Maxwellian in the co-moving frames, a good approximation for small fields. Another simplification may be obtained if one considers that the ionic thermal speed is much smaller than that of the electrons, hence for drift velocities much greater than the ion thermal velocity the ion distribution function may be represented by a δ function. An observer moving with the electron drift velocity observes a stationary thermal electron gas and ions drifting by with the velocity v_D. In equilibrium, the electric, and dynamical friction forces cancel, hence

$$qE = -m_a\langle\Delta v_\|\rangle = A_D l_b{}^2\left(1 + \frac{m_a}{m_b}\right)m_a G(l_b v_D) \qquad (11\text{-}81)$$

where a refers to the ions (here the test particles) and b to electrons. This is a transcendental equation to be solved for v_D. If the drift velocity is much smaller than the electron thermal velocity, $l_b v_D \ll 1$, and one may write

$$G(l_b v_D) \approx \left.\frac{dG}{dx}\right|_0 l_b v_D = \frac{2}{3\sqrt{\pi}} l_b v_D \qquad (11\text{-}82)$$

(see Exercise 11-14). Consequently in this case the crift velocity is directly proportional to the field and one may define the conductivity

$$\sigma = \frac{qnv_D}{E} = \frac{3(2\pi)^{3/2}\varepsilon_0{}^2 m_-}{q^2 \ln \Lambda}\left(\frac{kT}{m_-}\right)^{3/2} \qquad (11\text{-}83)$$

where we made use of $m_+/m_- \gg 1$. A more precise treatment, taking into account electron-electron collisions, changes the numerical factors slightly, but not the dependence on physical parameters.

One may note that the conductivity rises as $T^{3/2}$ and does not depend on the plasma density. If there are more particles, the friction is correspondingly larger and v_D proportionally smaller, but since the number of charge carriers has also proportionally increased, the current remains the same.

For large values of the electric field the approximation (11-82) breaks down and the full form of (11-81) is to be used. Since $G(x)$ never exceeds $G(1) = 0.214$, this equation has no solution for

$$E > 0.214\,\frac{A_D}{q}\,l_b{}^2\left(\frac{m_+}{m_-} + 1\right)m_+ = E_0 \qquad (11\text{-}84)$$

We have seen that the dynamical friction decreases for large test-particle velocities. If $E > E_0$, the electric force can no longer be balanced by the ever-decreasing friction force. The particles keep accelerating and a runaway current arises, to be limited only by collective processes such as the two-stream instability.

Consider now another aspect of electric conduction. If $E < E_0$, the field imparts energy to the plasma which turns into heat as a result of friction between electrons and ions. As the temperature increases, the conductivity rises rapidly, leading to ever-increasing drift velocity. At one point E_0 becomes smaller than the electric field (which is kept constant) and the conduction evolves into a runaway process.

E_0 can be given a simple physical interpretation. If the electric field is big enough, so that an average electron picks up roughly as much speed in a self-collision time as its own thermal speed, runaway current arises. The proof is left to the reader (Exercise 11-17).

How correct was our assumption that the two particle species retain their Maxwellian distribution in their respective frames? It is easy to see that there are always particles, far out in the Maxwellian tail, which run away no matter how small the field. A fast electron is subjected to a dynamical friction force proportional to $1/v^2$ pointing toward $v = 0$ in velocity space, and a constant electric force. The resulting acceleration and the flow lines in velocity space are shown in Fig. 11-7.

The electrostatic analogue of these flow lines is the superposition of a uniform field [the acceleration $(q/m)\mathbf{E}$] and the field of a point charge ($\langle \Delta v_{\parallel} \rangle$ produced by the bulk of the plasma distributed about $v = 0$). Clearly Fig. 11-7 gives a good representation of flow lines only for $l_b v \gg 1$. Another, more

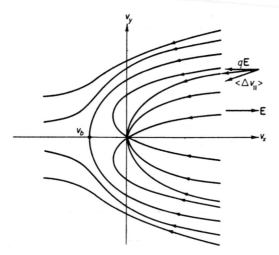

FIG. 11-7. Flow lines in velocity space of electrons subject to a uniform electric field and dynamical friction. Valid only for $l_b v \gg 1$. [From H. Dreicer, *Phys. Rev.* **117**, 329 (1960).]

serious limitation is the neglect of diffusion terms, where we know that $\langle \Delta v_\perp^2 \rangle$ dominates for fast particles. Nevertheless, the results contained in Fig. 11-7 can be regarded as an indication of the existence of runaways in weak fields, a finding confirmed by a more thorough analysis of the full F-P equation. The line (or rather surface obtained by rotation about v_x) separating the runaway region from the inner conduction region moves outward as E decreases. The stagnation point v_b can be calculated equating $(q/m)E$ with $\langle \Delta v_\parallel \rangle$:

$$\frac{q}{m} E = A_D \frac{1}{v_b^2} \tag{11-85}$$

where (11-69) and (11-59) have been used. Consequently,

$$v_b = \left(\frac{m_- A_D}{qE} \right)^{1/2} \sim \frac{1}{E^{1/2}} \tag{11-86}$$

Consider now a plasma in the presence of a uniform magnetic field. The component of the electric field along $\mathbf{B}$ results in the same conduction and runaway phenomena as in the absence of $\mathbf{B}$, while $\mathbf{E}_\perp$ gives rise to a drift of the plasma as a whole. Since in the latter case there is no relative drift of the electronic and ionic components, conduction or runaway currents do not arise.

It is of considerable interest, on the other hand, to investigate diffusion phenomena across the magnetic field caused by density and temperature gradients, since in the absence of instabilities, they present the final limitations to magnetic confinement. It will be assumed that the characteristic length scales of the gradients are far larger than the cyclotron radii, hence the guiding-center plasma model may be applied.

Consider first the effect of a binary collision on the positions of the guiding centers of the participating particles (Fig. 11-8). Prior to the collision the guiding center of a particle is located at

$$\mathbf{R} = \frac{\mathbf{p} \times \mathbf{B}}{qB^2} \tag{11-87}$$

as measured from the particle. During the collision the particle does not change its position appreciably, but the momentum changes by $\Delta \mathbf{p}$, hence

$$\Delta \mathbf{R} = \frac{\Delta \mathbf{p} \times \mathbf{B}}{qB^2} \tag{11-88}$$

Since the other particle changes its momentum by $-\Delta \mathbf{p}$, two interesting conclusions may be drawn immediately.

1. In collisions between particles of the same species,

$$\Delta \mathbf{R}_a + \Delta \mathbf{R}_b = 0 \tag{11-89}$$

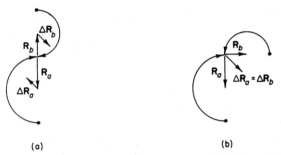

Fɪɢ. 11-8. Collision between particles gyrating in a magnetic field: (a) particles of the same species, (b) electron and singly charged positive ion.

It would be erroneous to conclude from this observation that the diffusion due to like particle collisions is zero (think, for example, of a situation where all guiding centers coincide initially). It is, however, an indication that the diffusion is slow, if all plasma parameters are nearly unchanged over the gyroradius scale. This guess is borne out by thorough investigation based on the F-P equation.

2. In collisions between electrons and singly charged ions, $q_A = -q_B$ and

$$\Delta \mathbf{R}_a = \Delta \mathbf{R}_b \tag{11-90}$$

Hence electronic and ionic guiding centers diffuse at the same rate. If there was no charge separation originally, the diffusion will not give rise to one.

Figure 11-9 shows the circular zero-order motion of an ion in a plasma with a density and temperature gradient in the $+y$ direction. Owing to collisions

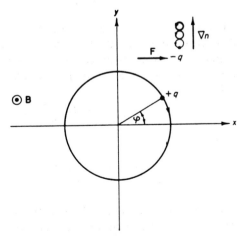

Fɪɢ. 11-9. Drag force acting on a gyrating ion from electrons with a density gradient.

with other particles, the test ion experiences a drag force **F**. Let us first set $\nabla T = 0$. To find **F** one has to consider the relative velocity of the test ion with the electron plasma. Since the guiding-center density increases with y, the average electron velocity is in the $+x$ direction (see Fig. 11-9). Because the ions can be considered motionless as compared to the electrons, one finds that **F** points in the $+x$ direction.

In the presence of a temperature gradient $dT/dy > 0$, **F** points in the $-x$ direction. Since the dynamical friction force is larger between particles with small relative velocity, the dominant effect is due to the colder electrons, whose guiding centers lie in the $-y$ direction. $\nabla n = 0$ was assumed.

In either case **F** gives rise to a guiding-center drift, with the velocity

$$\mathbf{u} = \frac{1}{q} \frac{\mathbf{F} \times \mathbf{B}}{B^2} \tag{11-91}$$

in the $-\nabla n$ and $+\nabla T$ directions, respectively.

Examining for simplicity the $\nabla T = 0$ case, one finds, from the pressure-balance equation,

$$\frac{B\nabla B}{\mu_0} + kT \nabla n = 0 \tag{11-92}$$

and from Ampere's law,

$$\frac{1}{\mu_0} \nabla B = qnv_r \tag{11-93}$$

where v_r is the average relative velocity between the species:

$$v_r = -\frac{kT}{qB} \frac{\nabla n}{n} \tag{11-94}$$

A quick result can be obtained if one approximates the electronic distribution function by a Maxwellian, moving with the velocity v_r with respect to the motionless ions, and assumes that v_r is small compared to the electron thermal speed. In this case one may use (11-81) and (11-82) to find

$$F = m\langle \Delta v_\parallel \rangle = \frac{2}{3\sqrt{\pi}} A_D l_b^3 \frac{m_+^2}{m_-} v_r \tag{11-95}$$

and the drift velocity

$$u = -\frac{1}{3(2\pi)^{1/2}} A_D \frac{1}{q^2 B^2} \frac{(m_-)^{1/2} m_+^2}{(kT)^{1/2}} \frac{\nabla n}{n} = -\frac{\nabla p_+}{\sigma B^2} \tag{11-96}$$

where σ is the conductivity defined by (11-83). Note that the diffusion velocity varies with B^{-2} and $T^{-3/2}$, both encouraging results, if one intends to confine a high-temperature plasma with large magnetic fields.

In the presence of ∇T the dominant process is heat transport, since energy transfer proceeds through the efficient process of like-particle collisions. Its magnitude can be estimated as follows. An average particle transfers energy during a self-collision time to a neighbor at a distance R_c, whose temperature is less by roughly $R_c \nabla T$. This results in a net energy flux for protons,

$$\mathbf{Q} = \tfrac{3}{2}k \, \nabla T \left\langle \frac{R_c^2}{t_c} \right\rangle n = \tfrac{3}{2}0.7 A_D \frac{k \, \nabla T \, m^{5/2}kTn}{(3kT)^{3/2}q^2B^2}$$

$$= \frac{0.7}{2\sqrt{3}} \frac{A_D}{q^2 B^2} \frac{\nabla(kT)m^{5/2}n}{(kT)^{1/2}} \tag{11-97}$$

A comparison with (9-96) shows that ionic heat flow is roughly $(m_+/m_-)^{1/2}$ times faster than particle flow by diffusion if the characteristic lengths in $\nabla T/T$ and $\nabla n/n$ are approximately equal. Consequently one expects the diffusion process to take place at approximately uniform temperature (except, of course, if a temperature gradient is artificially maintained).

For a particle gyrating in a uniform $\mathbf{B}$ field, the characteristic step size is the cyclotron radius, i.e., a particle (or guiding center) can be displaced by this distance in some collision time, or can transmit its energy to a location a cyclotron radius away. Since diffusion coefficients are of the form $D = \langle \Delta x^2 \rangle / t_c$, where $\langle \Delta x^2 \rangle$ is the mean square displacement in a collision and t_c the characteristic collision time, it is not surprising that both (11-96) and (11-97) scale with the cyclotron radius square.

If, for some reason, the step size is larger, one expects a corresponding increase in the diffusion velocities. A notable example is the tokomak geometry. There, as mentioned earlier, the magnetic field is made up of a toroidal component, produced by field coils, and a weaker poloidal field produced by a toroidal plasma current. The resulting field lines form toroidal helices. The surfaces containing these field lines are the magnetic surfaces, and they form nested toroids. Figure 11-10 shows such a surface, and a portion

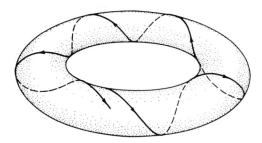

FIG. 11-10. Magnetic surface of a tokomak field. Poloidal field component has been exaggerated.

of a field line. Note that the field lines are seldom closed; a single field line usually covers the surface uniformly. Surfaces where field lines close after going around the long way n times (n integer) are the *mode rational surfaces*.

The intersections of magnetic surfaces with a plane, containing the axis of symmetry, are nested loops, represented by the concentric circles of Fig. 11-11. To the lowest order in the guiding-center approximation, a particle would follow a field line, with the guiding center sliding on a magnetic surface. To the next order, the ∇B and centrifugal drifts emerge, carrying the particle say upwards on Fig. 11-11. We recall that it was just this drift that prohibited particle confinement in a toroidal field without a poloidal component. In our case, however, the drift will carry the particle to inner magnetic surfaces in the lower half, and to outer surfaces in the upper half of the figure. The resulting surface may map onto a line on the cross section as shown in Fig. 11-11a; hence the particle is confined.

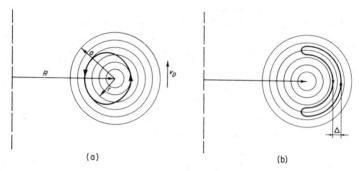

Fig. 11-11. Magnetic surfaces and projections of guiding center orbits in tokomaks: (a) untrapped orbit, (b) trapped (banana) orbit.

Most particles, however, behave quite differently. The field is stronger closer to the axis than further away, forming magnetic mirrors. Consequently, many particles are prevented from going around the toroid the short way and are trapped. Such an orbit is indicated on Fig. 11-11b. Because of its shape, this has been named *banana orbit*. Since the ratio of the largest to the smallest magnetic field, the mirror ratio, is $(R + r)/(R - r)$, the fraction of trapped particles is $\sqrt{2\varepsilon/(1 + \varepsilon)}$, where $\varepsilon = r/R$ (see Exercise 11-21). Since typically $a/R \approx \frac{1}{3}$, the majority of particles in a tokomak are trapped.

The point that interests us here is the excursion length in the direction of the density and temperature gradient. Since the magnetic surfaces are constant pressure surfaces [see (4-55)], the important scale length is the banana width Δ, which is usually much bigger than a cyclotron radius. This length replaces the cyclotron orbit scale length in calculating the diffusion coefficients.

This is called *neoclassical diffusion* and the diffusion coefficients are typically two orders of magnitudes larger than the classical ones calculated for a uniform magnetic field. Even with these large diffusion coefficients, one arrives at a manageable size for a tokomak thermonuclear reactor.

More dangerous for confinement is turbulent or *anomalous diffusion*. Instabilities, when saturated due to nonlinear processes, give rise to field fluctuations. Magnetic field fluctuations can lead to large excursions, and so do electric field fluctuations by giving rise to $\mathbf{E} \times \mathbf{B}$ drifts. We do not know yet how to calculate the fluctuation amplitudes, and what are the parameters that govern them.

11-5. Summary

The motion of a charged particle in a plasma is influenced, in addition to the "smooth" macroscopic electromagnetic field, by the microfields of adjacent particles within a Debye radius. While a complete description of these processes, called collisions, would require the solution of a practically insoluble many-body problem, the large number of particles in a Debye sphere facilitates transformation to a manageable statistical problem.

A particle suffers a large number of simultaneous small-angle binary collisions, whose effects may be linearly superposed. The cumulative effect of these collisions exceeds the effect of the very infrequent large-angle collisions, which are therefore neglected. The approximate equation used to describe the change of the distribution functions due to these small-angle collisions is the Fokker-Planck equation (11-49), which depends on the dynamical friction term $\langle \Delta \mathbf{v} \rangle$ and the diffusion tensor (in velocity space) $\langle \Delta \mathbf{v} \, \Delta \mathbf{v} \rangle$. These quantities, as well as a number of relaxation times—giving only order-of-magnitude estimates anyhow—may be used to make educated guesses about nearly Maxwellian or non-Maxwellian plasmas. It turns out, for example, that a non-Maxwellian plasma if left alone relaxes to thermal equilibrium in the following way. First, the electrons thermalize among themselves on a time scale given by the self-collision time (11-75). Subsequently [about $(m_+/m_-)^{1/2}$ times later] the slower ion component thermalizes, reaching in general a temperature different from that of the electrons. Finally [again $\approx (m_+/m_-)^{1/2}$ times later] a common equilibrium is established through the relatively slow ion-electron energy-exchange process (11-80).

The electrical conductivity for a nearly thermal plasma in a uniform d.c. electric field can be estimated by the simple consideration that the electrical force on the particles is counterbalanced by the dynamical friction force in the steady-state conduction process. The resulting conductivity is (11-83). However, since the dynamical friction decreases rapidly for large particle velocities, there are always some particles for which such an equilibrium is impossible. These particles are constantly accelerated by the field; they run away. Above a certain value of the electric field, described by (11-84), practically the entire plasma runs way, and one may no longer speak of a conduction process. Indeed this is the fate of any plasma in a constant electric field, however small, since joule heating pushes particles continuously above the runaway limit, until the entire plasma gets hot enough to run away.

A spatially nonuniform plasma is subject to diffusion processes. If the gyroradii are small compared to the scale of nonuniformities, the dominant diffusion is caused by electron-ion collisions. The rate of diffusion in a magnetized plasma with straight parallel field lines, due to a pressure gradient perpendicular to the magnetic field is equal for both species and its velocity is given by (11-96). In case a thermal gradient is also present, heat conduction takes place by the more effective process of ion-ion collisions. The heat flux is given by (11-97).

The diffusion coefficients scale as the square of the excursion length of a guiding center in a collision time along the density or temperature gradient. Often these excursion lengths are considerably longer than the gyroradius, as in the banana orbits in tokomaks. This leads to neoclassical diffusion, while turbulence caused excursions result in anomalous diffusion.

EXERCISES

11-1. Prove that the basic plasma characteristics ω_p, λ_D, $\bar{\kappa}$, β, ω_c are invariant against "smoothing out" the plasma microfields, by $q \to 0$, $m \to 0$ while q/m and qn are held constant.

11-2. Derive (11-1), using the same reasoning that led to the Boltzmann equation in Chapter III.

11-3. Prove that the substitution $f^{(2)} = f^{(1)}(\mathbf{r},\mathbf{v})f^{(1)}(\mathbf{r}_2, \mathbf{v}_2)$ in (11-3) leads to the Vlasov equation for $f^{(1)}$ ($N \gg 1$).

11-4. Derive (11-41).

11-5. Show that the potentials h and g are related through (11-48).

11-6. Prove that the momentum of a one-component plasma is conserved, in Fokker-Planck-type collisions, by multiplying (11-49) by $\mathbf{v}$ and integrating over velocity space. (*Hint*: Use the electrostatic analogy in the first term.)

11-7. Prove that a one-component plasma conserves energy in F-P collisions. [*Hint*: Integrate by parts, using (11-48). By electrostatic analogy the first term will correspond to the virial of forces, the second to the potential energy.]

11-8. Derive (11-62) and (11-63).

11-9. Compute the self-collision times (electrons and ions) for a plasma of density $n = 10^{20} m^{-3}$ if the kinetic temperature is (a) 100 eV, (b) 1 keV, (c) 10 keV.

11-10. Derive (11-78) and (11-79).

11-11. Compute d, λ_D, p_0, l, and l_σ (definitions at the end of Sec. 11-3) for
(a) $T = 10$ keV, $\quad n = 10^{21} m^{-3}$.
(b) $T = 100$ eV, $\quad n = 10^{17} m^{-3}$.

11-12. Use the Boltzmann equation with the Fokker-Planck terms on the right-hand side for the two particle species of the plasma to prove that the conduction current does not depend on the diffusion terms in the F-P equation (*Hint*: Take the first moment in velocity space.)

11-13. Use the moment equations derived in the previous exercise to find the energy-balance equation for a plasma in an electric field. Demonstrate that the power fed into the plasma by the electric field is divided between increasing the kinetic (drift) energy of the components and the "thermal" energy gain of the constituents by dynamical friction between the two species.

11-14. Show that $dG/dx|_0 = 2/3\sqrt{\pi}$.

11-15. At what temperature does the conductivity of a hydrogen plasma equal that of copper? [Use the low-field conductivity (11-83).]

11-16. What is the critical runaway field for a deuterium plasma at $T_1 = 10^6$ °K and $T_2 = 10^8$ °K?

11-17. Prove that the electric field E_0 accelerates an average electron during a self-collision time roughly by its average thermal speed.

11-18. Estimate the proportion of runaway electrons in a hydrogen plasma, as a function of E, using (11-86).

11-19. Estimate the diffusion velocity of a high-β deuterium plasma of temperature $T = 10^8$ °K confined by a magnetic field of $B = 10$ webers m^{-2} if the characteristic radius of the device is 1 m.

11-20. Estimate the characteristic cooling time of the device described in the previous exercise and compare it to its diffusion time (time needed to lose an appreciable amount of plasma by diffusion).

11-21. If $\varepsilon = r/R$, where R is the major tokomak radius and r is the minor radius of a toroidal magnetic surface, prove that the fraction of trapped particles is given by $\sqrt{2\varepsilon(1 + \varepsilon)^{-1}}$.

Appendix I

Useful Vector Relations

$$\mathbf{A} \cdot \mathbf{B} = A_x B_x + A_y B_y + A_z B_z \tag{A-1}$$

$$\mathbf{A} \times \mathbf{B} = \begin{vmatrix} \mathbf{e}_1 & \mathbf{e}_2 & \mathbf{e}_3 \\ A_x & A_y & A_z \\ B_x & B_y & B_z \end{vmatrix} \tag{A-2}$$

$$(\mathbf{A} \times \mathbf{B}) \times \mathbf{C} = (\mathbf{A} \cdot \mathbf{C})\mathbf{B} - (\mathbf{B} \cdot \mathbf{C})\mathbf{A} \tag{A-3}$$

$$\mathbf{A} \times (\mathbf{B} \times \mathbf{C}) = (\mathbf{A} \cdot \mathbf{C})\mathbf{B} - (\mathbf{A} \cdot \mathbf{B})\mathbf{C} \tag{A-4}$$

$$(\mathbf{A} \times \mathbf{B}) \cdot (\mathbf{C} \times \mathbf{D}) = (\mathbf{A} \cdot \mathbf{C})(\mathbf{B} \cdot \mathbf{D}) - (\mathbf{A} \cdot \mathbf{D})(\mathbf{B} \cdot \mathbf{C}) \tag{A-5}$$

$$(\mathbf{A} \times \mathbf{B}) \times (\mathbf{C} \times \mathbf{D}) = [\mathbf{A} \cdot (\mathbf{B} \times \mathbf{D})]\mathbf{C} - [\mathbf{A} \cdot (\mathbf{B} \times \mathbf{C})]\mathbf{D}$$
$$= [\mathbf{A} \cdot (\mathbf{C} \times \mathbf{D})]\mathbf{B} - [\mathbf{B} \cdot (\mathbf{C} \times \mathbf{D})]\mathbf{A} \tag{A-6}$$

$$\nabla\varphi\psi = \varphi\,\nabla\psi + \psi\,\nabla\varphi \tag{A-7}$$

$$\nabla \cdot (\varphi\mathbf{a}) = \varphi\,\nabla \cdot \mathbf{a} + (\nabla\varphi) \cdot \mathbf{a} \tag{A-8}$$

$$\nabla \times (\varphi\mathbf{a}) = \varphi(\nabla \times \mathbf{a}) + \nabla\varphi \times \mathbf{a} \tag{A-9}$$

$$\nabla \cdot (\mathbf{a} \times \mathbf{b}) = \mathbf{b} \cdot (\nabla \times \mathbf{a}) - \mathbf{a} \cdot (\nabla \times \mathbf{b}) \tag{A-10}$$

$$\nabla \times (\mathbf{a} \times \mathbf{b}) = \mathbf{a}(\nabla \cdot \mathbf{b}) + (\mathbf{b} \cdot \nabla)\mathbf{a} - \mathbf{b}(\nabla \cdot \mathbf{a}) - (\mathbf{a} \cdot \nabla)\mathbf{b} \tag{A-11}$$

$$\nabla(\mathbf{a} \cdot \mathbf{b}) = (\mathbf{a} \cdot \nabla)\mathbf{b} + (\mathbf{b} \cdot \nabla)\mathbf{a} + \mathbf{a} \times (\nabla \times \mathbf{b}) + \mathbf{b} \times (\nabla \times \mathbf{a}) \tag{A-12}$$

$$\nabla \times (\nabla \times \mathbf{a}) = \nabla(\nabla \cdot \mathbf{a}) - \nabla^2\mathbf{a} \tag{A-13}$$

$$\nabla \cdot (\nabla \times \mathbf{a}) = 0 \tag{A-14}$$

$$\nabla \times (\nabla\varphi) = 0 \tag{A-15}$$

Appendix II

Some Relations in
Curvilinear Coordinates

The line element in orthogonal coordinates is

$$ds^2 = (h_1\,dx_1)^2 + (h_2\,dx_2)^2 + (h_3\,dx_3)^2$$

The ∇ operations are

$$\nabla\psi = \frac{1}{h_1}\frac{\partial\psi}{\partial x_1}\,\mathbf{e}_1 + \frac{1}{h_2}\frac{\partial\psi}{\partial x_2}\,\mathbf{e}_2 + \frac{1}{h_3}\frac{\partial\psi}{\partial x_3}\,\mathbf{e}_3$$

$$\nabla\cdot\mathbf{a} = \frac{1}{h_1 h_2 h_3}\left[\frac{\partial}{\partial x_1}(h_2 h_3 a_1) + \frac{\partial}{\partial x_2}(h_1 h_3 a_2) + \frac{\partial}{\partial x_3}(h_1 h_2 a_3)\right]$$

$$\nabla\times\mathbf{a} = \frac{1}{h_2 h_3}\left[\frac{\partial}{\partial x_2}(h_3 a_3) - \frac{\partial}{\partial x_3}(h_2 a_2)\right]\mathbf{e}_1$$

$$+ \frac{1}{h_1 h_3}\left[\frac{\partial}{\partial x_3}(h_1 a_1) - \frac{\partial}{\partial x_1}(h_3 a_3)\right]\mathbf{e}_2$$

$$+ \frac{1}{h_1 h_2}\left[\frac{\partial}{\partial x_1}(h_2 a_2) - \frac{\partial}{\partial x_2}(h_1 a_1)\right]\mathbf{e}_3$$

$$\nabla^2\psi = \frac{1}{h_1 h_2 h_3}\left[\frac{\partial}{\partial x_1}\left(\frac{h_3 h_2}{h_1}\frac{\partial\psi}{\partial x_1}\right) + \frac{\partial}{\partial x_2}\left(\frac{h_3 h_1}{h_2}\frac{\partial\psi}{\partial x_2}\right) + \frac{\partial}{\partial x_3}\left(\frac{h_1 h_2}{h_3}\frac{\partial\psi}{\partial x_3}\right)\right]$$

Appendix III

Some Important Plasma Characteristics

Cyclotron frequency:
$$\omega_c = \frac{qB}{m} \qquad (2\text{-}10)$$

Magnetic moment:
$$\mu_m = \frac{\frac{1}{2}mv_\perp^{\,2}}{B} \qquad (2\text{-}14)$$

Plasma frequency:
$$\omega_p = \left(\frac{q^2 n}{m\varepsilon_0}\right)^{1/2} \qquad (7\text{-}6)$$

Debye length:
$$\lambda_D = \left(\frac{\varepsilon_0 kT}{q^2 n}\right)^{1/2} \qquad (6\text{-}6)$$

Ion sound velocity:
$$c_s = \omega_{pi}\lambda_D = \left(\frac{T}{M}\right)^{1/2} \qquad (7\text{-}102)$$

Alfven speed:
$$V = \frac{B}{(\mu_0 v)^{1/2}} \qquad (4\text{-}79)$$

Dielectric permeability:
$$\kappa_\omega = 1 - \left(\frac{\omega_p}{\omega}\right)^2 \qquad (6\text{-}138)$$

Beta:
$$\beta = \frac{P}{B^2/(2\mu_0)} \qquad (6\text{-}75a)$$

Dielectric tensor of cold plasma in magnetic field:

$$\overset{\leftrightarrow}{\kappa} = \begin{pmatrix} \kappa_T & i\kappa_H & 0 \\ -i\kappa_H & \kappa_T & 0 \\ 0 & 0 & \kappa_\parallel \end{pmatrix} \qquad (8\text{-}10)$$

$$\kappa_T = \frac{\kappa_R + \kappa_L}{2} \tag{8-11}$$

$$\kappa_H = \frac{\kappa_R - \kappa_L}{2} \tag{8-12}$$

$$\kappa_{R,L} = 1 - \frac{\Pi_p^2}{(1 \pm \beta^+)(1 \mp \beta^-)} \tag{8-4 and 8-5}$$

$$\kappa_\parallel = 1 - \Pi_p^2 \tag{8-6}$$

$$\beta^\pm = \left| \frac{\omega_c^\pm}{\omega} \right| \tag{8-8 and 8-9}$$

$$\Pi_p^2 = \frac{(\omega_p^+)^2 + (\omega_p^-)^2}{\omega^2} \tag{8-7}$$

Low-frequency dielectric permeability perpendicular to magnetic field:

$$\kappa_\perp \approx \frac{m^+ n}{\varepsilon_0 B^2} \tag{6-21}$$

Bounce frequency: $\quad \omega_b = \left(\frac{qkE}{m} \right)^{1/2} \tag{9-12}$

Conductivity: $\quad \sigma \approx \frac{3(2\pi)^{3/2} \varepsilon_0^2 m_-}{q^2 \ln \Lambda} \left(\frac{kT}{m_-} \right)^{3/2} \tag{11-83}$

Self-collision time: $\quad t_c \approx \frac{T^{3/2}}{Z^4 n} \left(\frac{m}{m_p} \right)^{1/2} \times 10^6 \tag{11-75}$

Suggested References

This list of references is not meant to be exhaustive. It is organized roughly in terms of difficulty and specialization, i.e., elementary and general works are listed first, followed by more advanced texts containing more specific topics.

F. Chen, "Introduction to Plasma Physics." Plenum, New York, 1974. An introductory text on the undergraduate level, including the description of some basic experiments and fusion devices.

M. O. Haglar and M. Kristiansen, "An Introduction to Controlled Thermonuclear Fusion." Heath, Indianapolis, Indiana, 1977. A good nonmathematical introduction to fusion, with description of principles and machine parameters as of 1977.

Lev A. Arzimovich, "Elementary Plasma Physics." Ginn (Blaisdell), Boston, Massachusetts, 1965. Translated from the Russian. A very readable introduction to plasma physics, concentrating on physics principles.

S. Gartenhaus, "Elements of Plasma Physics." Holt, New York, 1964. This book uses more mathematics than the previous references. It contains a good description of many-particle dynamics and the relationship of the Vlasov equation to statistical mechanics.

H. Alfven and C. G. Fälthammar, "Cosmical Electrodynamics." Oxford Univ. Press, London and New York, 1963. A classic work on guiding-center theory and magnetohydrodynamics with applications to cosmic plasma physics.

L. Spitzer Jr, "Physics of Fully Ionized Gases." Wiley (Interscience), New York, 1962. A classic treatment of two-fluid theory, collisions, and relaxation times.

N. A. Krall and A. W. Trivelpiece, "Principles of Plasma Physics." McGraw-Hill, New York, 1973. A general textbook on plasma physics on the graduate level.

T. H. Stix, "The Theory of Plasma Waves." McGraw-Hill, New York, 1962. This first comprehensive treatment of plasma waves in homogeneous and inhomogeneous media is still valuable reading.

G. Bekefi, "Radiation Processes in Plasmas." Wiley, New York, 1966. Discusses waves, their emission, absorption, scattering, and propagation in plasmas.

A. B. Mikhailovskii, "Theory of Plasma Instabilities." Consultants Bureau, New York, 1974. Translated from the Russian. An exhaustive description of plasma instabilities: Vol. I Instabilities of a Homogeneous Plasma; Vol. II. Instabilities in an Inhomogeneous Plasma.

T. G. Northrop, "The Adiabatic Motion of Charged Particles." Wiley (Interscience), New York, 1963. A thorough and rather mathematical treatment of guiding-center theory.

D. A. Tidman and N. A. Krall, "Shock Waves in Collisionless Plasmas." Pergamon, Oxford, 1974. A clear treatment of nonlinear waves and shocks in plasmas.

A. Hasegawa, "Plasma Instabilities and Nonlinear Effects." Springer-Verlag, Berlin and New York, 1975. A reference book, with an emphasis on space plasma instabilities.

R. Z. Sagdeev and A. A. Galeev, "Nonlinear Plasma Theory" (T. M. O'Neil and D. L. Book, eds.). Benjamin, Reading, Massachusetts, 1969. This compact book contains a wealth of information on nonlinear effects. It is not easy reading, but worth the effort.

R. C. Davidson, "Methods in Nonlinear Plasma Theory." Academic Press, New York, 1972. A rather formal but rigorous treatment of nonlinear waves and instabilities in plasmas.

V. I. Karpman, "Nonlinear Waves in Dispersive Media." Pergamon, Oxford, 1974. Translated from the Russian. This modern and strongly mathematical treatment of nonlinear waves is not restricted to waves in plasmas. Effects arising from strong nonlinearities, like solitons, are emphasized.

"Advances in Plasma Physics" (A. Simon and W. B. Thompson, eds.). Wiley, New York. This ongoing series contains review papers on various topics of current interest.

"Reviews of Plasma Physics" (M. A. Leontovich, ed.). Consultants Bureau, New York. Translated from the Russian. The Soviet equivalent of "Advances in Plasma Physics."

Index